NEUTRON AND X-RAY SCATTERING IN MATERIALS SCIENCE AND BIOLOGY

To learn more about AIP Conference Proceedings, including the
Conference Proceedings Series, please visit the webpage
http://proceedings.aip.org/proceedings

NEUTRON AND X-RAY SCATTERING IN MATERIALS SCIENCE AND BIOLOGY

Proceedings of the
International Conference on Neutron and X-Ray Scattering 2007

Serpong and Bandung, Indonesia   23 – 31 July 2007

EDITORS

Abarrul Ikram
Agus Purwanto
Sutiarso
National Nuclear Energy Agency of Indonesia, BATAN

Anne Zulfia
The University of Indonesia, UI

Sunit Hendrana
Indonesian Institute of Sciences, LIPI

Zeily Nurachman
Bandung Institute of Technology, ITB

All papers have been peer reviewed.

SPONSORING ORGANIZATIONS
National Nuclear Energy Agency of Indonesia (BATAN)
Directorate of Higher Education of the Ministry of National Education of Indonesia (DIKTI)
International Atomic Energy Agency (IAEA)
International Union of Crystallography (IUCr)
Abdus Salam International Centre for Theoretical Physics (ICTP)
Deutsche Forschungsgemeinschaft (DFG)

Melville, New York, 2008
AIP CONFERENCE PROCEEDINGS ■ VOLUME 989

CONTENTS

PLENARY LECTURES

PARTICIPANT CONTRIBUTIONS

ALLOYS

INSTRUMENTS AND METHODS

MODELING AND SIMULATIONS

MATERIALS CHEMISTRY

POLYMERS AND COLLOIDS

Committees

International Advisory Board Members

Dr. A. Aziz Bin Mohamed	MNA, Malaysia
Prof. A. Furrer	ETH Zurich & PSI, Switzerland
Dr. A. J. Hurd	LANSCE, USA
Prof. Bohari	UKM, Malaysia
Dr. C. J. Carlile	ILL, France
Prof. D. Chen	CIAE, China
Dr. J. Mesot	PSI, Switzerland
Dr. J. Root	Chalk River Laboratories, Canada
Prof. K. Mortensen	Risø National Laboratory, Denmark
Prof. M. Arai	JAEA, Japan
Prof. M. Shibayama	University of Tokyo, Japan
Dr. P. S. Goyal	IUC-DAEF, BARC, India
Dr. T. E. Mason	ORNL, USA
Dr. Y. Morii	JAEA, Japan

National Advisory Board Members

Dr. H. Hastowo	National Nuclear Energy Agency (BATAN)
Dr. S. Soentono	National Nuclear Energy Agency (BATAN)
Dr. P. Anggraita	National Nuclear Energy Agency (BATAN)
Dr. Ridwan	National Nuclear Energy Agency (BATAN)
Dr. A. Ikram	National Nuclear Energy Agency (BATAN)
Dr. A. Purwanto	National Nuclear Energy Agency (BATAN)
Dr. E. Giri Rachman Putra	National Nuclear Energy Agency (BATAN)
Prof. Dr. Marsongkohadi	Bandung Insitute of Technology (ITB)
Dr. Ismunandar	Bandung Insitute of Technology (ITB)
Dr. Z. Nurachman	Bandung Insitute of Technology (ITB)
Prof. Dr. Effendy	State University of Malang (UM)
Prof. Dr. A. K. Prodjosantoso	State University of Yogyakarta (UNY)
Dr. Kgs. Dahlan	Bogor Agricultural University (IPB)
Dr. M. Hikam	University of Indonesia (UI)
Dr. I N. Jujur	Indonesia Agency for Assessment and Application of Technology (BPPT)
Dr. Soetijoso	Padjadjaran University (UNPAD)

Organising Committees

Chairman(s)	Dr. E. Giri Rachman Putra	BATAN
	Dr. Ismunandar	ITB
Secretaries	Dr. Sutiarso	BATAN
	Dr. Z. Nurachman	ITB
Treasurer(s)	Ms. A. Insani	BATAN
	Dr. E. Sustini	ITB
Workshop Coordinator	Dr. A. Ikram	BATAN
Symposium Coordinator	Dr. A. Nugroho	ITB
Poster & Exhibitions	Dr. I N. Marsih	ITB
	Dr. F. Haryanto	ITB
Publication, Documentation & Proceeding	Dr. A. Fajar	BATAN
	Dr. B. Prijamboedi	ITB
Secretariat	Mr. Bharoto	BATAN
	Dr. I. Mulyani	ITB

Preface

The International Conference on Neutron and X-ray Scattering 2007 (ICNX2007) was held from July 23rd to 31st in Serpong and Bandung, Indonesia. This event is a continuation of the National Neutron and X-ray Scattering Seminar held annually since 1998 to promote neutron scattering techniques and applications in Indonesia by utilizing neutron scattering instruments established by BATAN (the National Nuclear Energy Agency of Indonesia) in 1992 in Serpong. This time, the conference focused on neutron and X-ray scattering in materials science and biology.

The conference comprises of five five-day workshops in Serpong and a two-day symposium in Bandung. The five-day workshops were held at BATAN nuclear center site in Serpong, 30 km south-west Jakarta, the capital city. The workshops were attended by 70 participants including 9 young scientists from abroad, i.e.: Algeria, India, Japan, Malaysia, Singapore and Turkey.

The two-day symposium activity took place at the Aula Barat (West Hall) of ITB, Bandung, 150 km south-east Jakarta. This year we began referring the event as "International Conference" rather than National Seminar, which has already taken place for six times, in order to encourage participation from abroad while elevating the confidence of the local participants. Prominent speakers from all corner of the world in the field of neutron and X-ray scattering were invited to present their papers. There were more than 150 participants coming from local, regional and international universities, research institutes and companies. There were more 110 abstracts for the poster sessions coming from Algeria, China, Czech, France, India, Iraq, Israel, Japan, Kenya, Malaysia, Pakistan, Poland, Republic of Korea, Russia, Turkey, Ukraine, United Kingdom, United States of America, Yemen and Indonesia. They were grouped into Alloys; Biology; Ceramics; Instruments and Methods; Modeling & Simulations; Materials Chemistry, Polymers & Colloids.

The Executive Committee dedicated the 2007 conference to Marsongkohadi, who turned 75 in May, for his pioneership in promoting, developing and establishing neutron scattering in Indonesia since 1960's. A conference dinner was held at the View Restaurant Dago Pakar located at the northern-side hill of Bandung neighborhood to wrap up the symposium first day. The magnificent scenery of Bandung city known as Parijs van Java during the night can be viewed from this conference dinner place.

Financial support for this conference was provided by the National Nuclear Energy Agency of Indonesia (BATAN), the Directorate of Higher Education of the Ministry of National Education of Indonesia (Dikti), the International Atomic Energy Agency (IAEA), the International Union of Crystallography (IUCr), the Abdus Salam International Centre for Theoretical Physics (ICTP), and the Deutsche Forschungsgemeinschaft (DFG). We would also like to acknowledge here the session Chairs and people who gave introductory remarks : Djoko Santoso, Brendan Kennedy, Abarrul Ikram, Abdul Aziz Mohamed, Agung Nugroho, Agus Purwanto, Vinod Aswal, Robert Knott, Ismunandar and Pudji Astuti.

We are very thankful to all members of the Organizing Committees, both in Serpong and in Bandung. The Executive Committee promoted and selected Abdul Aziz Mohamed from Malaysia Nuclear Agency as the Chair for the next International Conference on Neutron and X-ray Scattering in 2009. Finally, all participation, support and contribution are gratefully appreciated, and we look forward to meeting you all in Malaysia in 2009.

Edy Giri Rachman Putra (Chair)
Abarrul Ikram (Advisory Committee)

November 2007
Serpong, Indonesia

Workshop Program

23 - 28 July 2007
Neutron and X-ray scattering in materials science and biology
BATAN Nuclear Center Site
Kawasan Puspiptek Serpong, Tangerang 15314
INDONESIA

The workshop programs are divided into five groups:
- Nanostructure and macromolecules studies using small angle scattering technique.
- Residual stress measurement in alloys using neutron diffraction technique.
- Crystal and magnetic structure using high-resolution powder diffractometer.
- Non-destructive test on industrial materials using neutron radiography technique.
- Crystallography studies in protein and biomacromolecules.

Lectures on the theoretical and experimental aspects of neutron and X-ray scattering as well as demonstrations and hands-on training using several instruments and also data analysis will be given during the workshop

- Nanostructure and macromolecules studies using small angle scattering technique group

	Monday July 23	Tuesday July 24	Wednesday July 25	Thursday July 26	Friday July 27	Saturday July 28
08.00	Opening & Group Photo					
08.30		SANS: Some Applications (Sung-Min Choi)	Experiment with SMARTer (Edy Giri R. Putra)	Experiment IV SANS Experiment (Vinod K. Aswal & Edy Giri R. Putra)	Data Reduction & Analysis IV (Sung-Min Choi & Edy Giri R. Putra)	Special group meeting*
09.30	Morning tea					
10.00	Introduction to Neutron & X-Ray Scattering (Agus Purwanto)	Small Angle Neutron Scattering: An Introduction (Vinod K. Aswal)	Experiment I Sample Preparation (Vinod K. Aswal & Edy Giri R. Putra)	Data Reduction & Analysis I (Sung-Min Choi & Edy Giri R. Putra)	Participant Presentation	Special group meeting*
11.00	Introduction to Wide-Angle Neutron & X-Ray Scattering (Zin Tun)	Contrast Factor & Variation (Vinod K. Aswal)				
12.00	Lunch					
13.00	Introduction to Wide-Angle Neutron & X-Ray Scattering (Zin Tun)	Data Reduction & Analysis I (Sung-Min Choi)	Experiment II SANS Experiment (Vinod K. Aswal & Edy Giri R. Putra)	Data Reduction & Analysis II (Sung-Min Choi & Edy Giri R. Putra)	Participant Presentation	
14.00	Introduction to Small-Angle Scattering (Vinod K. Aswal)	Data Reduction & Analysis II (Sung- Min Choi)			Summary	
15.00	Afternoon tea					
15.30	Applications X-Ray Scattering in Biology (Bauke W. Dijkstra)	Designing a SANS Experiment (Vinod K. Aswal)	Experiment III SANS Experiment (Vinod K. Aswal & Edy Giri R. Putra)	Data Reduction & Analysis III (Sung-Min Choi & Edy Giri R. Putra)	Conclusions	
16.30					Closing	

General lectures/activities
Specific lectures/activities

◆ Residual stress measurement in alloys using neutron diffraction technique group

	Monday July 23	Tuesday July 24	Wednesday July 25	Thursday July 26	Friday July 27	Saturday July 28
08.15						
08.30	Opening & Group Photo	Residual Stress Measurement using X-ray Diffraction I (Takao Hanabusa)	Experiment Preparation of Residual Stress Measurement using Neutron Diffraction (Refai Musliih)	Experiment (Refai Muslih)	Data Analysis	Special group meeting*
09.30	Morning tea					
10.00	Introduction to Neutron & X-Ray Scattering (Agus Purwanto)	Residual Stress Measurement using X-ray Diffraction II (Takao Hanabusa)	Experiment Preparation of Residual Stress Measurement using Neutron Diffraction (Refai Musliih)	Experiment & Data Analysis	Uncertainty Calculation Residual Stress Measurement (Iwan Sumirat)	Special group meeting*
11.00	Introduction to Wide-Angle Neutron & X-Ray Scattering (Zin Tun)				Data Analysis	
12.00	Lunch					
13.00	Introduction to Wide-Angle Neutron & X-Ray Scattering (Zin Tun)	Residual Stress Measurement using Neutron Diffraction I (Masayuki Nishida)	Experiment (Refai Muslih)	Experiment & Data Analysis	Discussion	
14.00	Introduction to Small-Angle Scattering (Vinod K. Aswal)					
15.00	Afternoon tea					
15.30	Applications X-Ray Scattering in Biology (Bauke W. Dijkstra)	Residual Stress Measurement using Neutron Diffraction I (Masayuki Nishida)	Experiment (Refai Muslih)	Experiment & Data Analysis	Conclusions	
16.30					Closing	

General lectures/activities

Specific lectures/activities

• Crystal and magnetic structure using high-resolution powder diffractometer group

	Monday July 23	Tuesday July 24	Wednesday July 25	Thursday July 26	Friday July 27	Saturday July 28
08.15	Opening & Group Photo					
08.30		Neutron diffraction (Zin Tun)	Single Crystal Diffraction on Crystalline Materials (Zin Tun)	Data Analysis	Spallation Source Neutron Powder Diffraction for Materials Science (Takashi Kamiyama)	Special group meeting*
09.30	Morning tea					
10.00	Introduction to Neutron & X-Ray Scattering (Agus Purwanto)	Powder diffraction (Agus Purwanto)	Rietveld Analysis (Andika Fajar)	Single crystal Diffraction on Magnetic Materials (Zin Tun)	Complementary Aspects of Neutron and Synchrotron Powder Diffraction on Oxide Based Materials (Brendan Kennedy)	Special group meeting*
11.00	Introduction to Wide-Angle Neutron & X-Ray Scattering (Zin Tun)	Experiment with HRPD (Andika fajar)	Introduction to FullProof (Andika Fajar)	Group Discussion	Group Discussion	
12.00	Lunch					
13.00	Introduction to Wide-Angle Neutron & X-Ray Scattering (Zin Tun)	Experiment	Data Analysis	Group Discussion	Participant Presentation	
14.00	Introduction to Small-Angle Scattering (Vinod K. Aswal)				Summary	
15.00	Afternoon tea					
15.30	Applications X-Ray Scattering in Biology (Bauke W. Dijkstra)	Experiment	Data Analysis	Group Discussion	Conclusions	
16.30					Closing	

☐ General lectures/activities
☐ Specific lectures/activities

◆ Non-destructive test on industrial materials using neutron radiography technique group

	Monday July 23	Tuesday July 24	Wednesday July 25	Thursday July 26	Friday July 27	Saturday July 28
08.15						
08.30	Opening & Group Photo	Introduction to the Workshop of Neutron Radiography (Sutiarso)	Practical I : Neutron Beam Quality (N. Raghu - Sutiarso)	Practical III Real Time Method (Sutiarso – Tjiptono)	Visit to the facilities (Sutiarso)	Special group meeting*
09.30	Morning tea					
10.00	Introduction to Neutron & X-Ray Scattering (Agus Purwanto)	Basic Principle and Applications (N. Raghu)	Practical II Inspection of Engineering/ Biological Objects (N. Raghu - Sutiarso)	Practical III Real Time Method (cont'd) (Sutiarso – Tjiptono)	Visit to the facilities (Sutiarso)	Special group meeting*
11.00	Introduction to Wide-Angle Neutron & X-Ray Scattering (Zin Tun)					
12.00	Lunch					
13.00	Introduction to Wide-Angle Neutron & X-Ray Scattering (Zin Tun)	Simulation and Design of Collimator (P. Ilham)	Practical II Inspection of Engineering/ Biological Objects (cont'd) (N. Raghu – Sutiarso)	Neutron Tomography (Mardiyanto)	Report from other Facilities (Participants)	
14.00	Introduction to Small-Angle Scattering (Vinod K. Aswal)				Summary	
15.00	Afternoon tea					
15.30	Applications X-Ray Scattering in Biology (Bauke W. Dijkstra)	Neutron and X-ray Complimentary (Sutiarso)	Discussion	Discussion	Conclusions	
16.30					Closing	

▢ General lectures/activities
▢ Specific lectures/activities

- Crystallography studies in protein and biomacromolecules group

	Monday July 23	Tuesday July 24	Wednesday July 25	Thursday July 26	Friday July 27	Saturday July 28
08.15	Opening & Group Photo					
08.30		Preparing Protein Crystal (Bauke W. Dijkstra)	Diffraction Data Evaluation (Bauke W. Dijkstra)	Methods for Solving Phase Problem (Bauke W. Dijkstra)	Model building (Zeily Nurachman)	Special group meeting*
09.30	Morning tea					
10.00	Introduction to Neutron & X-Ray Scattering (Agus Purwanto)	Preparing Protein Crystal (Bauke W. Dijkstra)	Methods for Solving Phase Problem (Bauke W. Dijkstra)	Methods for Solving Phase Problem (Bauke W. Dijkstra)	Model building (Zeily Nurachman)	Special group meeting*
11.00	Introduction to Wide-Angle Neutron & X-Ray Scattering (Zin Tun)			Phase Improvement (Bauke W. Dijkstra)	Model Building and Crystallographic Refinement (Bauke W. Dijkstra)	
12.00	Lunch					
13.00	Introduction to Wide-Angle Neutron & X-Ray Scattering (Zin Tun)	X-Ray Techniques (Bauke W. Dijkstra)	Methods for Solving Phase Problem (Bauke W. Dijkstra)	Phase Improvement (Bauke W. Dijkstra)	Model Building and Crystallographic Refinement (Bauke W. Dijkstra)	
14.00	Introduction to Small-Angle Scattering (Vinod K. Aswal)				Analysis solved structure (Bauke W. Dijkstra)	
15.00	Afternoon tea					
15.30	Applications X-Ray Scattering in Biology (Bauke W. Dijkstra)	Crystallization of Lysozyme (Zeily Nurachman)	Rigid Body Refinement (Zeily Nurachman)	Phase Improvement and Generating Electron Density Map (Zeily Nurachman)	Conclusions	
16.30					Closing	

▨ General lectures/activities
☐ Specific lectures/activities

Symposium Program

29 - 31 July 2007
Aula Barat (West Hall) ITB
Ganesa No. 10, Bandung Institute of Technology
Bandung 40132
INDONESIA

The symposium programs are divided into several activities:
- Plenary lectures.
- Poster presentations.
  Classified into: Alloys, Biology, Ceramics, Instruments & Methods, Modeling & Simulations, Materials Chemistry, Polymers & Colloids groups.
- Exhibitions.

The conference dinner was held at the View Restaurant Dago Pakar Resort located on the hill at the northern side of Bandung neighborhood.

PLENARY LECTURES

Application of X-ray and Neutron Scattering Techniques in Materials Research: Lithium Batteries and Electronic Ceramics

A. R. West

University of Sheffield, Department of Engineering Materials,
Sir Robert Hadfield Building, Mappin Street, Sheffield, S1 3JD, United Kingdom

Abstract. X-ray and neutron powder diffraction provide complementary information on the structures of inorganic complex oxides, primarily because of the different dependence of atomic scattering power, or scattering length, on atomic number. Neutron diffraction is particularly useful for characterising novel lithium transition metal oxides which have applications in prototype advanced lithium battery systems. Thus, it is possible to establish conduction pathways for mobile Li^+ ions and to distinguish between transition metal ions in ordered spinel structures such as $Li_2NiMn_3O_8$, which has a charge-discharge potential of 4.7V. An additional feature of such materials is oxygen non-stoichiometry in which their oxygen content depends on sample preparation conditions and temperature. This oxygen non-stoichiometry may be analysed by thermogravimetry and the transition metal oxidation states determined, in the solid state, by the X-ray absorption technique, XANES. Structural changes as a function of temperature may be followed by high temperature diffraction methods and as a function of lithium content during charging/discharging of lithium batteries by in situ synchrotron XRD. A range of examples of the applications of these techniques will be presented.

References
1. Inorganic functional materials: Optimization of properties by structural and compositional control, A R West, The Chemical Record, 6, 206-216 (2006)
2. Crystallography of Ni-doped $Zn_7Sb_2O_{12}$ and phase equilibria in the system $ZnO-Sb_2O_5-NiO$, R Harrington, G C Miles and A R West, J. Eur. Ceram. Soc. 26, 2307-2311 (2006)
3. Structural characterisation of $REBaCo_2O_{6-\delta}$ phases (RE-Pr, Nd, Sm, Eu, Gd, Tb, Dy, Ho), P S Anderson, C A Kirk, J Knudsen, I M Reaney and A R West, Solid State Sciences, 7, 1149-1156 (2003)
4. Temperature-dependent crystal structure of ferroelectric $Ba_2LaTi_2Nb_3O_{15}$, G C Miles, M C Stennett, I M Reaney and A R West, J. Mater. Chem. 15, 798-802 (2005)
5. Oxygen content and electrochemical activity of $LiCoMnO_{4-\delta}$, D Pasero, S de Souza, N Reeves and A R West, J. Mat. Chem. 15, 4435-4440 (2005)

Keywords: Powder diffraction, synchrotron, lithium battery

CP989, *Neutron and X-ray Scattering in Materials Science and Biology, International Conference on Neutron and X-ray Scattering 2007,* edited by A. Ikram, A. Purwanto, Sutiarso, A. Zulfia, S. Hendrana, and Z. Nurachman
© 2008 American Institute of Physics 978-0-7354-0508-0/08/$23.00

Early Years of Neutron Scattering and Its Manpower Development in Indonesia

Marsongkohadi[1,2]

[1] Former Professor of Physics, ITB, Bandung, Indonesia
[2] Director of Materials Science Research Centre, BATAN, Kawasan Puspiptek Serpong, Indonesia, 1987-1996

Abstract. In this paper I shall give a short history of the development of neutron scattering at the Research Centre for Nuclear Techniques (PPTN), in Bandung, and the early development of a more advanced facilities at the Neutron Scattering Laboratory (NSL BATAN), Centre of Technology for Nuclear Industrial Materials, in Serpong. The first research reactor in Indonesia was the TRIGA MARK II in Bandung, which became operational in 1965, with a power of 250 KW, upgraded to 1 MW in 1971, and to 2 MW in 2000. The neutron scattering activities was started in 1967, with the design and construction of the first powder diffractometer, and put in operation in 1970. It was followed by the second instrument, the filter detector spectrometer built in 1975 in collaboration with the Bhabha Atomic Research Centre (BARC), India. A powder diffractometer for magnetic studies was built in 1980, and finally, a modification of the filter detector spectrometer to measure textures was made in 1986. A brief description of the design and construction of the instruments, and a highlight of some research topics will be presented. Early developments of neutron scattering activities at the 30 MW, RSG-GAS reactor in Serpong in choosing suitable research program, which will be mainly centred around materials testing/characterization, and materials/condensed matter researches has been agreed. Instrument planning and layout which were appropriate to carry out the program had been decided. Manpower development for the neutron scattering laboratory is a severe problem. The efforts to overcome this problem has been solved. International Cooperation through workshops and on the job trainings also support the supply of qualified manpower.

Keywords: Neutron scattering, powder diffractometer, magnetic studies
PACS: 61.12.Ex

INTRODUCTION

The first research reactor in Indonesia was a light water cooled, 250 KW TRIGA MARK II reactor which became operational in 1965, at the Research Centre for Nuclear Techniques (formerly the Bandung Reactor Centre) in Bandung with the thermal flux of ~ 10^{12} n/cm^2 sec. This reactor consists of 4 beam ports (1 tangential, 1 piercing, and 2 radials) was upgraded to 1 MW in 1971 (thermal flux 6 x 10^{12} n/cm^2 sec) and to 2 MW in 2000 (thermal flux 1.6 x 10^{13} n/cm^2 sec).

The neutron scattering group in Bandung started in 1966, when a training course on Neutron Diffraction organized by IAEA, namely, the India–Philippines-Agency (IPA) agreement was held in Manila in 1964.

As a first participant of this regional cooperation, I was in charge to build the first powder diffractometer in Bandung.

Design and Construction of Neutron Scattering Instrument in Bandung

FIGURE 1. Research Centre for Nuclear Techniques in Bandung operates the first reactor in Indonesia, TRIGA MARK II.

CP989, *Neutron and X-ray Scattering in Materials Science and Biology, International Conference on Neutron and X-ray Scattering 2007,* edited by A. Ikram, A. Purwanto, Sutiarso, A. Zulfia, S. Hendrana, and Z. Nurachman
© 2008 American Institute of Physics 978-0-7354-0508-0/08/$23.00

The design and construction of the first powder diffractometer was started in 1967 and became operational in 1970, Fig. 2. Except the spectrometer and the counting system, the other parts were locally made. The parts of the spectrometer, for example the gear system were made with high precision by Picker Nuclear Instrument, Co, USA.

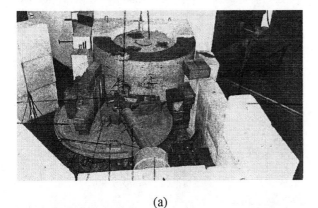

(a)

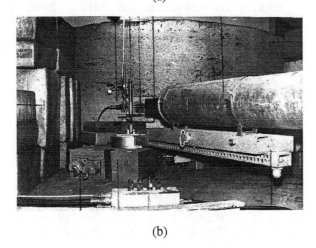

(b)

FIGURE 2. The first neutron diffractometer in Indonesia was built in 1967 (a) Top view (b) Side view.

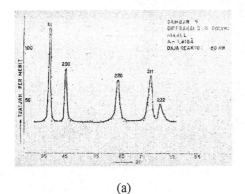

(a)

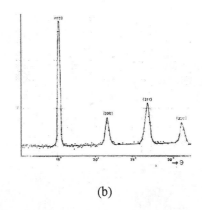

(b)

FIGURE 3. The neutron data were collected from (a) Ni polycrystal sample with λ = 1.318 Å at 60 kW of reactor power (b) FeCr polycrystal sample with λ = 1.063 Å at 1 MW of reactor power

The diffractometer was used mostly for educational purposes, by postgraduate students for their thesis, Fig. 3. It was dismantled in 1980, to be replaced by a new carefully designed diffractometer for magnetic studies.

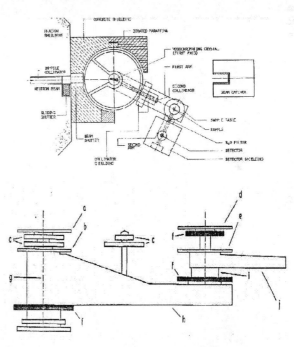

FIGURE 4. Schematic layout of design and construction of a filter-detector spectrometer (FDS) installed at the piercing beam port of TRIGA MARK II reactor.

In collaboration with Bhabha Atomic Research Centre (BARC) a filter detector spectrometer (FDS) was installed in 1975, Fig. 4. The spectrometer parts were made at BARC and the other parts were made locally. Dr. C. L. Thaper from BARC helped to setup the FDS and performed some experiments

An inelastic scattering experiment can also be performed by a FDS, instead of the usual Triple-Axis Spectrometer (TAS). By a FDS is meant an instrument in which the scattered neutrons pass through a beryllium oxide filter as fixed energy selection device (constant window analyzer), before being detected.

The neutrons scattered at 90° by the sample were viewed with a constant window analyzer comprised of a BeO filter and a BF_3 detector placed vertically.

A powder diffractometer designed by BARC for magnetic studies, necessitating the use of magnet and cryostat, was constructed at the local workshop in 1979 and put in operation in late 1980. A special table was constructed which provided independent support of the heavy magnet and cryostat. The magnetic field should be maintained parallel to the scattering vector, while the magnet rotates in unison with the rotation of the arm. A pulley and belt system was installed to make the 2:1 coupling between the detector and the sample drives. For step scanning operation a cam and micro switch was used to stop the drive at certain intervals.

The magnet provides magnetic field of 1.0 T at a gap of 3.4 cm with magnetizing current of 25 Amp, while the cryostat could maintain a temperature of 80 K continuously during 240 hours by refilling liquid nitrogen every 24 hours.

The magnet as well as the cryostat were specially designed for neutron diffraction and were locally made.

Due to technical reasons, the TRIGA MARK II reactor was operated below full power, viz. at 700 KW, Consequently, the FDS could not efficiently utilized for inelastic scattering experiment.

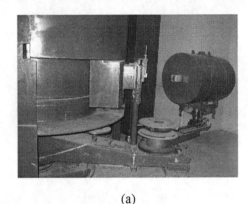

(a)

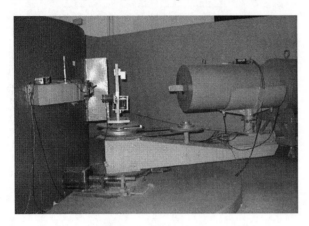

(b)

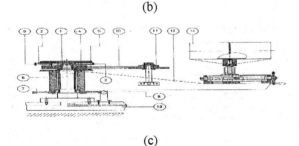

(c)

(b)

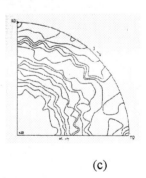

(c)

FIGURE 5. The photograph of a modified powder diffractometer (a) for magnetic studies and its schematic layout of design and construction (b) and (c) installed at the tangential beam port of TRIGA MARK II reactor. The soller slit was redesigned to obtain maximal resolution and intensity.

FIGURE 6. The photograph of a modified filter-detector spectrometer (FDS) for texture diffractometer (a), a texture goniometer (b) was made by Institute of Nuclear Physics and Techniques and the pole figure (c). The soller collimator was replaced by coarse collimation to obtain good luminosity.

To make use of the existing spectrometer, it has been decided to modify the FDS to be used as a texture diffractometer (TD), Fig. 6. This has been carried out with the help of Prof. Szpunar, IAEA expert from the Institute of Nuclear Physics and Techniques, Cracow, Poland.

First of all the FDS can be utilized as a diffractometer by rotating the arm around 90° (the original position). To make maximal luminosity, neutrons with wavelength around the peak of the reactor spectrum ($\lambda = 1.23$ Å) and coarse collimation were used.

Texture is determined by setting the TD for Bragg reflection and rotating the sample in the beam to obtain pole figures. To do this the sample is mounted on a texture goniometer, which for the present TD was made at the Institute of Nuclear Physics and techniques, Cracow, Poland.

Automation of the data collection has been made, The TD can be operated in three modes viz., manual, off-line and on-line. In the on-line mode, the TD is connected by a minicomputer PDP 11/34 A and the Canberra Datanim Counting System. Communication between these systems are controlled by a telecomputer interface through four subroutines.

Highlights of Some Research Topics

Since the operation of the first diffractometer, in 1970 several minerals and alloys have been investigated. An interesting work on MnO_2 mineral from Karangnunggal, West Java is presented bellow.

MnO_2 is used as a depolarizer in dry-cell manufacturing. There are seven phases of MnO_2 mineral, but only the γ-MnO_2 is suitable for battery making. By neutron diffraction it was found that the mineral from West Java were mostly

β-MnO_2 or pyrolusite. γ-MnO_2 is considered as a mixture of β-MnO_2 and R-MnO_2 (ramsdellite) to form a random layer structure. The degree of the mixture is characterized by the pyrolucite concentration, p. A sample of MnO_2 mineral from West Java was analyzed by the method introduced by de Wolff in 1959 for X-ray diffraction.

Based on this work a method to analyze MnO_2 has been established and transformation of the β-phase ore into γ-phase ore has been evolved.

Since 1975, when the FDS was installed some interesting experiments have been performed. Lattice dynamics of CaO is an example how phonon dispersion curve can be measured by a FDS.

All the optical branches and the high frequency part of the acoustic branches were measured on a FDS. The relatively low frequency acoustic branches however were measured on a TAS.

Another interesting experiment shows that inelastic neutron scattering provides a useful compliment to infrared, Raman and NMR techniques in determining low-energy molecular motions (librational motions).

Within the experimental resolution, it was possible to observe the librational motions of the water molecules in hydrates, $Ba(ClO_3)_2.H_2O$ and $BaCl_2.2H_2O$

The three kinds of librational motions are twisting vibration (motion about the B-axis), rocking vibration (motion about the C-axis) and waving vibration (motion about the A-axis). There was a good agreement between the librational frequencies observed in the neutron and infrared measurements.

However, due to the random orientation of the molecules with respect to the scattering vector, it has not been possible to observe distinctly the three kinds of librational motion. This can be overcome, however by using a single crystal sample oriented parallel to the scattering vector.

Since the operation of the magnetic powder diffractometer in 1980, research on magnetic materials such as ferrites, oxides and nitrides have been accomplished. One of an interesting result is: The magnetic structures and magnetic from factors of transition metal nitrides, Fe_4N and Mn_4N, is presented below,

The fact that the corner and face-centre metal atoms in both nitrides have significantly different environments leads to a difference in their magnetic moments. It should be immediately follow that the atoms must display different magnetic form factors. Our measurement clearly showed this.

The form factor of the face centre iron atom is significantly expanded as compared to that of the corner iron atom, implying a contracted moment density for the face centre atom. Furthermore, the form factor of the face centre manganese atom showed to be sharp. This suggests that the moment density of the face centre Mn atom is more spread out that of the corner atom.

The experimentally determined magnetic structures of Fe_4N and Mn_4N have been explained qualitatively from the shapes of the magnetic form factors, on the basis of Jardin and Labbe models.

The textures of some metals and alloys have been analyzed since the TD was available for operation. An interesting case was the zircaloy-4 because it was usually applied as a nuclear fuel cladding material.

Some specific properties of the alloy are suitable for cladding material, such as, low thermal neutron absorption, corrosion resistance and good strength and ductility some pole figures have been measured to obtain the crystallite orientation by means of a qualitative method. For the present sample the ideal orientation direction is (0001)[10-10].

A better and precise procedure is to use quantitative method by deriving the crystallite orientation distribution function (ODF), which can be calculated on the basis of pole figures. For materials with a cubic structure, the ODF has been calculated successfully, but since Zircalloy-4 has hexagonal structure, the calculation became complicated and was not successful.

Early Years of Neutron Scattering in Serpong

A small reactor such as the TRIGA MARK II has a limited range of research and development activities. Hence, the National Atomic Energy Agency of Indonesia (BATAN) built a bigger one, the 30 MW, RSG – GAS reactor in Serpong, and will considerably enhance the scope of neutron scattering in Indonesia.

In 1987, some research staffs, technicians, electronics, and myself were transferred to Serpong, where I was in charge of the newly formed "Materials Science Research Centre", MSRC (now Centre of Technology for Nuclear Industrial Materials). Physics graduates from the Bandung Institute of Technology (ITB), University of Indonesia (UI), and Gadjahmada University (UGM) were eager to joint the MSRC.

The research program and the neutron scattering facilities planning have been made and agreed by Prof. Schmatz of Karlsruhe, Dr. K. Werner and Dr. A. Djaloeis, both from Juelich and myself.

The procurement of the instruments depends on the scope of activities, which will be mainly centered around materials testing/characterization and applied materials/condensed matter researches.

Testing and characterization of materials are performed by a diffractometer for residual stress measurement (RSM), four-circle and texture diffractometers (FCD/TD), and high-resolution powder diffractometer (HRPD). Applied materials/ condensed matter researches are carried out by a small angle neutron spectrometer (SANS), high-resolution small angle neutron spectrometer (HRSANS), Triple Axis spectrometer (TAS) and high-resolution powder diffractometer (HRPD).

The layout and detailed specification will not be discussed here, it will be presented in the inviting papers by Dr. A. Ikram and Dr. A Purwanto in this conference.

The detailed specifications were made by Prof. Schmatz group and followed by an International tender. In total there were six diffractometers/ spectrometers, and a neutron radiography facility and two neutron guides to be constructed within four years.

Manpower Development

When I setup the neutron scattering group at PPTN, Bandung in 1966, only two staffs and four graduate students were available to construct the first diffractometer and, when the diffractometer become operational, a number of graduate students carried out their thesis research using this instrument. Consequently, they became interested in neutron scattering and joint the group as soon as they graduated. The group now having nine members was expanding to form a neutron laboratory.

The Condition in The MSRC, Serpong however was quite different from PPTN in Bandung. To get qualified manpower to handle the advanced instruments installed in the neutron scattering laboratory in Serpong, was the most crucial problem. This problem could be solved as follows:

According to the contract, the company which built the instrument was obliged to send abroad at least one research staff for every instruments they had made, to be trained in the operation and utilization of the instrument, furthermore, the BATAN-IAEA and BATAN-JAERI cooperations had granted fellowships to research staffs to be trained in neutron scattering abroad.

Starting in 1987 about thirty young research staffs and technicians were sent successively to institutions/ universities in Japan and other countries. After six or twelve months, most of the staffs and technicians returned to MSRC, but some eager and dynamic staffs were able to get scholarships to study towards PhD degree. The MSRC has at present fifteen PhD's, seven of them work in the neutron scattering laboratory and the rest are using neutron scattering technique for their research in materials science. The neutron scattering laboratory is expecting four PhD's more, who will be graduated in the near future.

International cooperation through workshops and on the job training also support the supply of qualified manpower BATAN has cooperation with IAEA, JAERI and some informal cooperation. BATAN-IAEA cooperation, commence in the 1960's when the neutron scattering activities began at PPTN in Bandung. The cooperation comprised of sending expert and equipments (technical assistance) to Bandung and granting fellowships to the staffs, to participate in the training course on neutron scattering, abroad. Two experts had come to Bandung to help setup and calibrated the spectrometers and doing some experiments. The implementation of the cooperation to Serpong consisted of sending equipments for SANS and expert to discuss and evaluate the research program and instruments planning, evaluate the performances of TAS and SANS and finally tested the initial data analysis for SANS. Some young research staffs were given opportunities to participate training course in neutron scattering abroad.

BATAN-JAERI cooperation, commenced in 1988 covered, residual stress measurement, neutron diffraction of industrial materials, basic research using neutron scattering and training on measurement in neutron scattering.

Under the BATAN-JAERI bilateral agreement the neutron powder diffractometer which was the first instrument installed at Serpong in 1987, was modified in 1995, for residual stress measurement in engineering components.

Since 1992 to 1995, four Workshops on the Utilization of Research Reactor in Neutron Scattering, sponsored by BATAN, STA (Science Technology Agency of Japan) and JAERI was held in Jakarta. Eight countries from the Asia-Pacific region, i.e. Australia, China, Indonesia, Japan, Korea, Malaysia and Thailand, participated in the workshop, which was attended by technical specialists.

High T_c superconductors are the worldwide interest and neutron scattering is an eminently suitable technique for examining the crystal structure. The new facilities at Serpong include HRPD which should be ideal for studies of this kind.

An On-the-Job Training (OJT) was held prior to the workshop which the aim to train participating scientists on the use of HRPD and in the analysis method (RIETAN). A three years program of work, starting from calibration/improvement of the HRPD, moving through studies of the simpler to more complicated high T_c superconductors, superionic conductors and magnetic materials.

To implement the three-years program, collaboration projects were established, between China-Indonesia on "Magnetic structure of RE_2Fe_{17} based intermetallic compounds", between Philippines-Indonesia on "Structure and transport properties of beta-alumina type superionic conductor" and between Thailand-Indonesia on "Li-doped Bismutth (2212) superconductor The research were so successful that seven papers have been published in international journals.

The new neutron scattering facilities in Serpong enhance international cooperation, which induce qualified manpower.

Opportunities for Materials Science and Biological Research at the OPAL Research Reactor

S. J. Kennedy

The Bragg Institute, Australian Nuclear Science and Technology Organisation,
Menai NSW 2234, Australia

Abstract. Neutron scattering techniques have evolved over more than ½ century into a powerful set of tools for determination of atomic and molecular structures. Modern facilities offer the possibility to determine complex structures over length scales from ~0.1 nm to ~500 nm. They can also provide information on atomic and molecular dynamics, on magnetic interactions and on the location and behaviour of hydrogen in a variety of materials. The OPAL Research Reactor is a 20 megawatt pool type reactor using low enriched uranium fuel, and cooled by water. OPAL is a multipurpose neutron factory with modern facilities for neutron beam research, radioisotope production and irradiation services. The neutron beam facility has been designed to compete with the best beam facilities in the world. After six years in construction, the reactor and neutron beam facilities are now being commissioned, and we will commence scientific experiments later this year. The presentation will include an outline of the strengths of neutron scattering and a description of the OPAL research reactor, with particular emphasis on it's scientific infrastructure. It will also provide an overview of the opportunities for research in materials science and biology that will be possible at OPAL, and mechanisms for accessing the facilities. The discussion will emphasize how researchers from around the world can utilize these exciting new facilities.

Keywords: Neutron beam facilities, research reactor, atomic and molecular dynamics

CP989, *Neutron and X-ray Scattering in Materials Science and Biology, International Conference on Neutron and X-ray Scattering 2007,* edited by A. Ikram, A. Purwanto, Sutiarso, A. Zulfia, S. Hendrana, and Z. Nurachman

J-PARC and Prospective Neutron Science

M. Arai

J-PARC Centre, Japan Atomic Energy Agency,
Tokai, Ibaraki 319-1195, Japan

Abstract. J-PARC is interdisciplinary facility with high power proton accelerator complex to be completed by 2008 after 7 years construction. Materials-Life Science Facility (MLF) will be very intensive pulsed neutron and muon facility at 1MW of the accelerated proton power. The neutron peak flux will be as high as several hundred times of existing high flux reactors. It is highly expected that new science will be opened up by using MLF. In the presentation I will explain the present status of J-PARC, strategy of user programme and prospective neutron science to be performed with it.

Keywords: Proton accelerator, pulsed neutron, muon.

CP989, *Neutron and X-ray Scattering in Materials Science and Biology, International Conference on Neutron and X-ray Scattering 2007,* edited by A. Ikram, A. Purwanto, Sutiarso, A. Zulfia, S. Hendrana, and Z. Nurachman
© 2008 American Institute of Physics 978-0-7354-0508-0/08/$23.00

Current Status and Future Works of Neutron Scattering Laboratory at BATAN in Serpong

A. Ikram

Center of Technology for Nuclear Industrial Materials, National Nuclear Energy Agency of Indonesia (BATAN)
Kawasan Puspiptek Serpong, Tangerang 15314, Indonesia

Abstract. Current status of neutron beam instruments using neutrons produced by the Multi Purpose Research Reactor – 30MWth (MPR 30, RSG GA Siwabessy) located in Serpong is presented. Description of the reactor as the neutron source is mentioned briefly. There are six neutron beam tubes coming from the beryllium reflector surrounding half of the reactor core providing neutrons in the experimental hall of the reactor (XHR). Four of them are dedicated to R&D in materials science using neutron scattering techniques. Neutron Radiography Facility (NRF), Triple Axis Spectrometer (TAS) and Residual Stress Measurement (RSM) Diffractometer are installed respectively at beam tubes S2, S4 and S6. The largest neutron beam tube (S5) is exploited to accommodate two neutron guide tubes that transfer the neutrons to a neighbouring building called neutron guide hall (NGH). There are three other neutron beam instruments installed in this building, namely Small Angle Neutron Scattering (SANS) Spectrometer (SMARTer), High Resolution SANS (HRSANS) Spectrometer and High Resolution Powder Diffractometer (HRPD). In the XHR, a Four Circle and Texture Diffractometer (FCD/TD) is attached to one of the neutron guide tubes. These seven instruments were installed to utilize the neutrons for materials science research, and recently the RSM diffractometer has shown its capabilities in identifying different amount of stress left due to different treatments of welding in fuel cladding, while the SANS spectrometer is now gaining capabilities in identifying different sizes and shapes of macromolecules in polymers as well as investigations of magnetic samples. In the mean time, non-destructive tests using the NRF is gathering more confidence from some latest real time measurements eventhough there are still some shortcomings in the components and their alignments. Future works including improvement of each facility and its components, even replacement of some parts are necessary and have to be carried out carefully. A plan for developing a neutron reflectometer at one of the neutron guide in the Neutron Guide Hall is also part of the near future activities

Keywords: Research reactor, neutron scattering, diffractometer, spectrometer.
PACS: 61.12.Ex

INTRODUCTION

As part of its program, the National Nuclear Energy Agency of Indonesia (BATAN), as the authority for implementing Indonesia national nuclear program built a Multi-Purpose Reactor in Serpong, 30 km south west of Jakarta, the capital city. This research reactor reached its criticality in 1987. In utilizing the neutrons produced by the reactor, some neutron beam instruments were installed and commissioned in August 1992 constructing a neutron scattering laboratory in Serpong. These instruments are four-circle/texture diffractometer, triple axis spectrometer, neutron radiography facility, small angle neutron scattering (SANS) spectrometer, high resolution SANS spectrometer and high resolution powder diffractometer (HRPD). The first three instruments together with a neutron powder diffractometer for residual stress measurement were installed in the reactor experimental hall (XHR) while the last three are located in the neutron guide hall (NGH). Those two halls are connected by a tunnel accommodating two neutron guides transferring neutrons from the reactor to the instruments in the neutron guide hall. Lay out of the neutron beam instruments in the reactor hall (XHR) and in the neutron guide hall (NGH) is presented in Fig. 1.

After fifteen years of its existence, full of not only obstacles and challenges, but also rigorous development, the current status and latest activities of the instruments, as well as some near future goals are presented in this paper hoping to gather some more responses and supports for further enhancements.

CP989, Neutron and X-ray Scattering in Materials Science and Biology, International Conference on Neutron and X-ray Scattering 2007, edited by A. Ikram, A. Purwanto, Sutiarso, A. Zulfia, S. Hendrana, and Z. Nurachman
© 2008 American Institute of Physics 978-0-7354-0508-0/08/$23.00

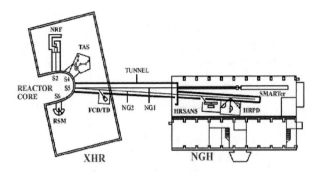

FIGURE 1. Lay out of the neutron beam instruments in the reactor experimental hall (XHR) and the neutron guide hall (NGH) [1].

THE NEUTRON SOURCE

The reactor as the neutron source is called RSG-GAS (Multipurpose Reactor - GA Siwabessy) in honor of the late Prof. GA Siwabessy for his contribution in promoting and developing nuclear techniques in Indonesia. The RSG-GAS has achieved its criticality in 1987 and reached its full power level of 30 MW thermal in 1992. However continuous regular operation was started not earlier than 1995. The delay was due to a long shutdown period owing to the installations of in-pile engineering loops, radioisotope production equipments, neutron scattering instruments, and also due to other technical problems. In 1996, the reactor was operated quite regularly for two cycles per month and nine days per cycle. In order to improve the radioisotope production, in 1997 the reactor operation was made weekly with 5 operation days with reactor power of 25MW thermal. This was not so convenience for the neutron scattering activities. From July 1998 and lasted until July 1999 the reactor operation mode has been compromised and changed to 12 days operation and followed by 9 days off. In return the reactor was only in operation at 15MW thermal to maintain the length of the operation with the available fuel elements. Due to some more problems in the financial sector, in the year 2000 up to 2003, the reactor had been in operation for 12 days followed by 16 days off. Finally, since 2004, the reactor has been run for 11 days followed by 10 days off, and in every three months 17 days were used for maintenance of the reactor. This means that we have neutron beams for about 15 days monthly which is about 150 days yearly. So far the neutron beams have been used for most of the scheduled operation days. These continuous and regular monthly scheduled operation days has given advantages in planning the experiments in advance and improve the reliability of the neutron instruments.

The reactor is a light water open pool reactor with designed maximum thermal power of 30 MW. This reactor is considered to be the first high-power

research reactor in the world, designed and constructed for the use of low enriched (less than 20%) uranium MTR-type fuel. It was designed to produce an average thermal neutron flux of 2.5×10^{14} n cm^{-2} sec^{-1} at the central irradiation position in its core. An L-shaped beryllium block reflector surrounds one half of the core. Six beam tubes - two tangentials and four radials - are available for neutron beam experiments. When they are not in use these beam tubes are flooded with water and closed by the insertion of concrete plugs. One of the beam tubes has a larger diameter ($\varnothing = 27$ cm) than the others ($\varnothing = 24$ cm), to accommodate two thermal neutron guides supplying neutron beam for the spectrometers in the external neutron guide hall (NGH). In addition, the reactor provides facilities for nuclear engineering experiments, neutron activation analysis and radioisotopes production. Those facilities include five in-core irradiation positions for materials/fuel testing and radioisotope production, seven irradiation holes in reflector area, five rabbit systems, one power ramp test facility and one wet neutron radiography. Those in-core and out-of-core irradiation positions are arranged to support commercial, research and development activities.

NEUTRON BEAM INSTRUMENTS

Various neutron beam instruments were installed at the Neutron Scattering Laboratory, NSLBATAN, Serpong – Indonesia. Those instruments are:
1) Powder Diffractometer (PD), DN-1
2) High-Resolution Powder Diffractometer (HRPD), DN-2
3) Four-Circle Diffractometer/Texture Diffractometer for single crystal structural studies and texture analysis (FCD/TD), DN-3
4) Neutron Radiography Facility (NRF) with real time measurement capability, RN-1
5) Triple Axis Spectrometer (TAS), SN-1
6) Small Angle Neutron Scattering (SANS) spectrometer, SMARTer, SN-2
7) High-Resolution Small Angle Neutron Scattering (HRSANS) spectrometer, SN-3

In 1995, the powder diffractometer was modified to accommodate residual stress measurements (DN-1M) and has been used for the measurements since. As seen in Fig. 1 the powder diffractometer, the triple axis spectrometer and the neutron radiography facility are installed at the beam ports S6 (tangential), S4 (radial), and S2 (tangential), respectively in the experimental hall of the reactor (XHR).

There are two neutron guides installed starting from S5 radial beam port, penetrating the reactor confinement building wall and going through a 35 m tunnel into the external neutron guide hall (NGH). These neutron guides comprise of four optically flat Borkron glasses glued together forming a 33 x 90 mm^2 cross section. The inner surfaces are coated with thin Ni^{58} deposit. The first neutron guide (NG1) is dedicated to the small angle neutron scattering spectrometer, which is installed at the end of the guide. The second neutron guide (NG2) has five beam ports labeled as NG2-1, NG2-2, NG2-3, NG2-4 and NG2-5 (end of the guide). The first three beam ports have already been used for the four-circle diffractometer, the high-resolution small angle neutron scattering spectrometer and the high-resolution powder diffractometer respectively. The last two ports are still available for any kind of neutron beam instruments. In order to shield the radiation coming from them, the neutron guides are confined inside a case mate having concrete walls along their whole length, except those parts inside the tunnel. The characteristics of the neutron guides are shown in Table 1 below.

TABLE 1. Characteristics of the neutron guides.

Characteristics	NG1, thermal	NG2, thermal
1. Coating	Ni^{58}	Ni^{58}
2. Cross section	33 x 90 mm^2	33 x 90 mm^2
3. Characteristic wavelength	2 Å	1.5 Å
4. Radius of curvature	3926 m	6979 m
5. Total length	50 m	78 m
6. Beam ports	end of the guide	NG2-1, NG2-2, NG2-3, NG2-4, end of the guide

Diffractometer for Residual Stress Measurement (RSM), DN-1M

This neutron diffractometer is situated in the reactor experimental hall (XHR) as the first instrument installed (1987). It is a standard two-axis type diffractometer with a Si (311) monochromator system that is bent horizontally and focused vertically to enhance the neutron flux at the sample position. The take-off angle of the monochromator can be varied from 0 to 90 degrees allowing some different neutron wavelengths to be used. Mylar films coated with gadolinium are used as collimator with angular divergences of 40' before the monochromator and 20' after the sample. Since 1995 the diffractometer has been used mostly for residual stress measurement after some modifications.

The modifications were made by installing a goniometer having three orthogonal translations as well as a turn table to rotate samples around the

incident beam with positioning accuracy of 0.025 mm. All sample movements and data collection are controlled by a personal computer. The size of both incident and scattered beams is defined by apertures having various sizes ranging from (0.1 x 10) mm^2 to 10 mm (in diameter). The volume sampled by the diffractometer is then defined by the intersection of the incident and scattered beams in the scattering plane. Residual stress in a standard specimen of a shrink-fit ring plug, similar to the VAMAS (Versailles project on Advanced Materials and Standards) standard sample has been measured to test the performance of the machine. The result shows that this machine has a comparable capability to RESA in JRR-3M at JAEA (Japan Atomic Energy Agency). This capability has provided high confidence in performing the measurements of residual stress in tungsten-copper composite materials and shot-peened materials.

FIGURE 2. Picture of the neutron diffractometer for residual stress measurement [1].

Development of this diffractometer includes detector movement mechanism which allows the detector to be moved closer to the sample table and increase the neutron intensity significantly. A full circle Eulerian goniometer has also been installed on the sample table allowing texture characterization of the sample prior to its residual stress measurements. Figure 2 presents the diffractometer while an experiment was running.

Four Circle Diffractometer / Texture Diffractometer (FCD/TD), DN-2

This diffractometer is installed at the NG2-1 beam port sited in the XHR. The dual purpose instrument was designed for both single crystal structural studies and texture measurements, and was constructed robustly to allow heavy attachments. The allowable load on the sample table is 300 kg. A monochromator of Cu (220) or bent Si(311) with a fixed take off angle of 46° is mounted inside the neutron guide case mate to obtain monochromatic beam having a wavelength of 0.997 Å or 1.271 Å, respectively. A beam narrower situated just outside the case mate provides a maximum beam size of 40 x 40 mm² at the sample position. The detector is BF₃ end window counter with 100 mm⌀ and 236 mm effective length. The center of sample table is 1500 mm from the front-end of first collimator and 350 mm from the detector. There are collimators after the monochromator and before the detector with angular divergences of 20' and 30' respectively.

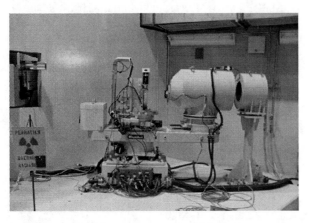

FIGURE 3. Picture of the four circle diffractometer / texture diffractometer [1].

This instrument is operated automatically by an IBM compatible computer using Windows Millennium Operating System. Visual Basic v6 is used for control system and data acquisition. This is part of our development in replacing the old control system. Ancillary equipments available are full and half circle Eulerian goniometers, a fiber specimen holder and a cryostat. Texture measurements of some samples have been performed using the new control system. Figure 3 shows the picture of the diffractometer. Future development will include fabrication of a funnel covering the neutron path from beam narrower to sample position.

High Resolution Powder Diffractometer (HRPD), DN-3

This instrument is a versatile diffractometer to study crystallographic and magnetic structure of powder samples. Since X-ray diffraction is insensitive to light atoms such as hydrogen and oxygen, neutron diffraction is indispensable for materials containing light atoms. Furthermore, neutron has inherent magnetic moment that enables in probing magnetic structure ranging from collinear commensurate to incommensurate configurations.

The diffractometer is installed at the neutron guide (NG2) in the Neutron Guide Hall (NGH) about 71 m away from the reactor core. A Ge(331) single crystal, pressed up to 70 kg cm⁻² during the heating at 850° C for one hour to increase the mosaic spread, is used as the monochromator. The monochromator drum has three beam exits corresponding to three different take off angles, $2\theta_M = 41.5°$, 89° and 130°. At present the instrument is set for $2\theta_M = 89°$. Although the Ge (331) monochromator system does not have focusing property, the performance however, is fairly good giving single wavelength of $\lambda = 1.822$ Å, with high intensity and good resolution.

FIGURE 4. Picture of the high resolution powder diffractometer.

There are three collimation systems : before and after monochromator, and before each detector. The collimation system consists of the guide tube (collimator 1) and gadolinium coated thin mylar films having angular divergences of 10' (collimator 2, before sample) and 6' (collimator 3, before detector). The detector system consists of 32 He3 detectors covering 160 degrees of scattering angle. This detector bank moves on air cushions to cover the 5 degree space in 100 steps. A picture of the High Resolution Powder diffractometer (HRPD) is shown in Fig. 4. Ancillary equipments for this instrument are a furnace with maximum temperatures of 900 K and a cryostat with closed cycle refrigerator down to 10 K.

Neutron Radiography Facility (NRF), RN-1

The thermal neutron radiography facility shown in Fig. 5 is placed at the S2 tangential beam port in the XHR and is used for non-destructive inspection of inactive bulk material using a direct method as well as real time examination. It comprises of an inner collimator inside the biological shielding of the reactor, an outer collimator just outside the biological shielding and a main beam shutter. The collimators are made of aluminum and produced L/D ratio of 83. The facility is equipped with a sample mounting mechanism and photographic recording as well as neutron television. An auxiliary shutter is placed just after the collimator exit, before the sample position. It is opened and closed by a pneumatic mechanism before and after each individual exposure. Gadolinium converter and X-ray film in an aluminum vacuum cassette is used for measurement with direct exposure technique while the real time method uses scintillator screen and CCD camera.

Neutron radiography as a non-destructive imaging technique is capable of visualizing the internal characteristics of a sample. Transmission of neutrons through a medium is determined by neutron cross-section for nuclei in the medium. Differential attenuation of neutrons through a medium can be measured, mapped and then visualized. The resulting image may be used to analyze the internal characteristics of the sample. Since neutron interaction probabilities depend on structure and stability of the nucleus and significantly different with X-ray's, neutron radiography has the ability to image light elements around the heavy ones in materials that do not attenuate X-ray.

FIGURE 5. Picture of the neutron radiography facility [1].

Neutron beam at sample position is almost uniform in the center covering a circle of 15 cm in diameter with flux above 10^7 n/cm^2.s. Outside this area a 7×10^6 n/cm^2.s of neutron flux can be maintained up to 13 cm from the center beam. Real time images from a neutron computed tomography system has been set up comprising of a neutron television system (NTV), step motor, motor driver and image frame grabber to capture the projected images of an object under the examination. Real time images from various kind of

dynamic objects have also been obtained and the results show a promising future for the facility.

Triple Axis Spectrometer (TAS), SN-1

This kind of neutron spectrometer yields valuable information on collective phenomena in condensed matter which include dispersion of phonons and magnons, translational motion in liquids and solids as well as rotational motion in molecules. The spectrometer is a conventional triple axis spectrometer installed at the S4 beam port in the XHR. It is capable of providing either unpolarized beam using PG (002), Ge (111) or Cu (220) monochromator or a polarized beam using Heusler monochromator Cu$_2$MnAl (111), magnetic guide and Mezei spin flipper. The PG (002) and Cu (220) monochromators can be rapidly interchanged by remote control. The monochromator angle $2\theta_M$ can be varied from 15° to 75° giving possibility of incident energy ranging from 4.9 meV for PG (002) to $\approx$ 100 meV for Cu (220) at low angle. Either the incident or the scattered neutron energy can be fixed in the energy analysis. Figure 6 shows picture of the spectrometer. In-house development in the last three years has overcome most of the problems related to instrument control and data acquisition. Software development is still in progress to ease the experiment and data collection.

FIGURE 6. Picture of the triple axis spectrometer [1].

Small Angle Neutron Scattering (SANS) Spectrometer (SMARTer), SN-2

This neutron spectrometer is installed at the end of the 50 m long neutron guide (NG1), situated in the neutron guide hall (NGH), to benefit from low background environment. The instrument applies small angle scattering technique suitable for investigation of various phenomenon in a wide range of materials such as alloys, polymers, colloids, liquid, crystals and biological systems having particle sizes or density fluctuations in the range of 1 - 100 nanometers. The incident beam is monochromatized by a slot-type mechanical velocity selector having a minimum rotational speed of 700 rpm and a maximum one of 7000 rpm. The selector's tilting angle is fixed at 0°. For experimental purpose, by varying these rotational speed 3500 – 7000 rpm produces neutron wavelengths of 2 – 6 Å and a effective Q range of ($0.002 < Q < 0.6$) Å^{-1} can be achieved.

The collimator is placed in an 18 m long tube, comprises of four sections of movable guide tube, and one section of a fixed collimator (non-reflecting) tube. Collimation is made by adjustable apertures (pinholes) at discrete distances of 1.5 m, 4 m, 8 m, 13 m and 18 m from the sample position. The detector, which can be moved continuously from 1.3 m to 18 m in another 18 m tube, is a 128 x 128 He-3 two dimensional position sensitive detector (2D-PSD) made by RISØ, with various beam stopper size that can be moved vertically and horizontally. The whole system, excluding the sample position is evacuated to 10^{-3} torr. Figure 7 shows picture of the instrument in the NGH.

FIGURE 7. Picture of the small angle neutron scattering (SANS) spectrometer [1].

In the last two years, different kinds of samples have been measured to characterize the instrument performance. These include wavelength calibration as well as its resolution, collimator and pin-hole setting and even applying magnetic field at sample position [2]. Results show that the instrument is capable of performing world class experiment even though rooms for improvement are still plenty.

High Resolution SANS Spectrometer (HRSANS), SN-3

This instrument is a neutron version of the Bonse-Hart type diffractometer for small angle X-ray scattering. It extends the Q range covered by SANS spectrometer (SMARTer) to lower Q. This increases the sensitivity of inhomogeneity on length up to some microns. The instrument has a PG (004) monochromator mounted on the neutron guide (NG2), 63 m away from the reactor core. Two nearly perfect Si (311) single crystals are used as monochromator ($\Delta\lambda/\lambda = 0.15$ %) and analyzer ($\Delta\theta = 0.0001°$), and installed on goniometers. These goniometers which provide rotation, translation and tilting movements are mounted on an optical bench moving on air cushions and can be controlled by a computer. Three detectors are installed for alignment, monitoring and detecting the diffracted beam.

Sample with a maximum size of 30x30 mm^2 is placed between the two Si (311) crystals whereas the whole system is installed in a thermostatic room to obtain constant sample temperature within 0.5 °C. Picture of the spectrometer can be seen in Fig. 8.

FIGURE 8. Picture of the high resolution small angle neutron scattering spectrometer (HRSANS) [1].

FUTURE DEVELOPMENT PROGRAM

There are still many rooms for improvement at different level for each of the neutron beam instrument mentioned above. Some needs electronics and software enhancements, others are ready to provide services for world class and high level as well as frontier research. In the mean time, there are also still some available facilities that should be utilized using neutron beam instruments such as beam port S-3 in the XHR as well as neutron gap NG2-4 and end of the guide in the NGH. Having those instruments covering a wide range of scales in materials, a neutron reflectometer could be an instrument that will extend the capability of the laboratory in providing neutron techniques.

Neutron reflectometry (NRy) probes variation in the neutron scattering length density (SLD) normal to a flat surface at depths of up to several thousand angstroms with a resolution of a few angstrom. This can be regarded as depth profiling of thin films and interfaces. This neutron SLD is determined by composition and density of the film of interest. Beyond the critical angle (θ_c), below which total reflection occurs, some of neutrons are absorbed and transmitted by the sample and the remainder is reflected. Reflectivity is measured by dividing the number of neutrons reflected from the sample by the number of neutron incident upon its surface, typically as a function of the incident angle of the neutron beam [3].

A proposal for building and installing a neutron reflectometer in Serpong is mostly based on the arguments of improving the utilization of the RSG GAS to its most capacity. The instrument will be positioned at the beam port NG2-4 which is still unused until now and will complete the coverage of range of materials research in Serpong from angstrom to micron in the size of interest. Table 2 shows the proposed specification of the instrument.

TABLE 2. Characteristics of proposed Neutron Reflectometer in Serpong

Sample	Vertical
Beam Tube	N G 2 - 4
Q range	0.003 – 0.4 Å
Wavelength	$\lambda = 2.4$ Å
Monochromator	PG(002), $\beta = 0.4$ deg
Filter	PG
Flux at sample	10^5 n/cm2.s
Detector	He-3 (single)

SUMMARY

There are more than 150 days of neutron beam yearly in Serpong. The three neutron diffractometers are running well and available for measurements of residual stress, texture and diffraction pattern for materials science research. Even though being developed for its new control system, the SANS machine is also running well and ready for any kind of measurements. These instruments measure features on the scale of 1 – 1000 nanometer. Investigation of bulk materials can also be served by the neutron radiography facility. Crystal dynamics using TAS is under preparation since the machine is starting to work again after almost 10 years in silent. Suggestions and contributions are welcome for the revitalization of the HRSANS machine as well as for the new proposed neutron reflectometer.

ACKNOWLEDGEMENTS

The author would like to thank all members of the Neutron Scattering Lab. for their hard and also smart work so the state of all neutron scattering instruments can be lifted to this point and the brighter future of each instrument has been spotted. Financial support from the Government of Indonesia through all the projects and schemes is also acknowledged. Special thank is also due to Mr. Yatno and Ms. AD Puspitasari for their help in preparing this paper.

REFERENCES

1. http://www.batan.go.id/ptbin/bsn/index.html, 27-7- 2007
2. E. Giri Rachman Putra, A. Ikram, Bharoto, E. Santoso, submitted to J. Nucl. Related Tech. 2007; E. Giri Rachman Putra, A. Ikram, submitted to PRAMANA J. Phys. 2008.
3. A. Ikram, Proceeding of the 6th National Seminar on Neutron and X-Ray Scattering, Serpong 2005, pp. 9 - 15

Magnetic Excitations in Transition-metal Oxides Studied by Inelastic Neutron Scattering

M. Braden

Institute of Physics, University of Cologne, Germany

Abstract. Inelastic neutron scattering using a triple axis spectrometer is a very efficient tool to analyze magnetic excitations. We will discuss several recent experiments on transition-metal oxides where orbital degrees of freedom play an important role. Different kinds of experimental techniques including longitudinal and spherical polarization analysis were used in order to determine not only magnon frequencies but also polarization vectors. In layered ruthenates bands of different orbital character contribute to the magnetic excitations which are of both, ferromagnetic and antiferromagnetic, character. The orbital dependent magnetic excitations seem to play different roles in the superconducting pairing as well as in the metamagnetism . In manganates the analysis of the magnon dispersion in the charge and orbital ordered phase yields direct insight into the microscopic coupling of orbital and magnetic degrees of freedom and helps understanding, how the switching between metallic and insulating phases in manganates may occur. In multiferroic $TbMnO_3$ the combination of our polarized neutron scattering results with the infrared measurements identifies a soft collective excitation of hybridized magnon-phonon character.

Keywords: Inelastic neutron scattering, magnetic excitations.

CP989, *Neutron and X-ray Scattering in Materials Science and Biology, International Conference on Neutron and X-ray Scattering 2007*, edited by A. Ikram, A. Purwanto, Sutiarso, A. Zulfia, S. Hendrana, and Z. Nurachman

Pulsed Neutron Powder Diffraction for Materials Science

T. Kamiyama

Materials and Life Science Facility, J-PARC Center, High Energy Accelerator Research Organization, Tsukuba, Ibaraki 305-0801 JAPAN

Abstract. The accelerator-based neutron diffraction began in the end of 60's at Tohoku University which was succeeded by the four spallation neutron facilities with proton accelerators at the High Energy Accelerator Research Organization (Japan), Argonne National Laboratory and Los Alamos Laboratory (USA), and Rutherford Appleton Laboratory (UK). Since then, the next generation source has been pursued for 20 years, and 1MW-class spallation neutron sources will be appeared in about three years at the three parts of the world: Japan, UK and USA. The joint proton accelerator project (J-PARC), a collaborative project between KEK and JAEA, is one of them. The aim of the talk is to describe about J-PARC and the neutron diffractometers being installed at the materials and life science facility of J-PARC. The materials and life science facility of J-PARC has 23 neutron beam ports and will start delivering the first neutron beam of 25 Hz from 2008 May. Until now, more than 20 proposals have been reviewed by the review committee, and accepted proposal groups have started to get fund. Those proposals include five polycrystalline diffractometers: a super high resolution powder diffractometer (SHRPD), a 0.2 %-resolution powder diffractometer of Ibaraki prefecture (IPD), an engineering diffractometers (Takumi), a high intensity $S(Q)$ diffractometer (VSD), and a high-pressure dedicated diffractometer. SHRPD, Takumi and IPD are being designed and constructed by the joint team of KEK, JAEA and Ibaraki University, whose member are originally from the KEK powder group. These three instruments are expected to start in 2008. VSD is a super high intensity diffractometer with the highest resolution of $\Delta d/d = 0.3\%$. VSD can measure rapid time-dependent phenomena of crystalline materials as well as glass, liquid and amorphous materials. The pair distribution function will be routinely obtained by the Fourier transformation of $S(Q)$ data. Q range of VSD will be as wide as $0.01\text{Å}^{-1} < Q < 100\text{Å}^{-1}$. IPD is fully funded by Ibaraki prefecture for the promotion of new industries based on advanced science and technologies. It is for the first time in neutron facilities in Japan that a prefecture owns neutron instruments as well as neutron beam will be provided widely to industrial users. To make it successful, the user system is quite important because those users are expected to use IPD like chemical analyzers in their materials development process. Based on questionnaire data to several hundreds industries, IPD is designed as a versatile diffractometer including texture measurement, small angle scattering and total scattering as well as usual powder diffraction. IPD covers d range $0.15 < d (\text{Å}) < 4$ with $\Delta d/d = 0.15\%$, and covers $4 < d (\text{Å}) < 60$ with gradually changing resolution. Q range of IPD will be as wide as $0.01\text{Å}^{-1} < Q < 50\text{Å}^{-1}$ to be utilized for varieties of structures: local structure, nano structure and crystal structure analyses. Typical measuring time for the typical 'Rietveld-quality' data is several minutes with the sample size of laboratory X-ray: 0.4 cc. SHRPD is designed to be the world highest resolution with $\Delta d/d = 0.03\%$ without sacrificing intensity. The combination of the high quality data from HRPD and their high-precision analysis gives us information on tiny structural changes which have been overlooked. After careful examination with the moderator group five years ago, we have decided to develop a high-resolution & good S/N moderator to achieve the 0.03 % resolution within 100 m flight path. This development was almost successful up to now. Instrumental simulation and radiation analysis were almost completed. The d range $0.5 < d (\text{Å}) < 4$ with $\Delta d/d = 0.03\%$, and covers $4 < d (\text{Å}) < 45$ with gradually changing resolution. Takumi is the first priority instrument in JAEA for stress mapping inside structure materials with the highest resolution of $\Delta d/d = 0.2\%$ (corresponding to 10^5 to 10^6 strain precision). The typical gauge volume will be 1 mm3. JED has transmission radiography detectors to support stress mapping. Software group is planning so that basic software to cover data acquisition and data treatment should be common. Since 1 Gbyte data are typically obtained for single experiment in an instrument, the basic software is quite important. International TV conference between ISIS, IPNS, SNS has been held every month to exchange information on each development. KEK developed manyo-lib to help basic analysis. Analysis software development including powder diffraction is strongly related with the activity of the software group. However, users of IPD will be from various field of science and their background is different. It should cover wide topics and help both beginners and well-trained users. We have started with neutron intensity database, peak-search software, peak-match software, pattern simulation, whole pattern fitting, PDF and RDF analysis, and now start coding Rietveld software.

Keywords: Pulsed neutron, powder diffractometer, crystalline, glass, amorphous.

CP989, Neutron and X-ray Scattering in Materials Science and Biology, International Conference on Neutron and X-ray Scattering 2007, edited by A. Ikram, A. Purwanto, Sutiarso, A. Zulfia, S. Hendrana, and Z. Nurachman

Thermal Stress Behavior of Micro- and Nano-Size Aluminum Films

T. Hanabusa[1], M. Nishida[2], K. Kusaka[3]

[1]Institute of Technology and Science, the University of Tokushima, Tokushima, Japan
[2]Department of Mechanical Engineering, Kobe City Collage of Technology, Kobe, Japan
[3]Institute of Technology and Science, the University of Tokushima, Tokushima, Japan

Abstract. In-situ observation of thermal stresses in thin films deposited on silicon substrate was made by X-ray and synchrotron radiation. Specimens prepared in this experiment were micro- and nano-size thin aluminum films with and without passivation film. The thickness of the film was 1 micrometer for micro-size films and 10, 20 and 50 nanometer for nano-size films. The stress measurement in micro-size films was made by X-ray radiation whereas the measurement of nano-size films was made by synchrotron radiation. Residual stress measurement revealed tensile stresses in all as-deposited films. Thermal stresses were measured in a series of heating- and cooling-stage. Thermal stress behavior of micro-size films revealed hysteresis loop during a heating and cooling process. The width of a hysteresis loop was larger in passivated film that unpassivated film. No hysteresis loops were observed in nano-size films with SiO_2 passivation. Strengthning mechanism in thin films was discussed on a passivation film and a film thickness.

Keywords: Thermal stress, in-situ observation, aluminum film, X-ray measurement, synchrotron radiation

INTRODUCTION

The phenomena of a stress migration and an electro-migration have been concentrated on nano-size lead lines [1-5] in large-scale integrated circuit (LSI). The electro-migration is a biased movement of atoms caused by the application of high electrical current densities [2-4], while stress migration damages are caused by the thermal stresses in the line, which develop in a manufacturing process of the LSI due to the different thermal expansion or shrinkage between lines and the substrate [5]. Temperature increase in a service condition may also arise in stress migration. Therefore, it is important to investigate the phenomena of stress migration in a nano-size structure.

If temperatures change in the film/substrate system, thermal stresses develop in the film because of the difference in the coefficient of thermal expansion of the film and the substrate. Large temperature change produces large stresses which may induce plastic deformation as well as creep deformation in the film. The strength of a film will be controlled by a passivation layer on a film surface and depend on a thickness of the film.

Usually, the mechanical properties of film are different from those of bulk material. Many investigations on the thermal stress development in micrometer-size film have been appeared. They found hysteresis loops in the thermal stress development during heating and cooling thermal cycles [6-8]. However, little data is available on the properties of nano-size thin films. The present study is a preliminary investigation of a mechanism of thermal stress in micro-size and nano-size thin films.

The strength of a film will be controlled by a passivation layer on a film surface and depend on a thickness of the film. A roll of passivation layer and the thickness of the film on strengthening of thin films was experimentally investigated. In-situ thermal stress behavior was observed by X-ray and synchrotron radiation.

EXPERIMENTAL PROCEDURE

Preparation of the specimen

Aluminum films were deposited on thermally oxidized silicon wafers by rf magnetron sputtering with an input power of 80 W. Argon was used as a reactive gas. Substrate temperature was maintained at room temperature. The thickness of the first type of micro-size film is 1 micrometer. Films with and

CP989, Neutron and X-ray Scattering in Materials Science and Biology, International Conference on Neutron and X-ray Scattering 2007, edited by A. Ikram, A. Purwanto, Sutiarso, A. Zulfia, S. Hendrana, and Z. Nurachman
© 2008 American Institute of Physics 978-0-7354-0508-0/08/$23.00

without passivation of AlN were prepared. The second type of sample is nano-size film of the thickness of 10, 20 and 50 nm. A silicon oxide (SiO_2) film with 600 nm in thickness was deposited by spin coating on the aluminum surface as a passivation layer.

Stress measurement

In-situ thermal stress measurement in aluminum micro-size films was carried out by laboratory X-ray system. On the other hand, measurements of nano-size films were carried out by synchrotron radiation (SR) at the BL13XU of the SPring-8 with the approval of the Japan Synchrotron Radiation Research Institute (JASRI). A multi-axis diffractometer shown in Fig. 1 was used in this investigation at SPring-8. Since this diffractometer has a very high rigidity, a precise diffraction angle measurement is available even in low diffraction angles. The SR energy was 12.398 keV which corresponds to the wave length of 0.099956 nm.

In order to measure thermal stresses in thermal cycles of heating and cooling processes, a specimen was attached on a ceramic heater plate mounted on the specimen holder. The temperature of the specimen holder surface was measured by a thermo-couple and controlled to the planned temperature level using a temperature controller.

FIGURE 1. Multi-axis diffractometer used in stress measurement.

The sample was heat-treated from room temperature to 200 °C, 300 °C or 400 °C at the rate of about 40 °C /min. The thermal stresses were measured at every 50 or 100 °C intervals in the heat cycles. The stress measurement was made after the temperature was equilibrated at each temperature.

Two-exposure measurement was used because of a strong 111 fiber texture in film structure.

Stress Analysis of {111}-oriented Film by Diffraction

The two-tilt method [9-10] was used to evaluate stresses because an adequate diffraction intensity could only be obtained at two ψ angles, i.e., $\psi_1 = 0°$ and $\psi_2 = 70.5°$, from the {111} fiber-textured film. The following equation [10] gives the in-plane biaxial stress in the film:

$$\sigma = \frac{2}{s_{44}} \frac{\varepsilon(\psi_1) - \varepsilon(\psi_2)}{\sin^2 \psi_1 - \sin^2 \psi_2} \qquad (1)$$

where s_{44} is the elastic compliance of the film, and ψ the angle between the normal of lattice planes and the normal of film surface. Expressing the lattice strain ε in terms of the diffraction peak angle 2θ, Eq. (1) becomes:

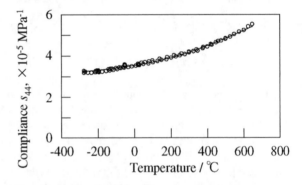

FIGURE 2. Effect of temperature on elastic compliance of single crystal aluminum.

$$\sigma = -\frac{1}{s_{44}} \frac{\pi}{180} \cot(\theta_0) \frac{2\theta(\psi_1) - 2\theta(\psi_2)}{\sin^2 \psi_1 - \sin^2 \psi_2} = K \cdot M \qquad (2)$$

where K is the stress constant, and M the slope in the 2θ-$\sin^2\psi$ relation. The elastic compliance s_{44} varies with the temperature as shown in Fig. 2 [11-13]. From this figure, the s_{44} of aluminum is expressed by the following equation as a function of temperature T:

$$s_{44} = 2.0 \times 10^{-6} T^2 + 1.7 \times 10^{-3} T + 3.50 \ (\times 10^{-5} \, \text{MPa}^{-1}) \ (3)$$

EXPERIMENTAL RESULTS

Thermal stress behavior in micro-size films

Structural evaluation of aluminum film

Figure 3 shows a diffraction pattern of aluminum film measured by CrK_α characteristic radiation at $\psi=0°$. It shows that only two diffraction peaks, i.e., 111 at 2θ = 58.6°, and 222 at 2θ = 156.7°, appear in the whole 2θ range. This means that the {111} plane of the aluminum crystal is arranged parallel to the surface of aluminum film. An additional observation at any orientation of the film surface revealed that the aluminum film is composed of small crystals and that each crystal has 2π rotational freedom about <111> orientation as has been assumed previously.

In-situ thermal stress measurement of aluminum film

Upper part of Figure 4 shows the results of thermal stress measurement on unpassivated aluminum film during a series of heating and cooling cycles. The first figure (a) shows thermal stress alterations between room temperature and 200 °C. The initial residual stress is about 50 MPa. As temperature increases, a small compressive stress develops from 100 °C to 200 °C. In the cooling stage, stresses due to a thermal mismatch increase almost linearly with decreasing temperatures. The specimen was then again heated to 300 °C (the second figure (b)). The initial high tensile stress of 160MPa rapidly drops to zero at 100 °C beyond which small compressive stresses remain almost constant.

The stress originating from the temperature change ΔT of the aluminum film/silicon substrate system is calculated by the following equation:

$$\Delta\sigma = \frac{E_{Al}}{(1-\nu_{Al})\Delta\alpha\Delta T} \qquad (4)$$

where $\Delta\alpha$ is the difference in the coefficient of thermal expansion between silicon (α_{Si}) and aluminum (α_{Al}). E_{Al} is Young's modulus and ν_{Al} is Poisson's ratio of aluminum. The values are summarized in Table 1.

TABLE 1. Coefficient of thermal expansion and elastic constants of aluminum and silicon.

Properties	Al	Si
Young's modulus (GPa)	70	112
Coefficient of thermal expansion (10^{-6}/deg)	23.9	9.6
Poisson's ratio	0.33	

According to Eq. (4), the gradient $\Delta\sigma/\Delta T$ is – 1.49MPa/°C. Although the measurement involves only two data points, i.e., room temperature and 100 °C, the initial change in measured stress with increasing temperature is linear in consistent with the calculated one. Above 100 °C, the data points deviate from this line, showing no significant increase in thermal stresses.

In the cooling stage, stress initially rapidly increases to about 50 MPa, then slowing down in the next stage. The increase rate again picks up below 100 °C. These changes in behavior in the heating and the cooling stages are qualitatively the same in the next heat cycle between room temperature and 400 °C as shown in the third figure (c).

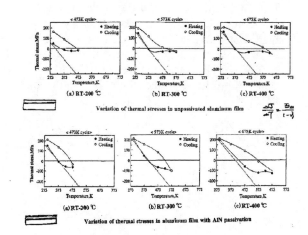

Variation of thermal stresses in unpassivated aluminum film

Variation of thermal stresses in aluminum film with AlN passivation

FIGURE 4. Variation of thermal stress in aluminum film with and without AlN passivation.

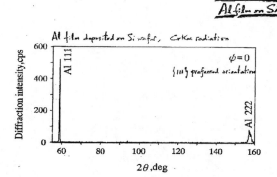

FIGURE 3. Diffraction pattern of aluminum film deposited on a thermally oxidized silicon substrate.

The lower part of Fig. 2 shows a thermal stress variation of AlN passivated film. As shown in the first figure (a), AlN passivation creates tensile residual stress of about 150 MPa in the aluminum film. As temperature increases the state of thermal stress varies from tension to compression. Initially, the change in thermal stress approximates the slope expected by Eq. (4). When the temperature rises above 100 °C, the rate of thermal stress change becomes smaller than has been expected. However, compressive stress continues to increase and the maximum stress value attained at the terminal point of the heating cycle becomes larger than that of unpassivated aluminum film. The behavior of the thermal stress in the cooling stage is rather smooth and does not appear complicated as is shown in the unpassivated film. The stress variation in aluminum film passivated by TiN film is almost the same as that in the case of AlN passivation.

We understand from these results that residual stresses after heat cycles increase and the development of compressive thermal stresses is evidently intensified by passivation.

Thermal stress behavior in nano-size films

Characteristics of diffraction profiles and accuracy in stress calculation

Figures 5, 6 and 7 show 111 diffraction patterns measured from aluminum films with the thickness of 50, 20 and 10 nm, respectively. Since the films deposited by rf-sputtering have a strong {111} texture, 111 diffraction profiles are available at two ψ angles of 0° and 70.5°. The profiles in these figures were measured at 300 °C in the first heating cycle and no significant change in the pattern was observed in any another stage. Although the intensities decreased with decreasing the film thickness, we had still clear intensity profiles from the film of 10 nm thickness as shown in Fig. 7.

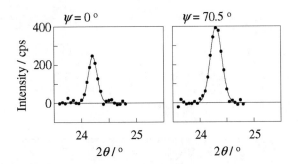

FIGURE 5. Diffraction patterns measured from 50 nm film.

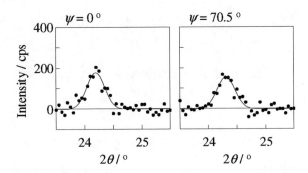

FIGURE 6. Diffraction patterns measured on 20 nm film.

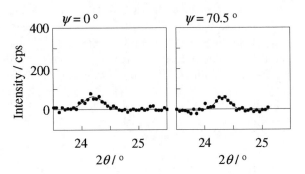

FIGURE 7. Diffraction patterns measured from 10 nm film.

Figure 8 shows 2θ -$\sin^2\psi$ diagrams measured at 300 °C in the first heating stage. Three times measurement made at each ψ angle revealed the good accuracy in stress calculation. It is obvious that the confidence limit of the stress calculation increases with decreasing the film thickness due to decreasing diffraction intensities.

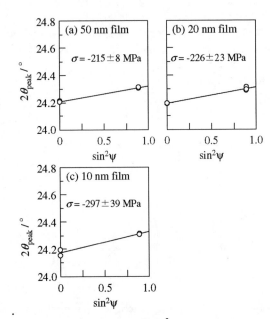

FIGURE 8. Examples of 2θ -$\sin^2\psi$ relation and stress

The full width at the half maximum (FWHM) and the intensity of the diffraction profile were also evaluated as a function of the film thickness as well as the heating temperature. The FWHM was 0.16° for the film of 50 nm thick and becomes large with decreasing in the film thickness, counting 0.44° for 10 nm but was independent of heating temperature. The intensity of 111 diffraction at $\psi=0°$ varied with the film thickness as shown in Figs. 5~7 but no significant change in the intensity was found with increasing temperatures calculation.

Behavior of thermal stresses in heat cycles

Thermal stress measurement of the aluminum film was repeated two times for the heat cycle between room temperature and 300 °C. Figure 7 shows the in-situ thermal stress measurement of 50 nm aluminum film covered by SiO_2 passivation. The initial residual stress in the aluminum film was 200 MPa in tension. In the first heating cycle, the stresses in the film linearly changed from tension to compression up to 250 °C. The thermal stress became compressive above 130 °C and reached to -200 MPa at 250 °C. Further increase in temperature up to 300 °C showed no change in compressive stresses. In the following cooling cycle, the thermal stresses linearly varied with decreasing temperatures, falling to 330 MPa in tension at room temperature. In the second heat cycle, the thermal stresses linearly traced the same pass as in the first cooling cycle.

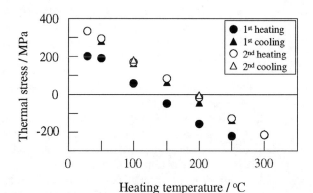

FIGURE 9. In-situ thermal stress measurement of 50 nm thick Al films with SiO_2 passivation.

Figure 8 shows the results of the stress measurement of 20 nm aluminum film with SiO_2 passivation. Since the thickness of the film was small to generate enough intensity for determining 2θ values, measuring errors were somewhat larger than those of the 50 nm film. It seems however that the thermal stress variations are almost linear with both increasing and decreasing temperatures. No hysteresis loop was

observed in the thermal history. The results for 10 nm film exhibited essentially the same figure as is found in Fig. 10 (see Fig. 11).

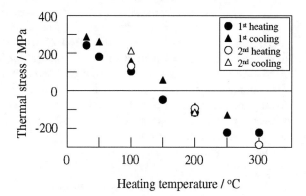

FIGURE 10. In-situ thermal stress measurement of 20 nm thick Al films with SiO_2 passivation.

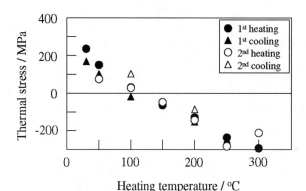

FIGURE 11. In-situ thermal stress measurement of 10 nm thick Al films with SiO_2 passivation.

DISCUSSIONS

We have observed non-linear variations of thermal stresses during heating and cooling cycles in a few micrometer thick aluminum [9,10] and copper films [14]. Furthermore, the thermal stress variation shows a clear hysteresis loop during the heat cycle. Variations in thermal stresses are also different between unpassivated and passivated films [9,10,15]. Figure 12 shows a model of thermal stress development in films with a thickness of the order of micrometer. Thermal stresses of the film without passivation behave in a linear way in the initial heating stage followed by small or no change in stresses after reaching a compressive state. This is because of a stress relaxation by forming hillocks on a film surface. In a heating stage, compressive stresses developed in the film at high temperatures tend to migrate atoms through a free surface and grain boundaries. The hillocks are then considered to be formed on the film surface at triple points of grain boundaries. In a cooling stage, on the other hand, reverse migration occurs to shrink and/or eliminate hillocks formed in the heating stage. Furthermore, the tensile stresses make voids at grain boundaries in the film. Due to these hillock and void formation, large stress relaxation occurs resulting in a formation of narrow hysteresis curve in the thermal history. Since a hard passivating film prevents these hillock and void formation, the hysteresis curve can be formed by tracing the elastic and yield lines as shown in the right side in Fig. 12. Compressive and tensile yield lines for the passivated film may become large compared with the unpassivated film.

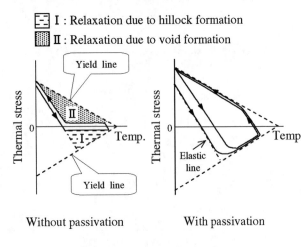

Without passivation   With passivation

FIGURE 12. Models of the thermal stress development in thick films with and without passivation film.

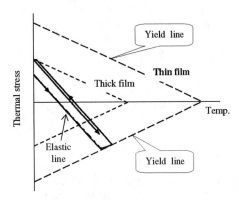

FIGURE 13. Models of the thermal stress development in thin films with the size of nano-meter order.

Figure 13 shows a model of the thermal stress development in thin films with the size of nano-meter order. Due to a very small cross section of the film, the number of Frank-Read sources may be small in addition with a small distance of column dislocations which intercept a slip plane. Therefore, the length of the dislocation which acts as a Frank-Read source is very short. Since the stress required to act Frank-Read source is inversely proportional to the column distance [16], the yield stress increases with decreasing the film thickness. In addition with this factor, the motion of the dislocation is inhibited by the existence of a passivation film. The passivation film acts as an image force in a way to repulse a moving dislocation from the interface [17]. Therefore, the yield stress and the flow stress are significantly increased with nano-scale structure and a hard passivation film. As shown in Fig. 13, the thermal stresses linearly change with increasing temperature in the first heating stage. If the thermal stress reaches the compressive yield line, plastic deformation causes a non-linear thermal stress development. In a cooling stage, the thermal stress varies along the linear elastic line to the terminal tensile residual stress state. In the second and the following thermal cycles, the stresses trace the same straight line as of the former cooling stage. These explanations are supported with the experimental facts that no hysteresis was observed in the present investigation.

In such a way, very thin films are strengthened by the hard passivation layer on the film and by the nature of a nano-scale thin film structure.

CONCLUSION

The thermal stress observation by X-ray and synchrotron radiation shows that:

1. In-situ thermal stress measurement of micro–size aluminum film revealed a hysteresis loop in a sequence of heating and cooling thermal cycle.
2. Passivation layer makes a large hysteresis loop and then strengthens a thin film.
3. Stresses in 10, 20 and 50 nm thick aluminum films could be observed by ultra-high intense synchrotron radiation.
4. In-situ thermal stress observation reveals that the thermal stresses in nano-size films behave linearly during the heating and cooling cycles between room temperature and $300\,^{\circ}C$.
5. Thin films are strengthened by a hard coating layer as well as by the nature of a nano-size thin film structure.

ACKNOWLEDGEMENT

The authors would like to thank Dr. H. Koike of the Japan Synchrotron Radiation Research Institute (JASRI) for his support. The synchrotron radiation experiments were performed using the SPring-8 with the approval of the JASRI (Proposal No. 2003A0305-ND1-np and 2003B0324-ND1d-np).

REFERENCES

1. International Technology Roadmap for Semiconductor, International SEMATECH, 1999.
2. P. R. Besser, M. C. Madden and P. A. Flinn: *J. Appl. Phys.* **72** (1992) 3792.
3. T. N. Marieb, E. D. Abratowski, J. C. Bravman: *AIP Conference Proceedings* **305**, Austin, TX, 1993, 1.
4. P.-C. Wang, G. S. Cargill III, I. C. Noyan and C.-K. Hu: *Appl. Phys. Lett.* **72** (1998) 1296.
5. P. S. Ho, I. –S. Yeo, and S. G. H. Anderson: *AIP Conference Proceedings* **305**, Austin, TX, 1993, 62.
6. R. Venkeatraman and J. C. Bravman: *Mat. Res. Soc. Symp. Proc.* **239** (1992) 127.
7. S. P. Baker, A. Kretschmann and E. Arzt: *Acta Mater.* **49** (2001) 2145.
8. D. Weiss, H. Gao and E. Arzt: *Acta Mater.* **49** (2001) 2395.
9. T. Hanabusa: *Mat. Sci. Res. Intl.* **5** (1999) 63.
10. K. Kusaka, T. Hanabusa, M. Nishida and F. Inoko: *Thin Solid Films* **290-291** (1996) 248.
11. P. M. Sutton: *Phys. Rev.* **91** (1953) 816.
12. G. N. Kamm and G. A. Alers: *J. Appl. Phys.* **35** (1964) 322.
13. J. Vallin, M. Mongy, K. Salama and O. Beckman: *J. Appl. Phys.* **35** (1964) 1822.
14. T. Hanabusa and M. Nishida: *Mat. Sci. Res. Intl.* **7** (2001) 54.
15. T. Hanabusa, K. Kusaka and O. Sakata: *Thin Solid Films* **459** (2004) 245.
16. F. C. Frank and W. T. Read: *Phys. Rev.* **79** (1950) 722.
17. D. Weeks, J. Dunders and M. Stippes: *J. Eng. Sci.* **6** (1968) 365.

Neutron Reflectometry as a Surface Probe:
A Personal Perspective

Z. Tun

Canadian Neutron Beam Centre, National Research Council Canada, Chalk River Laboratories
Chalk River, Ontario, Canada K0J 1J0

Abstract. Development of neutron reflectometry has enabled neutron scattering laboratories worldwide to make important contributions to the study of surfaces, interfaces and thin-films. As a result, neutron scattering has become an invaluable research tool for the scientific disciplines that did not traditionally use neutrons for research as recently as 20 years ago. At Chalk River (Canada), one discipline with which we have formed a close affiliation is electrochemistry. Our decision in the early 1990s to reach out to this potential user community was based on the fact that Canada has many researchers active in corrosion science. The virtue of this affiliation is best demonstrated in our experiments where reflectometry is performed simultaneously on the sample being investigated with electro-impedance spectroscopy, a standard electrochemical technique. The two methods in combination have led to the results that would have otherwise been missed or wrongly interpreted.

Keywords: Neutron reflectometry, electro-impedance spectroscopy, corrosion

CP989, *Neutron and X-ray Scattering in Materials Science and Biology, International Conference on Neutron and X-ray Scattering 2007*, edited by A. Ikram, A. Purwanto, Sutiarso, A. Zulfia, S. Hendrana, and Z. Nurachman
2008 American Institute of Physics 978-0-7354-0508-0/08/$23.00

Dynamical Scaling, Fractal Morphology and Small-angle Scattering

S. Mazumder

Solid State Physics Division
Bhabha Atomic Research Centre, Trombay, Mumbai 4000 85, India

Abstract. When a system with continuous symmetry is quenched instantly to a broken symmetry state, new phases of topological defects appear in an otherwise homogeneous medium of continuous symmetry. The phenomenon of new phase formation is a representative example of first order transition. The phenomenon is of immense interest as an example of a highly nonlinear process far from equilibrium. The second phase grows with time and in late stages all domain sizes are much larger than all microscopic lengths. In the large time limit, the new phase forming systems exhibit self-similar growth pattern with dilation symmetry, with time dependent scale, and scaling phenomenon. Extensive investigations on dynamical scaling phenomenon have been carried out so far for Euclidean systems. The question arises about the validity of the scaling laws for dynamical systems in non-Euclidean fractal geometry. Some of the questions, arising purely because of the geometrical constraints in the physical systems and others on experimental observations, are posed here.

Keywords: Dynamical scaling, fractals, small-angle scattering
PACS: 64.75.+g, 64.90.+b, 61.43.Hv, 61.50.Ks

INTRODUCTION

If a system with continuous symmetry is quenched instantly to a broken symmetry state, new phases of topological defects appear in an otherwise homogeneous medium of continuous symmetry. The further growth of the topological defects are of continuous nature such that the time evolution of the system can be described [1], both for conserved and non-conserved order-parameter field, by Ginzburg-Landau free energy functional.

The phenomenon of new phase formation is a representative example of first order transition. The phenomenon is fundamental and of immense interest as an example of a highly nonlinear process far from equilibrium. The second phase grows with time and in late stages all domain sizes are much larger than all microscopic lengths. In the large time limit, the new phase forming systems exhibit self-similar growth pattern with dilation or scaling symmetry, with time dependent scale, and scaling phenomenon. The phenomenon is indicative of the emergence of a morphological pattern of the domains at earlier times looking statistically similar to a pattern at later times apart from the global change of scale implied by the growth of time dependent characteristic length scale

$L(t)$ – a measure of the time dependent domain size of the new phase.

The scaling hypothesis assumes the existence of a single characteristic length scale $L(t)$ such that the domain sizes and their spatial correlation are time invariant when the lengths are scaled by $L(t)$. Quantitatively, for isotropic systems, the equal-time spatio-temporal composition modulation auto-correlation function $g(r,t)$, reflects the way in which the mean density of the medium varies as a function of distance from a given point, should exhibit the scaling form with time-dependent dilation symmetry

$$q(r,t) = f(r/L(t)) \qquad (1)$$

The scaling function $f(r/L(t))$ is universal in the sense that it is independent of initial conditions and also on interactions as long as they are short ranged. However, form of $f(r/L(t))$ depends non-trivially on n, the number of components in the vector order-parameter field exhibiting the scaling behavior, and d, the dimensionality of the system. It is important to note that the scaling hypothesis has not been proved conclusively so far except for some model systems.

The Fourier transform of $g(r,t)$, the structure factor or scattering function $S(q,t)$ for a d dimensional

CP989, *Neutron and X-ray Scattering in Materials Science and Biology, International Conference on Neutron and X-ray Scattering 2007*, edited by A. Ikram, A. Purwanto, Sutiarso, A. Zulfia, S. Hendrana, and Z. Nurachman
© 2008 American Institute of Physics 978-0-7354-0508-0/08/$23.00

Euclidean system, obeys simple scaling ansatz at late times,

$$S(q,t) = L(t)^d F(qL(t)) \qquad (2)$$

Extensive investigations on dynamical scaling phenomenon have been carried out so far for Euclidean systems. The question [2] arises about the validity of the scaling laws for dynamical systems in non-Euclidean fractal geometry. We will pose some of the questions arising either purely because of the geometrical constraints in the physical systems or due to some noteworthy experimental observations

INCOMPREHENSIBILTY: BACKGROUND

For last several years now, there has been strong interest in the phenomenon of dynamics of new phase formation in condensed systems of continuous symmetry. The field has been receiving considerable experimental attention due to the relevance of this phenomenon in a wide range of materials including metallic alloys, polymers, glasses, liquid mixtures, binary gases, ceramics, xerogels etc. The field has also been enriched by a series of theoretical and computational contributions. Very early stage of this phenomenon can be rigorously described by a linear theory, based on diffusion equation. The linear theory implies that the Fourier amplitude $A(q,t)$ of the composition modulation $C(r,t)$, follows a linear temporal relation

$$\frac{d[A(q,t)]}{dt} = \alpha(q) A(q,t). \qquad (3)$$

where, $\alpha(q)$ is the time t independent proportionality constant; r denotes spatial coordinate of the system with $|r|=r$; q is the scattering vector with modulus $|q|=q$. Accordingly, time dependent isotropic structure factor $S(q,t)$ should exhibit an exponential growth

$$S(q,t) = S(q,0)exp[2\alpha(q)t] \qquad (4)$$

The phenomenon of new phase formation has immense practical significance. For example, the kinetics of phase separation and the microscopic structures of the phase separating phases determine the suitability of an alloy for it's technological end use.

Dynamical scaling phenomenon in a multi-component systems may arise even when topological defects are poly-disperse in nature. The polydispersity may arise due to varying sizes, shapes and different contrasts. For a new phase forming system, the scaling function $F(qL(t))$ in Fourier space in the domain of large q $(qL(t)>>1)$ asymptotically approaches (denoting $qL(t)$ by x) a form

$$F(x) \sim x^{-(d+n)} \qquad (5)$$

It is noteworthy that Porod Law, applied for systems with scalar order-parameter and recognised as arising from defect configurations with sharp domain walls, $S(q) \sim q^{-4}$ is recovered from the asymptotic relation as the special case for $d=3$ and $n=1$ (scalar field). For $n=d$, the topological defects are point defects. For $n < d$, the defects are spatially extended where the field of order parameter varies only in n dimensions orthogonal to the defect core and is uniform in the remaining $d-n$ dimensions parallel to the core. The domain walls are the surfaces of dimension $d-n$. For $n=1$, the defects are interfaces or domain boundaries. For $n=2$, the defects are vortices/anti-vortices and for $n=3$, the defects are monopoles. There exist substantial differences between the $n=1$ and $n>1$ cases. For $n=1$, the interfaces are sharp while for $n>1$, the interfaces are smooth because of the symmetry obeyed by the order parameter.

For systems where a new phase is poly-disperse in nature, there is no universal form for $L(t)$. In cases where scaling phenomenon has been observed, $L(t)$ has been taken to be the reciprocal of the first moment of $S(q,t)$ in q space, i.e.,

$$L(t) = [I \int qS(q,t)dq]^{-1} \qquad (6)$$

In some other cases, the phenomenon has been observed where $L(t) = [q_m(t)]^{-1}$ and $q_m(t)$ is the value of q at which $S(q,t)$ has its maximum. For a model system where free energy functional is invariant under global rotation of order parameter, defined by n-component vector field, with $n=\infty$ and arbitrary d, it is found that scaling ansatz (2) breaks down because of the existence of two marginally different length scales - $L(t) \sim t^{1/4}$ and $[q_m(t)]^{-1} \sim [t/\ln(t)]^{1/4}$. In the scaling regime $L(t) \sim t^\beta$, where β depends on the conservation laws governing the dynamics and the dimensions of n and d. The exponent β is also, in general, model dependent and a wide range of values has been predicted for it. It is known that the elastic effect is responsible for the deviation from one-third law of Lifshitz-Slyozov. Most of the features predicted have been corroborated well by scattering experiments in binary alloys. For a multi-component alloy, it has been observed that β is temperature dependent and at lower temperature β is not uniform over the entire time range. At higher temperature, the growth of second phase is driven by the diffusion of atoms while at lower temperature the growth is due to the development of coherence between the clusters. The investigation hinted at the possibility of non-unique characteristic length.

Extensive investigations on scaling phenomenon have been carried out for Euclidean systems having three and two dimensions and having new phase for which the number of components of the order parameter, $n \leq d$, the spatial dimension. The phenomenon has been investigated even for multi-component alloys. Investigations have also established the validity of the scaling laws for some systems which do not support stable topological defects, having $n > d$, like liquid crystals with complicated order parameter. The question arises about the validity of the scaling laws for new phase formation in the case of non-Euclidean fractal systems. The validity of the dynamical scaling laws for new phase formation in the case of non-Euclidean fractal systems is still an open question.

Objects exhibiting scale invariance property are called fractal objects. Scale invariance property implies that the autocorrelation function remains invariant under change of scale upto a multiplicative factor when the scale of observation is changed by some factor b. Hence, $g(br) \sim g(r)$ indicating the validity of dilation or scaling symmetry. These kind of objects are called self similar under dilation symmetry. This is possible only when $g(r)$ follows a power law correlation of the form $g(r) \sim r^{-\gamma}$. Mass fractal objects follow long-range power-law correlation in density, i.e. mass $M(R)$ within a sphere of radius R is represented as, $M(R) \sim R^{D_m}$, where D_m is generally a fractional number and is called mass fractal dimension of the object.

For fractal objects, the scale b is isotropic or directionally independent in nature. When b is not isotropic that is directionally dependent, objects are not termed self-similar but self-affined. Many real so called fractal objects observed in the nature or in synthesized materials are not strictly self-similar but self-affined in nature.

For ramified structures, embeded in a three-dimensional space, the value of D_m lies in the range $1 < D_m < 3$. For a ramified rod and disk, the value of D_m lies in the range $0 < D_m < 1$ and $1 < D_m < 2$, respectively. The smaller the value of D_m, the more ramified is the object. Nanoparticles suspended in a liquid, under certain conditions, can aggregate together forming self-similar mass fractal clusters. For example, the structures of porous colloidal aggregates (gold, silica, latex) formed by Brownian motion exhibit dilation symmetry and are well described as mass fractals. For a mass fractal aggregate, the following relation holds good:

$$R = L[M(R)]^{(1/D_m)} \qquad (7)$$

where L is termed as lacunarity constant. Lacunarity is a counterpart to the fractal dimension that describes the texture of a fractal. A fractal object having large gaps or holes has high lacunarity. On the otherhand, a fractal object having low lacunarity is almost translationally invariant. Fractal objects having same fractal dimensions can look widely different because of having different lacunarity.

Like mass fractal objects, there are objects with uniform internal density but outer surface exhibiting self-similar geometric properties. Fractal surfaces represent topographical irregularities of the complex surfaces. For a surface fractal object, surface area $S(R)$ within a circle of radius R is represented as $S(R) \sim R^{D_s}$, where D_s is generally a fractional number lies in the range $2 < D_s < 3$ and is called surface fractal dimension of the object. Many deposited surfaces produced by a variety of chemical and ballistic procedures are surface fractal in nature. Further, all mass fractal objects are surface fractal in nature but surface fractal objects need not be mass fractal. In the limit of diffusion-limited aggregation, particles are more likely adhere to the outer branches of the growing agglomerate rather than penetrate to the core and produce surface fractal object. When the mass fractal becomes space filling both mass and surface fractal dimensions tend to the value 3 and $S(q,t)$ asymptotically approaches a form $S(q,t) \sim q^{-3}$. Then when mass consolidates itself on length scale larger than q^{-1} and the surface stays rough, structures undergo transition to a surface fractal regime and Porod exponent η becomes greater than 3. For a mass fractal cluster with size ξ and fractal dimension D_m, R_G (the Guinier radius) and ξ are related by $(R_G/\xi)^2 = 1/2 D_m(D_m-1)(D_m+1)$.

For objects whose volume or mass is fractal such as cluster aggregates, $S(q,t)$ asymptotically approaches a form $S(q,t) \sim q^{-\eta}$ where the exponent η reflects directly the mass fractal dimension D_m. For a mass fractal object, $\eta = D_m$ with $1 < \eta < 3$ and $1 < D_m < 3$. The expected Porod exponent for an ideal smooth surface is 4. There are also materials like pore fractals where the pore space and the material-pore interfaces are fractal in nature. Fractal objects are thermodynamically metastable because of higher surface energy and gradually should transform to objects with regular Euclidean dimension. But in reality, that transformation could be so sluggish that many fractal objects do not undergo perceptible changes in measurable time.

New phase formation is a diffusion controlled process but aspects like diffusion or random walk in non-Euclidean geometry are not satisfactorily comprehensible yet. Diffusion in non-Euclidean fractal geometry is similar, in physical terms, to random walk of drunken messenger on a road system designed by another drunken engineer. In Euclidean geometry, it is established that the curvilinear distance $<r(t)>$ covered in time t for a random walk is given by $\sim t^{1/2}$ and the exponent ½ is independent of dimensionality. If it is conjectured that the curvilinear distance covered for a random walk in fractal geometry also behaves as a power law given by $\sim t^{(1/2\nu)}$ - the exact dependence of ν on the fractal geometry is also yet to be established. For two-dimensional Sierpinski gasket, ν is ratio of fracton and fractal dimensions. Monte carlo simulations for diffusion on the fractal structure generated by random walk on a two-dimensional lattice shows $<r(t)> \sim t^{\alpha} (\ln t)^{\beta}$ with $\alpha = 0.325 \pm 0.01$ and $\beta = 0.35 \pm 0.03$. For Sierpinski gasket, both fracton dimension $\left(2\dfrac{\ln(d+1)}{\ln(d+3)} \right)$ and fractal dimension $\left(\dfrac{\ln(d+1)}{\ln 2} \right)$ are uniquely defined by the dimension of embedding Euclidean space – well characterized by one space dimension d. Objects in Euclidean geometry require only the dimension of the embedding Euclidean space for description of their physical properties. However in general, self-similar objects in non-Euclidean geometry require at least three dimensions namely, embedding Euclidean dimension d, fractal dimension and spectral or fracton dimension τ for complete description of various physical quantities.

The mass fractal dimension is associated with the heterogeneity and space-filling properties of an object- smaller the value of fractal dimension the more ramified is the object, surface fractal dimension reflects the roughness of the surface - higher the value of fractal dimension the more microporous surface is and can be interpreted as the tendency of the surface to spread out into the volume in which it resides while the fracton dimension τ reflects the connectedness of a fractal network. More connectivity between the subunits will lead to higher fracton dimension. For a Euclidean system, the vibrational density of states $\rho(\omega) \sim \omega^{d-1}$ depends on embedding Euclidean dimension d only. But for a fractal object, $\rho(\omega) \sim \omega^{\tau-1}$ does not depend on either d or fractal dimension but depends on fracton dimension τ.

Diffusion co-efficient in Euclidean geometry is time independent and is expressed as ratio of mean squared displacement $<r^2(t)>$ and time. In non-Euclidean geometry, mean squared displacement should increase more slowly and diffusion co-efficient will no longer be time independent, but rather a decreasing function of time as it leads a drunken messenger into regions of fewer and fewer roads.

Exact time dependence of diffusion co-efficient in non-Euclidean geometry, in general, is yet to be established. Since the growth of new phase in a phase separating system is a diffusion limited process and diffusion in non-Euclidean geometry is yet an open problem, the related phenomenon of dynamical scaling of structure factor in non-Euclidean geometry is far from comprehensible.

EXPERIMENTAL OBSERVATIONS

Phase separation kinetics of multi-component 350-grade maraging steel has been investigated [3,4] at two different temperatures, viz. 430 °C and 510 °C, for different aging times in the recent past. Unlike previous observations, at both the temperatures, dynamical scaling behavior is observed at the early stages of phase separation accompanied by the diffuse interface of the secondary phases. Dynamical scaling phenomenon has been observed when $L(t)$ is the reciprocal of the first moment of $S(q,t)$ in q space. The investigation also hints at the possibility of existence of a time window within which scaling phenomenon holds good. For significantly large values of the two bounds of the time window, the phenomenon appears to be valid only in the limit $t \rightarrow \infty$. But when the upper bound is not far off, then breakdown of scaling is appreciated beyond some finite time. When the lower bound is close to $t=0$, scaling is observed in the initial stage of new phase formation. It is the position of the two bounds of the time window which determines the stages at which the scaling laws are observed.

The examination, involving real time kinetics of hydration of silicates with light and heavy water, on validity of the dynamical scaling phenomenon in non-Euclidean geometry have been reported [5-7] only recently.

It has been observed that the kinetics, of hydration of silicates with light and heavy water, are of nonlinear nature even at the initial time. The scattering data could not be interpreted in terms of a linear theory based on the diffusion equation even at the very initial stage of hydration. In the case of hydration of silicates with light water, the hydrating mass exhibits a mass fractal nature throughout hydration, with the mass fractal dimension increasing with time and reaching a plateau. The second phase appears to grow with time initially. Subsequently, the domain size of the second phase saturates. It has also been demonstrated that light water hydration of silicates exhibits a scaling phenomenon for a characteristic length with a different measure. The exhibition of scaling phenomenon for non-Euclidean geometry has been observed for the first time. The temporal behavior of the characteristic length has been observed to be far from a power law.

32

As far as chemistry is concerned, the hydration of silicates with light and heavy water is expected to be quite similar except for kinetics. Due to different molecular masses of H_2O and D_2O, diffusion is expected to be more sluggish for D_2O. However, some incomprehensible contrasting behavior has been observed in the case of hydration of silicates with heavy water as far as the kinetics of new phase formation is concerned. The domain size of the density fluctuations grows in the beginning for a while and subsequently appears to shrink with time, reaching saturation ultimately. In the case of hydration of silicates with heavy water, the microstructure of the hydrating mass undergoes a transition from mass fractal to surface fractal and subsequently to mass fractal. The scaling phenomenon, with all possible measures of the characteristic length, has not been established for the hydration kinetics with heavy water. It has been conjectured that the different rates of diffusion of light and heavy water in forming gel structure in silicates lead to formation of different structural networks with different scattering contrast.

Not going beyond what has already been observed for hydration of silicates with light and heavy water, it can be inferred that dynamical scaling hypothesis is valid in non-Euclidean geometry if structure remained topologically same. When structure changes topologically, the dynamical scaling hypothesis is invalidated. But such an important conclusion needed further scrutiny. So examining about the uniqueness of silicates in cementitious material to exhibit this contrasting behavior of hydration was considered important.

With that point in mind, temporal evolution of mesoscopic structure and hydration kinetics of sulphates with light and heavy water have been investigated. For hydration of sulphates with light and heavy water, contrasting behavior of hydration, as observed in the case of hydration of silicates, has not been observed. Hydration of sulphates with light and heavy water does not have distinct characteristic as far as the agreement with dynamical scaling hypothesis is concerned. There is no change of the topographical mesoscopic structure for hydrating sulphates, unlike silicates, with light and heavy water. A disagreement has been observed, for the first time, with the hypothesis of dynamical scaling of the structure factor of non-Euclidean system maintaining same topographical morphology during temporal evolution of the structure. The observations are completely incomprehensible.

The experimental observations show that both for light and heavy water hydration of sulphates, time evolution of the scattering functions do not exhibit scaling phenomenon for a characteristic length with any possible measure although there is no topographical change of the hydrating mass as a function of time. Hydrating mass remains mass fractal throughout. The present investigation clearly indicates disagreement, for the first time, with the hypothesis of dynamical scaling of structure factor for fractal systems without temporal topolographical change

It has been conjectured that effect like hydrogen bonding may be playing some role. The spectacular observations also points to the grey area of our understanding of the hydration of cements. The poor understanding of both the hydration kinetics and the physical nature of hydration products was glaringly evident. It is only pertinent to note here that intracellular water dynamics in biological systems also yields fascinating surprises of different kind.

It is also inferred that temporal evolution of mesoscopic structure of hydrating cementitious material is far from understood and deserves intense scrutiny. The present manuscript only highlights some grey areas in our understanding of the phenomenon of dynamical scaling of structure factor in non-Euclidean Geometry and otherwise.

CONCLUSIONS

Some incomprehensible issues in dynamical scaling phenomenon have been brought to light. It is recently discussed that a system, with conserved-order-parameter and $n = \infty$, should exhibit multi-scaling. This is based on theoretical work. The question remains to be answered that whether this multi-scaling behavior persists to smaller values of n and related defect dynamics clearly deserves further investigations.

To the best of our knowledge, investigations for examining behavior of the scaling laws under confined geometry, for quasi-crystalline matrix and phase separating systems under random fields have not been touched upon so far and investigations on these lines are strongly recommended in the future. In general, it is felt that the phenomenon of dynamical scaling should be brought into closer scrutiny for different systems in the future with more objective and extensive scattering experiments

REFERENCES

1. A. J. Bray, *Adv. Phys.* **43**, 357-459 (1994) and references therein.
2. S. Mazumder, *Physica B.* **385-386**, 7-10 (2006) and references therein.
3. S. Mazumder, D. Sen, I.S. Batra, R. Tewari, G.K. Dey, S. Banerjee, A. Sequeira, H. Amenitsch and S. Bernstorff, *Phys. Rev.* B **60**, 822-830 (1999) and references therein.

4. R. Tewari, S. Mazumder, I.S. Batra, G.K. Dey and S. Banerjee, *Acta. Metall.* **48**, 1187-1200 (2000) and references therein.

5. S. Mazumder, D. Sen, A.K. Patra, S.A. Khadilkar, R.M. Cursetji, R. Loidl, M. Baron and H. Rauch, *Phys. Rev. Lett.* **93**, 255704-1-255704-4 (2004) and references therein.

6. S. Mazumder, D. Sen, A.K. Patra, S.A. Khadilkar, R.M. Cursetji, R. Loidl, M. Baron and H. Rauch, *Phys. Rev. B.***72**, 224208-1-224208-10 (2005) and references therein.

7. S. Mazumder, R. Loidl and H. Rauch, *Phys. Rev.* B **76**, 064205-1-064205-8 (2007) and references therein.

Deuterium Labeling and Neutron Scattering for Structural Biology

Peter Timmins

Institut Laue-Langevin, BP 156, 6 rue Jules Horowitz
38042 Grenoble Cedex 9, France

Abstract. Neutron scattering applications in structural biology are strongly enhanced by deuterium labeling either of the aqueous solvent or of the macromolecules themselves, or both. In protein crystallography deuteration of the solvent and protein allows a clear visualization of hydrogen sites (deuterium atoms) as well as increasing signal/noise in the diffraction data. This allows crystals as small as $0.1mm^3$ to be used. In small angle scattering and reflectometry deuteration of the aqueous solvent allows one to exploit manipulate the contrast between different components complexes such as proteins, nucleic acids or lipids. In the case of multi-protein complexes the in vivo deuteration of individual protein sub-units allows an even more sophisticated contrast variation to be performed.

Keywords: Structural biology, neutron scattering, deuterium labeling, contrast variation
PACS: 87.15.-v

INTRODUCTION

Neutron scattering is a term which in fact encompasses many different techniques from atomic resolution crystallography, through reflectometry for the study of interfaces to small angle scattering for the study of macromolecular complexes in solution. In addition neutron spectroscopic techniques exist which can give information on the dynamics of macromolecules. For studies of biological structure the most important advantage of neutron scattering over other techniques is the substitution of hydrogen by its heavier isotope deuterium. In this paper we will review the use of deuterium labeling and show its impact on structural studies of biological macromolecules and macromolecular complexes.

SCATTERING LENGTHS AND SCATTERING LENGTH DENSITIES

The neutron scattering lengths of atoms constituting biological materials are shown in Table 1 along with the equivalent values for X-rays. Note that these are in units of cm^{-1}. For X-rays these are frequently quoted simply as the number of electrons (e.g. 1 for hydrogen) as it is the electrons from which are scattered the X-rays. The scattering of neutrons is more complex and is from the nucleus but is not a simple function of atomic mass. Hence it may be noted that the neutron scattering lengths (i) vary randomly amongst the elements (ii) are different for different isotopes of the same element (iii) may have negative values (corresponding to a phase change of the scattered with respect to the incoming wave). For biological studies the most striking difference property is the scattering length of hydrogen which is similar in magnitude to the other elements and which has a different sign for hydrogen (^1H) and deuterium (^2H).

High resolution – Deuterium labelling

When hydrogen is replaced by deuterium in high resolution studies (e.g in crystallography) the deuterium atoms are seen as strong positive density, similar to that of carbon. Unexchanged hydrogen atoms are seen as negative density. By high resolution we mean here approximately 2Å resolution, i.e. a resolution at which adjacent atoms may be distinguished. This has a great advantage in protein crystallography and to some extent in reflectometry in that deuterium can be clearly visualized whereas in X-ray experiments because of its very low scattering power hydrogen (or deuterium) can only be seen at ultra-high resolution, <1Å, which is usually not possible in X-ray crystallography. The replacement of hydrogen has two major advantages for neutrons (i) the removal of hydrogen reduces the incoherent

CP989, *Neutron and X-ray Scattering in Materials Science and Biology, International Conference on Neutron and X-ray Scattering 2007,* edited by A. Ikram, A. Purwanto, Sutiarso, A. Zulfia, S. Hendrana, and Z. Nurachman

TABLE 1. Scattering Lengths of Atoms Found in Biological Molecules

Atom	Neutron Scattering Length b_{coh} (x 10^{-12} cm-1)	X-ray Scattering Length $f(2\theta=0)$ (x 10^{-12} cm-1)	No. of electrons
^{1}H	-0.3472	0.28	1
^{2}H	0.6671	0.28	1
C	0.6651	1.69	6
N	0.9400	1.97	7
O	0.5804	2.25	8
P	0.5170	4.23	15
S	0.2847	4.50	16

scattering and hence improves the signal/noise ratio of the data and (ii) at slightly lower than atomic resolution, 2 – 2.8Å, the scattering length density due to hydrogen atoms tends to be cancelled by the positive density from adjacent carbon nitrogen or oxygen atoms. Deuteration removes this effect.

Low resolution – Contrast Variation

At lower resolution, where individual atoms are not visualized then the relevant quantity to be determined is the scattering length density of a group of atoms which may, for example be part or all of a protein molecule. This quantity is calculated simply by taking the sum of the scattering lengths and dividing by the volume of the molecule or part of a molecule. It will of course depend on whether or not the hydrogen atoms have been replaced by deuterium. When a biological macromolecule is placed in a D_2O-containing solution there will be an exchange of hydrogen atoms between the macromolecule and the water. This will only concern atoms of hydrogen connected to oxygen or nitrogen atoms; the carbon-hydrogen bond is generally too strong to allow spontaneous exchange. This exchange means that the scattering length density of a macromolecule will vary depending on the deuterium content of the aqueous solvent in which it is dissolved. The exchange of hydrogen and deuterium between macromolecule and solvent is not an instantaneous process. Indeed in the case of globular proteins some peptide protons in the interior of the protein may take years to exchange. However in practice it has been found that ~70% of labile protons exchange within 12-24 hours and further exchange takes place on a much longer time-scale. Figure 1 shows the scattering length density of water and a number of biological macromolecules as a function of the H_2O/D_2O content of the solvent. These are only examples and the case for individual proteins, nucleic acids and in particular lipids may vary somewhat depending on their exact composition. From this plot we may draw a number of conclusions (i) proteins have the same scattering length density as water when in a mixture of ~40% D_2O/60% H_2O – we call this the contrast match point or isopicnic point (ii) nucleic acids have a contrast match point at 65 – 70% D_2O (iii) lipids have a contrast match point at ~10% D_2O but this can vary

quite widely depending on the nature of the head group. Detergents, which are used for solubilising membrane proteins behave in a similar manner to lipids but are even more variable in match point (iv) per-deuterated macromolecules have a scattering length density which is greater that water at all deuterium contents and can therefore not be completely matched. From the behaviour of per-deuterated and non-deuterated proteins and DNA it can be seen that these molecules can indeed be matched out in 100% D_2O if they are only partially deuterated. In practice a 75% deuterated protein is contrast matched in 100% D_2O.

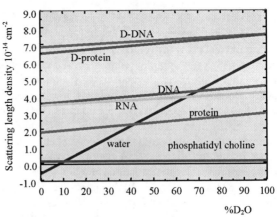

FIGURE 1. Neutron scattering length densities of components of biological complexes. D-DNA and D-protein stand for per-deuterated DNA and protein respectively.

In a low resolution experiment the scattering of the neutrons derives from the contrast between the dissolved particle and its solvent environment, i.e. the difference in scattering length density between the particle and its environment. Examination of figure 1 therefore shows that in certain H_2O/D_2O mixtures the contrast for a particular kind of macromolecule may be zero – at the contrast match point. We may therefore measure the scattering from a macromolecular complex where one of the components is effectively invisible. So for example if a protein DNA complex is examined with neutrons in a solvent containing ~70% D_2O/30% H_2O then the DNA will be invisible. Conversely in a mixture containing 40% D_2O then the protein will become invisible. We may therefore determine the shape of individual components in situ. This technique is known as contrast variation and it has been widely exploited to examine protein nucleic acid complexes and protein/lipid or protein/detergent complexes. Another class of biological complexes of major importance is multi-protein complexes (with or without associated nucleic acid). In this case the individual proteins will each have the same scattering length density and cannot therefore be distinguished from each other. This difficulty can however be overcome by deuteration of specific proteins in the complex. As we see in Figure 1 perdeuterated or partially perdeuterated proteins have a very different scattering length density from hydrogenated proteins and can therefore be easily distinguished. Indeed perdeuterated proteins have a scattering length density which exceeds even that of 100% D_2O. If however proteins are deuterated to a level of about 75%, which can be attained by growing bacteria on 85% D_2O without deuterated carbon source (P. Callow, private communication) then they can be contrast matched in 100% D_2O.

SAMPLE DEUTERATION

The simplest way of deuterating a sample is by dissolving it in D_2O. In this way all labile protons will exchange with deuterons in the water so long as there is contact between the water and the proton containing groups. As explained above in a protein all hydrogens linked to oxygen or nitrogen can be considered as labile and therefore potentially able to exchange with the solvent. The time for this exchange is however very variable and depends on the exposure of different parts of the protein to the solvent. Thus surface protons will exchange on a time scale of seconds whereas protons in the center of a tightly folded protein may take weeks or even months. On average we consider that ~ 80% of labile protons exchange within 24 hours. On average only about 25% of the hydrogen atoms in a protein molecules are labile. Hydrogen atoms attached to carbon atoms are much more tightly bound and are usually not exchanged with the solvent. If we therefore wish to replace them by deuterium atoms this must be done during synthesis of the macromolecule. Chemically such a process is only possible for very small proteins or peptides and the method of choice is via biosynthetic pathways. Molecules in which all, exchangeable and non- exchangeable hydrogen atoms have been replaced by deuterium of referred to as per-deuterated. The work on biosynthetic deuteration was pioneered by Crespi & Katz[1] some 40 years ago. It was first used extensively in neutron scattering experiments in the work on the 30S ribosome by Engelman and Moore[2]. The early work was carried out

before modern molecular biology methods were available and were therefore restricted to proteins which could be produced in organisms able to grow in a fully deuterated environment. Higher organisms can only tolerate low levels of deuteration and only bacteria, algae or yeasts can grow in 100% D_2O.

In the last 20 years methods have been developed to produce large amounts of deuterated protein in robust E. Coli expression systems that allow elevated levels of expression in deuterated media[3,4,5].

E. Coli cells are transformed using a plasmid containing the gene for the required protein. These cells are then grown in high density culture using a fully deuterated minimal media containing a deuterated carbon source such a glycerol or succinic acid. The cells must first be adapted to the deuterated medium which may require typically four to six cycles of dilution and growth and then final growth in a computer controlled fermenter to optimize supply of nutrients, oxygen etc.

Protein purification is then carried out by te same procedure as used for hydrogenated protein. The level of isotopic incorporation can be determined by electrospray ionization mass spectrometry.

An alternative to using a deuterated carbon source has been described by Liu et al.[6] based on the earlier work of Crespi and Katz. Here, algae which are capable of autotrophic growth with only CO_2 as a carbon source are grown in D_2O to produce fully deuterated organisms which are then themselves a source of deuterated nutrients for subsequent cultivation of bacteria.

APPLICATIONS

Deuteration is of major importance in all applications of neutron scattering in biology. Here we shall describe examples of its application in just two fields, that of high resolution protein crystallography and in small angle scattering from solutions of biological macromolecules.

Neutron Protein Crystallography

X-ray protein crystallography is the foremost technique used in structural biology and has been extraordinarily successful in solving many protein structures in particular. However, due to the fact that X-rays are scattered by electrons, as explained earlier, then the scattering from hydrogen atoms is very weak and in practice only when very well ordered crystals diffracting to better than 1 Å resolution are available can hydrogen atoms be observed in protein crystals. Neutron diffraction would therefore indeed be the method of choice but for the fact that neutron sources are generally much weaker than X-ray sources and data consequently if lower quality. For many years this limited the use of neutrons diffraction to just a handful of proteins where crystals of several cubic millimeters could be grown to compensate the low neutron flux. Now, however, with the advent of a new generation of neutron diffractometers, such as LADI3 at the ILL, and the development of deuteration techniques to reduce the incoherent background, crystals of less than 1mm^3 or less can be used for data collection. The smallest crystal used to date has been one of aldose reductase[7], a 45Kd protein, where a 0.15mm^3 sample of perdeuterated protein produced data to 2.3Å resolution. At this resolution (compared to <1 Å for X-rays) it was possible to clearly visualize hydrogen atoms. Figure 2 shows a histidine residue of the aldose reductase molecule with hydrogen bond contact to a water molecule. The blue density represents the neutron scattering density at a resolution of 2.3A whereas the red density is from the X-ray crystallographic data at 0.66A. The X-ray data show very clearly the carbon and hydrogen atoms but only the neutron data (at much lower resolution) show the hydrogen atoms. From this it can be seen that only one of the nitrogen atoms is protonated, the second one being a hydrogen bond acceptor from the well-defined water molecule.

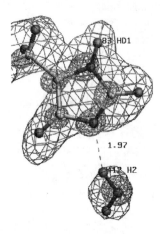

FIGURE 2. Scattering length density from a histidine residue in aldose reductase. Blue density from the 2.2Å neutron structure and red from the 0.66Å X-ray structure.

Small Angle Neutron Scattering (SANS)

Deuteration of biological macromolecules for neutron scattering was first developed for SANS studies of the ribosome[2]. In this case single proteins or pairs of deuterated proteins were reconstituted into otherwise hydrogenated ribosomal subunits in order to determine the distances between the constituent proteins and then by triangulation determine the overall quaternary structure of the particle. No attempt was made in the early work to describe the shapes of the sub-units.

More recently deuteration has been used to determine the positions and interactions of proteins in multi-protein complexes. Such studies have benefited not only from advances made in deuteration techniques but also in the striking progress made in the determination of molecular shapes in solution from small angle scattering. Methods exist now to reconstruct molecular shapes either *ab initio*[8] or using partial data from X-ray crystallographic studies[9]. One example of the latter technique is the structure of the trimeric protein troponin from skeletal muscle[10]. Troponin is a Ca^{2+}-sensitive switch that regulates the contraction of vertebrate striated muscle by participating in a series of conformational events within the actin-based thin filament. Troponin is a heterotrimeric complex consisting of a Ca^{2+}-binding subunit (TnC), an inhibitory subunit (TnI), and a tropomyosin-binding subunit (TnT). Ternary troponin complexes were produced by assembling recombinant chicken skeletal muscle TnC, TnI and the C-terminal portion of TnT known as TnT2. A full set of small-angle neutron scattering data was collected from TnC-TnI-TnT2 ternary complexes, in which all possible combinations (7) of the subunits were deuterated, in both the calcium bound (+Ca^{2+}) and the calcium free (-Ca^{2+}) states. These were made from samples dissolved in 41.6% D$_2$O/58.4% H$_2$O such that the hydrogenated sub-units were contrast matched. Small-angle X-ray scattering data were also collected from the same troponin TnC-TnI-TnT2 complex. Analysis of the lowest angle data showed that the complex was monomeric in solution and that there was a large change in the radius of gyration of the TnI sub-unit when it went from the +Ca^{2+} to the -Ca^{2+} state. A rigid-body Monte Carlo optimization procedure was then used starting with a model based on the human cardiac troponin crystal structure, to yield models of chicken skeletal muscle troponin, in solution, in the presence and in the absence of regulatory calcium. It was assumed that the tightly folded parts of the structure were conserved between cardiac and skeletal troponins but that all domains could move relative to each other. The optimization was carried out simultaneously against all of the scattering data sets.

Figure 3 shows the molecular models obtained for the troponin molecule in both the Ca-bound and Ca-free states and Figure 4, as an example, the scattering curves for the calcium-free state.

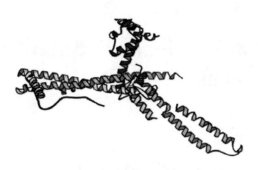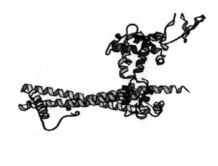

FIGURE 3. Molecular models of Troponin without (left) and with (right) bound calcium. TnC - red, TnT – yellow and TnI-blue. Note the large conformational change in TnI on binding of calcium.

A key feature is that TnC adopts a dumbbell conformation in both the $+Ca^{2+}$ and $-Ca^{2+}$ states. The most significant feature however is the change in conformation of the I sub-unit when Ca^{2+} is bound.

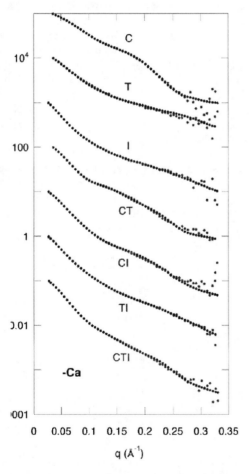

Figure 4. Fit of measured scattering curves (red) to scattering calculated from the model for Troponin (blue) in the absence of Ca^{2+}. The letters on each scattering curve indicate the "visible" sub-units, i.e. those not contrast matched.

CONCLUSIONS

Deuteration is playing an increasingly important part in neutron scattering studies of biological systems. Both the reduction in incoherent background and the increase in scattering at atomic resolution or in contrast variation contribute to this increase.

ACKNOWLEDGEMENTS

I thank M. Blakeley for Fig.2. Sincere thanks are due to the members of the EMBL/ILL Deuteration laboratory including M. Hartlein, T. Forsyth, M. Moulin, M. Blakeley, P. Callow and others for discussions concerning deuteration.

REFERENCES

1. H.L. Crespi, J.J. Katz, *Methods Enzymol.* **26**, 627-637 (1972)
2. D.M. Engelman, P.B. Moore, *Proc. Natl. Acad. Sci. USA* **69**, 1997-9 (1972)
3. K. Vanatalu, T. Paalme, R. Vilu, N. Burkhardt, R. Junemann, R.P. May, M. Ruhl, J. Wadzack, K.H. Nierhaus, *Eur. J. Biochem.* **216**, 315-321 (1993)
4. F. Meilleur, J. Contzen, D.A. Myles, C. Jung, *Biochemistry*, **43**, 8744-8753 (2004)
5. J-B. Artero, M. Hartlein, S. McSweeney, S., P.A. Timmins, *Acta Cryst.* **D61**, 1541-1549 (2005)
6. X. Liu, B.L. Hanson, P. Langan, R.E. Viola, *Acta Cryst.* **D63**, 1000-1008, (2007)
7. M. Blakeley, F. Ruiz, R. Cachau, I. Hazemann, F. Meilleur, A. Mitschler, S. Ginell, P. Afonine, O.N. Ventura, A. Cousido-Siah, M. Haertlein, A. Joachimiak, D. Myles, A. Podjarny, *Proc. Natl. Acad. Sci. USA* in press (2008)
8. D.I. Svergun, *Biophys J.*, **76**, 2879-86, (1999)
9. M.V. Petoukhov, D.I. Svergun, *Eur Biophys J.* **35**, 567-576 (2006)
10. W.A. King, D.B. Stone, P.A. Timmins, T. Narayanan, A.A.M. von Brasch, R.A. Mendelson, P.M.G. Curmi, *J. Mol. Biol.* **345**, 797-815, (2005)

Membrane Structure Studies by Means of Small-Angle Neutron Scattering (SANS)

R. B. Knott[1,2]

[1]Bragg Institute, ANSTO, Private Mail Bag, Menai NSW 2234, Australia
[2]CSIRO Minerals, Box 312, Clayton South VIC 3169, Australia

Abstract. The basic model for membrane structure – a lipid bilayer with imbedded proteins – was formulated 35 years ago, however the detailed structure is still under active investigation using a variety of physical, chemical and computational techniques. Every biologically active cell is encapsulated by a plasma membrane with most cells also equipped with an extensive intracellular membrane system. The plasma membrane is an important boundary between the cytoplasm of the cell and the external environment, and selectively isolates the cell from that environment. Passive diffusion and/or active transport mechanisms are provided for water, ions, substrates etc. which are vital for cell metabolism and viability. Membranes also facilitate excretion of substances either as useful cellular products or as waste. Despite their complexity and diverse function, plasma membranes from quite different cells have surprisingly similar compositions. A typical membrane structure consists of a phospholipid bilayer with a number of proteins scattered throughout, along with carbohydrates (glycoproteins), glycolipids and sterols. The plasma membranes of most eukaryotic cells contain approximately equal weights of lipid and protein, which corresponds to about 100 lipid molecules per protein molecule. Clearly, lipids are a major constituent and the study of their structure and function in isolation provides valuable insight into the more complex intact multicomponent membrane. The membrane bound protein is the other major constituent and is a very active area of research for a number of reasons including the fact that over 60% of modern drugs act on their receptor sites. The interaction between the protein and the supporting lipid bilayer is clearly of major importance. Neutron scattering is a powerful technique for exploring the structure of membranes, either as reconstituted membranes formed from well characterised lipids, or as intact membranes isolated from selected biological systems. A brief summary of membrane structure will be followed by an outline of the neutron scattering techniques used to understand membrane structure and dynamics. The emphasis will be on the small angle neutron scattering technique since there is a very powerful instrument at Serpong, however brief mention of other techniques will be included to demonstrate how a multidisciplinary approach is usually required

Keywords: neutron scattering, membrane structure, biology, SANS

CP989, *Neutron and X-ray Scattering in Materials Science and Biology, International Conference on Neutron and X-ray Scattering 2007*, edited by A. Ikram, A. Purwanto, Sutiarso, A. Zulfia, S. Hendrana, and Z. Nurachman
© 2008 American Institute of Physics 978-0-7354-0508-0/08/$23.00

Structure and Function of Glucansucrases

B. W. Dijkstra and A. Vujičić-Žagar

Laboratory of Biophysical Chemistry, University of Groningen
Nijenborgh 4, 9747 AG Groningen, The Netherlands

Abstract. Glucansucrases are relatively large (~160 kDa) extracellular enzymes produced by lactic acid bacteria. Using sucrose as a substrate they synthesize high molecular mass glucose polymers, called α-glucans, which allow the bacteria to adhere to surfaces and create a biofilm. The glucan polymers are of importance for the food and dairy industry as thickening and jellying agents. An overview is given of the current insights into the structure and functioning of these and related enzymes

Keywords: glucansucrase; amylosucrase; crystal structure; catalysis; catalytic mechanism.
PACS: 87.17.Uv

INTRODUCTION

In 1861 Louis Pasteur inferred that the jellifying and thickening texture of sugar cane and sugar beet syrups was of microbial origin. Twenty years later, in the 1880s, lactic acid bacteria of the species *Leuconostoc mesenteroides* were identified as the organisms that were responsible for the thickening by producing a viscous, high molecular-weight glucan polysaccharide [1-3]. The polymer was named dextran (from dextrorotatory), but how it was produced by the bacteria remained obscure until the 1950s, when it was discovered that it was produced by an enzyme. This enzyme converts sucrose (β-D-fructofuranosyl-(2,1)-α-D-glucopyranoside) into the dextran polymer with fructose monomers as a side product. Hence, the name dextransucrase was coined for the enzyme, which was later generalized to glucansucrase.

The glucans produced by glucansucrases have vast potential application. They are already used in the food industry in the production of dairy products because of their thickening and water-binding properties. They are also used as prebiotics[a] and as bioactive agents such as immune-modulators, and anti-ulcer and cholesterol-lowering agents. Chemically or physically modified glucans are promising as metal-complexing compounds, emulsifiers, detergents, bleaching assistants, strengthening additives, and in waste-water purification [4,5]. It has also been discovered that some glucans have anti-corrosive properties (see *e.g.*

---

[a] Prebiotics are non-digestible food additives that stimulate the growth of beneficial intestinal bacteria.

http://www.bioprimer.nl/pdf/2005002.pdf)[6].This opens up a vast perspective of replacing common anti-corrosive treatments that make use of heavy metal-containing alloys by an environment-friendly alternative for corrosion prevention or delay

CLASSES OF GLUCANSUCRASES

The properties of the high molecular weight glucans are determined by the glucansucrases that produce them. Glucansucrases are relatively large, extracellular enzymes (~160 kDa) that are secreted by lactic acid bacteria of the genera *Streptococcus, Leuconostoc, Weissella, Lactobacillus* and *Oenococcus*[4]. Their amino acid sequences are ~30 to ~80% identical, which led to their classification in a single homology family (GH70) in the Glycoside Hydrolase section of the Carbohydrate-Active Enzymes database (see http://www.cazy.org/).

On the basis of the size and glycosidic linkage type of their glucan products glucansucrases have been categorized in four different groups. The first group comprises dextransucrases. These enzymes are mainly found in *Leuconostoc* spp. and make glucans mainly having α(1→6) linkages [1,7]. Dextrans are known for their thickening and jellifying properties, and are nowadays used as chromatographic column material and in microsurgery [1,8].

The second group consists of mutansucrases, which are mainly found in oral *Streptococcus* spp. These enzymes produce water-insoluble mutan, a mainly α(1→3) linked glucose polymer. They play an

CP989, Neutron and X-ray Scattering in Materials Science and Biology, International Conference on Neutron and X-ray Scattering 2007, edited by A. Ikram, A. Purwanto, Sutiarso, A. Zulfia, S. Hendrana, and Z. Nurachman
© 2008 American Institute of Physics 978-0-7354-0508-0/08/$23.00

important role in the formation of dental plaque and induction of dental caries [9].

The third group is made up of alternansucrase, an enzyme only reported for *Leuconostoc mesenteroides* NRRL B-1355. It alternates the formation of α(1→3) and α(1→6) bonds, and therefore the synthesized α-glucan is named alternan [10].

Finally, the fourth group contains reuteransucrases. These enzymes, which include GTFA and GTFO isolated from *Lactobacillus reuteri* strains 121 and ATCC55730, make reuteran, a glucose polymer linked by α(1→4) and α(1→6) glycosidic bonds [11-14].

DOMAINS IN GLUCANSUCRASE

Analysis of the amino acid sequences of GH70 glucansucrases led to the hypothesis that these proteins have three putative domains: *1)* an N-terminal variable domain (~120-700 residues), preceded by a secretion peptide (36-40 residues), *2)* a conserved catalytic domain (~900 residues), and *3)* a C-terminal domain (~300-400 residues) [4].

N-terminal Variable Domain

This domain is variable in both length and composition. Its primary structure contains different repeating units not found before (see Table 1 for consensus sequences of these sequence motifs). The "RDV" motif occurs in *L. reuteri* 121 GTFA, *L. reuteri* 180 GTF180, and others [15]; the "motif-T" was found in *L. mesenteroides* DSRT [16], the "motif-S" in *L. mesenteroides* DSRE [17], while the "TTQ" motif was identified in *Lactobacillus parabuchneri* 33 GTF33 [15]. A-repeats are usually found at the C-terminus of GH70 glucansucrases, but the motif occurs

at the N- terminus of *L. mesenteroides* DSRB, DSRS and ASR [18]. The role of the N domain and its repeats remain unclear to date. Its deletion in *L. reuteri* GTFA and *L. parabuchneri* GTF33 had only minor effects on the glucan product [15], and deletion of the domain in the *S. sobrinus* glucansucrase also had no effect on the enzyme's activity [19]. Similar results were found for *S. downei* GTFI [20] and *L. mesenteroides* DSRS [21]. One member of the GH70 family, the *L. mesenteroides* DSRA dextransucrase even completely lacks this N-terminal domain [22]. Thus, its function is apparently not essential.

Catalytic Domain

This domain contains amino acid stretches that show high sequence similarity to four sequence motifs (I-IV; see Table 2) identified in the catalytic (β/α)$_8$-barrel domain of amylolytic enzymes from Glycoside Hydrolase families 13 and 77 (http://www.cazy.org). The (β/α)$_8$-barrel fold, in which eight parallel β-strands are surrounded by eight α-helices, was first found in chicken muscle triosephosphate isomerase and is also known as the TIM-barrel fold. It is one of the most ubiquitous folds and is found in all kingdoms of life. Proteins with a (β/α)$_8$-barrel fold are almost always enzymes involved in energy and molecular metabolism [23]. The strictly conserved amino acid residues in the four sequence motifs (Table 2) are involved in catalysis and substrate binding. Therefore, the GH13, GH70, and GH77 enzyme families are believed to share the same catalytic mechanism and are grouped together in the α-amylase super-family or GH-H clan [3].

TABLE 1. Sequence repeats found in the N- and C-terminal ends of GH70 glucansucrases

Repeat Units	Consensus Sequence
RDV	R(P/N)DV-x$_{12}$-SGF-x$_{19-22}$-R(Y/F/D)S
Motif T	TDDKA(A/T)TTA(A/D)TS
Motif S	PA(A/T)DKAVDTTP(A/T)T
TTQ	TTTQN(A/T)(P/A)NN(S/G)N(D/G)PQS
CW_binding_1^b	NGWIKDNGNWYYFDSDGKM
A	WYYFNxDGQAATGLQTIDGQTVFDDNGxQVG
B	VNGKTYYFGSDGTAQTQANPKGQTFKDGSVLRFYNLEGQYVSGSGWY
C	GKIFFDPDSGEVVKNRFV
D	GGVKNADGTYSKY
YG	NDGYYFxxxGxxH^{O}x(G/N)XH^{O}H^{O}H^O
ASR-repeats	DGx$_4$APY
KYQ	AVK(T/A)A(K/Q)(A/T)(Q/K)(L/V)(A/N)K(T/A)KAQ(I/V)(A/T)KYQ-KALKKAKTTKAK(A/T)QARK(S/N)LKKA(E/N)(T/S)S(F/L)(S/T)KA

Sequence repeats "RDV" to "TTQ" are found at the N-terminus while repeats A to "KYQ" are found at the C-terminus of the polypeptide chain. In some GH70 GTFs the A repeats may also be found at the C-terminus, see text. Legend: x = non-conserved amino acid residue, H^O = hydrophobic residue. ^{b}Pfam database [2]. For references see text

The order of the four conserved sequence motifs in the primary structure of GH70 glucansucrases is different from that found in the GH13 and GH77 families due to a "circular permutation" event that must have occurred during evolution of the GH70 glucansucrases[24]. In the GH13/GH77 families the $(\beta/\alpha)_8$-barrel starts with strand β1, followed by helix α1, β2α2, etc. and it ends with strand β8 followed by helix α8 (β1α1, β2α2, … β8α8 arrangement, Fig. 1). The GH70 "circularly permuted" $(\beta/\alpha)_8$-barrel starts with an α-helix which corresponds to helix α3 of the GH13/GH77 barrel and ends with strand β3 (α3, β4α4,…β8α8, β1α1, β2α2, β3 arrangement, Fig. 1). The four conserved motifs in the GH13/GH77 families are found at the C-terminal ends of β-strands β3 (motif I), β4 (motif II), β5 (motif III) and β7 (motif IV). Instead, in GH70 glucansucrases, sequence motif I is found C-terminally of motifs II, III and IV. Thus the sequence motifs I, II, III and IV are located at the C-terminal ends of β-strands β8 (motif I), β1 (motif II), β2 (motif III), and β4 (motif IV). These β-strands are equivalent to strands β3, β4, β5, and β7 of the GH13 family $(\beta/\alpha)_8$-barrel, respectively.

C-Terminal Domain

This domain, which consists of ~300-400 residues, follows the catalytic domain in all GH70 glucansucrases. It contains tandem sequence repeats that vary in composition, length and number, and that are specific for each glucansucrase enzyme [7]. Sequence repeats A and C (see Table 1) are found in *Streptococcus* and *Leuconostoc* species [7]. Repeats B are found in *S. downei* GTFI [25]; D repeats have solely been identified in *Streptococcus salivarius* ATCC 25975, while N-repeats have so far been only identified in *L. mesenteroides* DSRS [26]. Other sequence repeats unique to *L. mesenteroides* ASR [10] and *L. parabuchneri* GTF33 [15], respectively, have also been identified (Table 1). Within the A, B, C, and D repeats, so-called YG-repeats could be discerned. They occur in *Streptococcus* and *Lactobacillus* glucansucrases. The 33 residues long A repeats resemble cell wall binding repeats (~20 residues long, CW_binding_1, Pfam database [2]). Both repeats consist of conserved aromatic and glycine residues. According to Janeček *et al.* [18], one and a half CW binding repeat would correspond to one A repeat, but Shah *et al.* [27] argue that the A repeats are different from the CW binding repeats. The CW binding repeats are found in *Clostridium difficile* and *Clostridium sordelli* toxins [28], and in choline-binding proteins of *Streptococcus pneumoniae* [29]. The C-terminal repetitive domain (CRD) of *Clostridium* toxins binds carbohydrates in the gut epithelium, while choline-binding proteins bind choline-containing teichoic acids found in *S. pneumoniae* cell walls. Because of the similarity of the A and CW repeats, *Streptococcus downei* GTFI was tested for the ability to bind choline molecules, but no evidence for such binding was obtained [27].

The C-terminal domains of the glucansucrases from oral streptococci, which contain A, B, and C repeats [7], bind glucans such as dextran and were thus called glucan binding domains (GBD) [27]. They are believed to be responsible for the sucrose-dependent attachment of the bacteria to tooth surfaces. Involvement of the domain in glucan binding is well documented for oral streptococcal glucansucrases, but the term glucan binding domain is also widely used for the C-terminal domains of other GH70 glucansucrases [7], even though there is no clear evidence that these domains are involved in glucan binding [27].

TABLE 2. Conserved sequence motifs of the catalytic domains of *L. reuteri* GTF180 glucansucrase (GH70) aligned with *N. polysaccharea* amylosucrase (GH13), *A. oryzae/niger* (TAKA) α-amylase (GH13) and *B. circulans* CGTase (GH13).

Motif I		
Glucansucrase	1503	ADWVPDQ
Amylosucrase	181	VDFIF**NH**
TAKA-amylase	116	VDVVA**NH**
CGTase	134	IDFAP**NH**
Consensus		XDXXX**NH**
Motif II		
Glucansucrase	1021	GI**R**VDAVDN
Amylosucrase	282	IL**R**MDAVAF
TAKA-amylase	202	GL**R**IDTVKH
CGTase	225	GI**R**MDAVKH
Consensus		GX**R**XDXXZZ
Motif III		
Glucansucrase	1059	INIL**E**DWG
Amylosucrase	324	FFKS**E**AIV
TAKA-amylase	226	YCIG**E**VLD
CGTase	253	FTFG**E**WFL
Consensus		XXXX**E**ZZZ
Motif IV		
Glucansucrase	1131	FVRA**HD**
Amylosucrase	388	YVRS**HD**
TAKA-amylase	292	FVEN**HD**
CGTase	323	FIDN**HD**
Consensus		XXBB**HD**

AMYLOSUCRASE

So far no crystal structure has been published of a family GH70 glucansucrase, which could confirm the above given domain organization and clarify the function of the sequence repeats. However, some of the properties of GH70 glucansucrases may be inferred from the crystal structure of the GH13 amylosucrase

(E.C. 2.4.1.4) from *Neisseria polysaccharea*. The latter enzyme catalyses the transfer of the sucrose glucosyl moiety to an acceptor molecule with formation of exclusively α(1→4) glucosidic bonds [30]. In this way an amylose-like insoluble glucose polymer is obtained.

The *in vivo* function of the enzyme is believed to be the extension of glycogen-like oligosaccharides, since a 98-fold increase in k_{cat} is observed in the presence of such oligosaccharides [31,32].

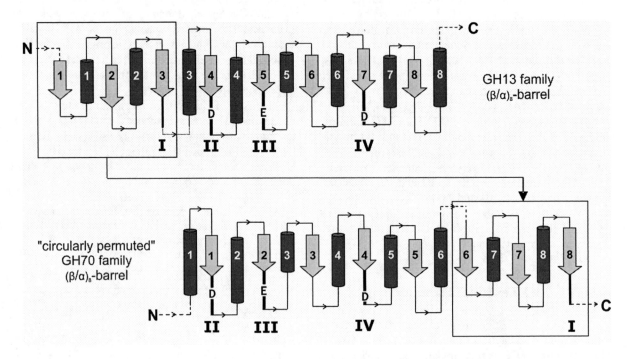

FIGURE 1. Order of the $(\beta/\alpha)_8$-barrel elements in the GH13 and GH70 family enzymes.

The *N. polysaccharea* amylosucrase is a 72 kDa enzyme, and thus significantly smaller than the GH70 glucansucrases. Its crystal structure revealed that the polypeptide chain is folded into five distinct domains. An α-helical domain N is located at the N-terminus. It is followed by the catalytic domain A, which is folded into a $(\beta/\alpha)_8$-barrel. Domains B and B' are inserted between strand β3 and helix α3, and strand β7 and helix α7 of the $(\beta/\alpha)_8$-barrel, respectively, while domain C, which has an eight-stranded β-sandwich fold, is located at the C-terminus of the polypeptide chain. Domains A, B and C are found in other GH13 family members such as *A. niger* α-amylase [33], cyclodextrin glycosyl-transferase, and oligo-1,6-glucosidase, but domains N and B' are specific to amylosucrase. Domain B in *N. polysaccharea* amylosucrase consists of two short antiparallel β-sheets, which are sandwiched between two α-helices. This domain, which contributes to substrate binding [34], varies considerably in both sequence and length among the GH13 enzymes [35]. In contrast, domain C, which is built up of β-strands in a Greek-key motif, is generally structurally conserved. This domain is involved in raw starch and oligosaccharide binding [36,37]. The amylosucrase-specific domain B' consists of three short α-helices and a short β-sheet; it is believed to assist in elongation of glycogen branches *via* conformational movements [31].

The active site of amylosucrase is located at the bottom of a deep and narrow pocket [38]. Sucrose binds with its glucosyl moiety in subsite -1 and with its fructosyl moiety in subsite +1. Co-crystallization experiments with maltoheptaose allowed mapping of subsites +2 to +5, and revealed also the presence of two other binding sites located at the surface of the enzyme in domains B' and C[37]. In contrast to the endo-acting enzymes from the GH13 family, such as TAKA-amylase, in which the active site is a long cleft containing subsites mapped from -3 to +5 [33,39], the active site of amylosucrase is a pocket closed off by a salt-bridge, which does not extend beyond subsite -1. Complexes with sucrose [38] and a covalently bound glucosyl enzyme intermediate [40] showed that the amino acid residues involved in binding the glucosyl moiety of sucrose and in cleavage of the sucrose glycosidic bond are strictly conserved, as expected for a GH13 family enzyme (Fig. 1).

Upon cleavage of the sucrose glycosidic bond and formation of the covalent intermediate, the fructosyl moiety should leave the active site making place for the non-reducing end of an acceptor molecule (a glycogen branch). After the transfer reaction, the acceptor molecule elongated by one glucose residue leaves the active site and a new sucrose molecule can enter. The B' sugar binding site is most probably the "anchoring platform" that captures the acceptor polymer (glycogen) and directs it towards the active site. Thus, a semi-processive elongation of the glycogen by *N. polysaccharea* amylosucrase has been proposed [31].

Albenne *et al.* [31] have suggested that binding of glycogen in proximity of the active site results in exclusion of water molecules, which could explain the absence of sucrose hydrolysis in the presence of glycogen. Additionally, as recently described for GH77 amylomaltases [41], the covalent intermediate of amylosucrase is protected from hydrolysis in three different ways. The -1 glucose is bound in its low energy $^4C_1$ chair conformation, the acid/base Glu328 is not productively positioned to activate a water molecule, and the ester bond between the nucleophile (Asp286) and the glucose ring is almost perpendicular to the plane of the sugar making the C1 atom less accessible to an incoming nucleophile. Thus, this transglycosylase from family GH13 protects its covalent intermediate in the same way as the related GH77 family enzymes.

GLUCANSUCRASE MECHANISM

Based on the sequence similarity to enzymes of the GH13 α-amylase family, including a glucansucrase (*N. polysaccharea* amylosucrase) a reaction mechanism similar to that of GH13 enzymes has been proposed for GH70 glucansucrases. The reaction is supposed to proceed *via* a double displacement mechanism, involving a covalent β-glucosyl-enzyme intermediate, and retaining the α-anomeric configuration of the substrate in the product [4,21]. However, how GH70 glucansucrases transfer the covalently linked glucosyl moiety to an acceptor during the second half of the reaction and, in particular to which end of the acceptor molecule, has been a matter of debate for several decades. Altogether, two possibilities have been proposed. The first one is analogous to the general mechanism of GH13 enzymes as described above, in which the covalently linked glucosyl moiety is transferred to the non-reducing end of the growing glucan chain[7]. The other proposed mechanism involves two catalytic centres and transfers the glucosyl-moiety to the reducing end of a polymer. Nowadays, the first reaction mechanism has found

wider acceptance, because of the successful trapping of a single β-glucosyl-enzyme intermediate in GH70 *Streptococcus sobrinus* glucansucrase by Mooser *et al.* [42] and the crystallographic analysis of the glucosyl-enzyme intermediate of the GH13 glucansucrase *N. polysaccharea* amylosucrase [40]. In addition, Moulis *et al.*[21] have recently reported detailed biochemical and mutagenesis experiments with the *Leuconostoc mesenteroides* alternansucrase and dextransucrase, which also support a single active site and a reaction mechanism with elongation of the polymer chain at the non-reducing end.

OUTLOOK AND PROSPECTS

GH70 glucansucrases are relatively large (~160 kDa) extracellular enzymes produced by lactic acid bacteria. Using sucrose as a substrate they synthesize high molecular mass glucose polymers, called α-glucans. Depending on whether the C2, C3, C4 or C6 hydroxyl group of the acceptor glucose points towards the covalently bound glucosyl intermediate, $\alpha(1\rightarrow2)$, $\alpha(1\rightarrow3)$, $\alpha(1\rightarrow4)$, or $\alpha(1\rightarrow6)$ glucosidic bonds may be formed. This broad product specificity results in products with markedly different physical and rheological properties, which make them interesting compounds for various industrial applications. To date, the lack of any structural information has hampered the in-depth understanding of the molecular factors that are responsible for the formation of the various products. However, with the recent crystallization of an N-terminally truncated, fully active glucansucrase from *Lactobacillus reuteri* 180[43] structural information will soon become available providing the molecular details for further investigations of how these intriguing enzymes convert sucrose to those high molecular weight polysaccharides with highly interesting properties.

ACKNOWLEDGMENTS

This work was supported by the Ministry of Economic Affairs of The Netherlands *via* the IOP Genomics program (project IGE01021).

REFERENCES

1. V. Crescenzi, *Biotechnol. Prog.* **11**, 251-259 (1995).
2. A. Bateman, L. Coin, R. Durbin, R. D. Finn, V. Hollich, S. Griffiths-Jones, A. Khanna, M. Marshall, S. Moxon, E. L. Sonnhammer, D. J. Studholme, C. Yeats, and S. R. Eddy, *Nucleic Acids Res.* **32**, D138-141 (2004).
3. E. A. MacGregor, S. Janeček, and B. Svensson, *Biochim. Biophys. Acta* **1546**, 1-20 (2001).

4. S. A. F. T. van Hijum, S. Kralj, L. K. Ozimek, L. Dijkhuizen, and G. H. van Geel-Schutten, *Microbiol. Mol. Biol. Rev.* **70**, 157-176 (2006).

5. G. H. Van Geel-Schutten, *United States Patent 20050059633 A1* (2005).

6. G. M. Ferrari and H. J. A. Breur, in *16th ICC Congress* (Beijing, China, 2005).

7. V. Monchois, R. M. Willemot, and P. Monsan, *FEMS Microbiol. Lett.* **23**, 131-151 (1999).

8. N. Jallali, *Microsurgery* **23**, 78-80 (2003).

9. M. Balakrishnan, R. S. Simmonds, and J. R. Tagg, *Aust. Dent. J.* **45**, 235-245 (2000).

10. M. A. Argüello-Morales, M. Remaud-Simeon, S. Pizzut, P. Sarçabal, and P. Monsan, *FEMS Microbiol. Lett.* **182**, 81-85 (2000).

11. G. H. van Geel-Schutten, E. J. Faber, E. Smit, K. Bonting, M. R. Smith, B. Ten Brink, J. P. Kamerling, J. F. Vliegenthart, and L. Dijkhuizen, *Appl. Environ. Microbiol.* **65**, 3008-3014 (1999).

12. S. Kralj, G. H. van Geel-Schutten, H. Rahaoui, R. J. Leer, E. J. Faber, M. J. van der Maarel, and L. Dijkhuizen, *Appl. Environ. Microbiol.* **68**, 4283-4291 (2002).

13. S. Kralj, G. H. van Geel-Schutten, M. J. van der Maarel, and L. Dijkhuizen, *Microbiol.* **150**, 2099-2112 (2004).

14. S. Kralj, E. Stripling, P. Sanders, G. H. van Geel-Schutten, and L. Dijkhuizen, *Appl. Environ. Microbiol.* **71**, 3942-3950 (2005).

15. S. Kralj, G. H. van Geel-Schutten, M. M. Dondorff, S. Kirsanovs, M. J. van der Maarel, and L. Dijkhuizen, *Microbiol.* **150**, 3681-3690 (2004).

16. K. Funane, K. Mizuno, H. Takahara, and M. Kobayashi, *Biosci. Biotechnol. Biochem.* **64**, 29-38 (2000).

17. S. Bozonnet, M. Dols-Laffargue, E. J. Faber, S. Pizzut, M. Remaud-Simeon, P. Monsan, and R. M. Willemot, *J. Bacteriol.* **184**, 5753-5761 (2002).

18. Š. Janeček, B. Svensson, and R. R. Russell, *FEMS Microbiol. Lett.* **192**, 53-57 (2000).

19. H. Abo, T. Matsumura, T. Kodama, H. Ohta, K. Fukui, K. Kato, and H. Kagawa, *J. Bacteriol.* **173**, 989-996 (1991).

20. V. Monchois, M. A. Argüello-Morales, and R. R. Russell, *J. Bacteriol.* **181**, 2290-2292 (1999).

21. C. Moulis, G. Joucla, D. Harrison, E. Fabre, G. Potocki-Veronese, P. Monsan, and M. Remaud-Simeon, *J. Biol. Chem.* **281**, 31254-31267 (2006).

22. V. Monchois, R. M. Willemot, M. Remaud-Simeon, C. Croux, and P. Monsan, *Gene* **182** (1996).

23. N. Nagano, C. A. Orengo, and J. M. Thornton, *J. Mol. Biol.* **321**, 741-765 (2002).

24. E. A. MacGregor, H. M. Jespersen, and B. Svensson, *FEBS Lett.* **378**, 263-266 (1996).

25. J. J. Ferretti, M. L. Gilpin, and R. R. Russell, *J. Bacteriol.* **169**, 4271-4278 (1987).

26. V. Monchois, A. Reverte, M. Remaud-Simeon, P. Monsan, and R. M. Willemot, *Appl. Environ. Microbiol.* **64**, 1644-1649 (1998).

27. D. S. Shah, G. Joucla, M. Remaud-Simeon, and R. R. Russell, *J. Bacteriol.* **186**, 8301-8308 (2004).

28. J. S. Moncrief and T. D. Wilkins, *Curr. Top. Microbiol. Immunol.* **250**, 35-54 (2000).

29. E. García, J. L. García, P. García, A. Arrarás, J. M. Sánches-Puelles, and R. López, *Proc. Natl. Acad. Sci. USA* **85**, 914-918 (1988).

30. G. Potocki de Montalk, M. Remaud-Simeon, R. M. Willemot, P. Sarçabal, V. Planchot, and V. Monchois, *FEBS Lett.* **471**, 219-223 (2000).

31. C. Albenne, L. K. Skov, V. Tran, M. Gajhede, P. Monsan, M. Remaud-Simeon, and G. André-Leroux, *Proteins* **66**, 118-126 (2007).

32. G. Potocki de Montalk, M. Remaud-Simeon, R. M. Willemot, and V. Monchois, *FEMS Microbiol. Lett.* **186**, 103-108 (2000).

33. A. Vujičić-Žagar and B. W. Dijkstra, *Acta Crystallogr.* **F62**, 716-721 (2006).

34. B. A. Van der Veen, J. C. Uitdehaag, B. W. Dijkstra, and L. Dijkhuizen, *Biochim. Biophys. Acta* **1543**, 336-360 (2000).

35. Š. Janeček, *Progr. Biophys. Mol. Biol.* **67**, 67-97 (1997).

36. D. Penninga, B. A. Van der Veen, R. M. A. Knegtel, S. A. F. T. Van Hijum, H. J. Rozeboom, K. H. Kalk, B. W. Dijkstra, and L. Dijkhuizen, *J. Biol. Chem.* **271**, 32777-32784 (1996).

37. L. K. Skov, O. Mirza, D. Sprogoe, I. Dar, M. Remaud-Simeon, C. Albenne, P. Monsan, and M. Gajhede, *J. Biol. Chem.* **277**, 47741-47747 (2002).

38. O. Mirza, L. K. Skov, M. Remaud-Simeon, G. Potocki de Montalk, C. Albenne, P. Monsan, and M. Gajhede, *Biochemistry* **40**, 9032-9039 (2001).

39. A. M. Brzozowski and G. J. Davies, *Biochemistry* **36**, 10837-10845 (1997).

40. M. H. Jensen, O. Mirza, C. Albenne, M. Remaud-Simeon, P. Monsan, M. Gajhede, and L. K. Skov, *Biochemistry* **43**, 3104-3110 (2004).

41. T. R. M. Barends, J. B. Bultema, T. Kaper, M. J. E. C. van der Maarel, L. Dijkhuizen, and B. W. Dijkstra, *J. Biol. Chem.* **282**, 17242-17249 (2007).

42. G. Mooser and K. R. Iwaoka, *Biochemistry* **28**, 443-449 (1989).

43. T. Pijning, A. Vujičić-Žagar, S. Kralj, W. Eeuwema, L. Dijkhuizen, and B. W. Dijkstra, *Biocatal. Biotransform.* **in press** (2008).

Neutron Protein Crystallography: Beyond the Folding Structure

N. Niimura

Institute of Applied Beam Science, Graduate School of Science and Engineering, Ibaraki University,
4-12-1 Naka-Narusawa, Hitachi, Ibaraki 316-8511, Japan

Abstract. Neutron diffraction provides an experimental method of directly locating hydrogen atoms in proteins, a technique complementary to ultra-high-resolution X-ray diffraction. A neutron diffractometers for biological macromolecules has been constructed in Japan, and it has been used to determine the crystal structures of proteins up to resolution limits of 1.5-2.5 Å. Results relating to hydrogen positions and hydration patterns in proteins have been obtained from these studies. Examples include the geometrical details of hydrogen bonds, the role of hydrogen atoms in enzymatic activity, CH_3 configuration, H/D exchange in proteins and oligonucleotides, and the dynamical behavior of hydration structures, all of which have been extracted from these structural results and reviewed.

Keywords: Neutron, protein, crystallography, hydrogen, hydration, J-PARC, SNS
PACS: 83.85.Hf, 87.15.-v

INTRODUCTION

The three dimensional structure determinations of biological macromolecules such as proteins and nucleic acids by X-ray crystallography have improved our understanding of many of the mysteries involved in biological processes. At the same time, these results have clearly reinforced the commonly-held belief that hydrogen atoms and water molecules around proteins and nucleic acids play a very important role in many physiological functions. However, since it is very hard to identify hydrogen atoms accurately in protein molecules using X-ray diffraction alone, a detailed discussion of protonation and hydration sites can only be speculated upon so far. In contrast, it is very well known that neutron diffraction provides an experimental method of directly locating hydrogen atoms.

The development of a neutron imaging plate (NIP) became a breakthrough event in neutron protein crystallography[1] for reactor neutron sources. The first application of the NIP was a structure determination of tetragonal hen-egg-white lysozyme[2] using a quasi-Laue diffractometer, LADI[3], at the Institut Laue Langevin (ILL) in Grenoble, France. After that, we have constructed high-resolution neutron diffractometers dedicated to biological macromolecules (BIX-3, BIX-4), which use a monochromatized neutron beam as well as the NIP at

the Japan Atomic Energy Agency (JAEA)[4,5]. More recently, the LANSCE time-of-flight ^{3}He detector for neutron protein crystallography has started producing results[6]. Thus these three technical developments have greatly improved the capability of neutron protein crystallography: namely, the time needed to measure data, the diffraction resolution reached and the molecular weight ceiling reachable. As a result, the field has enjoyed a significant resurge in recent years.

Unfortunately, neutron protein crystallography still remains to this day a severely limited technique, but hopefully things will improve substantially in the coming future. The current development of "next generation" spallation neutron sources, such as J-PARC (the Japanese Proton Accelerator Research Complex) in Japan and SNS (the Spallation Neutron Source) in the USA, as well as new developments and improvements at existing sources (e.g., LADI-3 at the ILL) will enable several more powerful protein crystallographic instruments to be installed. In some of the afore-mentioned new spallation sources, a gain in neutron intensity of almost two orders of magnitude is expected. At that point, the use of neutron diffraction is expected to greatly expand the field of structural biology.

The general subject of neutron protein crystallography has been reviewed earlier by several authors[7-13]. Also of potential interest to readers are articles describing the synergy and complementarity between neutron diffraction and ultra-high-resolution

CP989, *Neutron and X-ray Scattering in Materials Science and Biology, International Conference on Neutron and X-ray Scattering 2007*, edited by A. Ikram, A. Purwanto, Sutiarso, A. Zulfia, S. Hendrana, and Z. Nurachman
© 2008 American Institute of Physics 978-0-7354-0508-0/08/$23.00

X-ray diffraction[14]. In this paper, we will summarize selected results regarding hydrogen positions and hydration in proteins, obtained mainly using the two BIX-type diffractometers in JAEA.

Diffractometer for neutron protein crystallography

The neutron imaging plate (NIP) consists of a neutron converter material, Gd, which captures neutrons and emits secondary charged particles, which are absorbed by a photostimulated luminescence (PSL) material ($BaFBr: Eu^{2+}$). The neutron capture efficiency at neutron wavelength 1Å is about 80%, and a cylindrical shape of the detector is easily obtained with the NIP since it is a large flexible (plastic) sheet. The high spatial resolution (less than 0.2 mm x 0.2 mm) of the NIP has allowed us to reduce the distance betwen the sample and the detector to 200 mm, which contributes to an overall decrease in size of the diffractometer[1]. After developing a novel practical neutron monochromator (a set of elastically bent perfect Si plates), two BIX-type diffractometers (BIX-3 and BIX-4)[4,5] dedicated for protein crystallography were constructed at Japan Atomic Energy Agency (JAEA). The most characteristic and novel design of BIX-type diffractometer is the overall vertical arrangement of the main components of the diffractometer (Fig. 1).

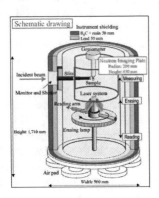

FIGURE 1. A schematic view of the instrument

This enables one to realize a compact design, and consequently a higher flux of neutrons is obtained because of the close proximity of the sample to the monochromator (i.e., to the neutron source). The BIX-3 and BIX-4 instruments are conceptually similar to the LADI diffractometer at the Institut Laue-Langevin (ILL) in Grenoble, France which uses a Laue geometry equipped with NIP and a broad spectral range to maximize the neutron flux on the sample[3,15], except that LADI is based on a horizontal configuration. In contrast, the PCS (Protein Crystallographic Station)

diffractometer[6] located at the LANSCE neutron spallation source in Los Alamos National Laboratory (New Mexico, U.S.A.) is unique in that it is the first instrument designed to use a pulsed neutron source for macromolecular crystallography.

Hydrogen and Hydration in Proteins Data Base (HHDB)

The results from neutron protein crystallography have been accumulating significantly in recent years and we have realized that it is becoming increasingly important to deposit this information in a publicly-accessible site so that this data can be retrieved more efficiently. We have created a 'Hydrogen and Hydration in Proteins' Data Base (HHDB) that catalogs all H atom positions in biological macromolecules and in hydration water molecules that have been determined thus far by neutron macromolecular crystallography. This allows important structural information such as the geometry of H-bonds to be abstracted and categorized for the first time in the field of structural biology.

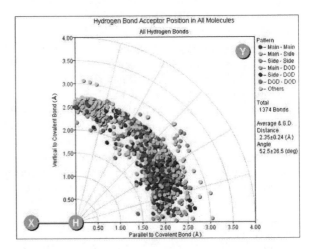

FIGURE 2. Schematic statistics of hydrogen bonds. Acceptorpositions (Y) are searched and plotted in the first zone. In thisexample, H-bond lengths are defined by X...Y values less than 2.7 Å, 2.6Å and 3.1Å for the acceptor atoms, N, O, and S,respectively; and H-bond angles are defined to be thesupplement of 180 degrees. These cut-off values (2.7 Å, 2.6Åand 3.1Å) are just default and users can modify them. Note thevery surprising result that in proteins there are very fewacceptor atoms which form truly co-linear X-H -----Y hydrogen bonds, in contrast to the situation in small inorganic and organic compounds in which linear H bonds are quite common. (Note:the word "others" on the right-hand side of this plot refers toentities other than protein residues or water molecules, such asheme groups, substrates, counter-ions, etc.)

The HHDB provides us with a graphic interface for visualizing all types of interactions involving H atoms, such as hydrogen bonds (defined by X–H...Y, where X is a hydrogen donor and Y is a hydrogen acceptor). For example, one type of plot (Fig. 2) allows the user to visualize the distribution of the acceptor atoms (Y) as a function of the H---Y distance and the supplementary X–H...Y angle (180° minus the actual angle). From these results, it is very surprising that in proteins there are very few acceptor atoms which form truly co-linear X-H---Y hydrogen bonds, in contrast to the situation in small inorganic and organic compounds in which linear H bonds are quite common. In the HHDB the user has the option of easily changing the cutoff H-bonding values (such as 2.7, 2.6 or 3.1 Å) if he or she wishes, and we found that the non-linearity of hydrogen bonds is not sensitive to what the cutoff values are. The strengths of H-bonds obtained in X-ray protein crystallography have often been assigned using the X---Y length as a criterion, but in our opinion this assumption may be somewhat risky, because this plot suggests that the usual argument that X-H---Y hydrogen bonds are approximately linear may not be valid in proteins. The HHDB depository has been developed by cooperation between the Japan Science and Technology Agency (JST) and the Japan Atomic Energy Agency (JAEA) under a database project funded by the JST. It is available via the website http://hhdb01.tokai-sc.jaea.go.jp/HHDB/. From the HHDB, we have provided an Internet link to the Protein Data Bank (PDB) and the protein code and the nomenclature used in both databases are the same.

RESULTS

Methyl CH_3 group configuration

In gaseous ethane, C-C bond rotates freely but a staggered conformation is most stable and an eclipsed conformation is most unstable. In proteins CH_3 group conformations in some amino acid residues such as alanine, valine, leucine, and isoleucine were not well discussed so far, because hydrogen atoms of these CH_3 groups were not identified well. Our high-resolution neutron diffraction analyses of myoglobin and other proteins have provided this information. As shown in Fig. 3, the hydrogen atoms of alanine and isoleucine in rubredoxin are clearly identified. These positions have been deposited in the HHDB. All the CH_3 group configurations in amino acid residues are extracted and discussed as follows: Fig. 4(a) shows CH_3 group

conformations as Newman projections and the dihedral angles of the CH_3 group are shown as curved arrows.

CH3 Configuration

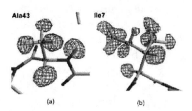

FIGURE 3. $|F_o|-|F_c|$ omit nuclear density map of the hydrogen atoms around the residues (a) Ala43 and (b) Ile7 of wild type rubredoxin.

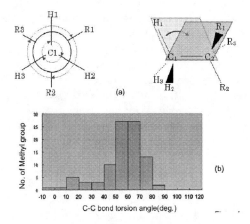

FIGURE 4. (a) CH_3 group conformations depicted by Newman projection formula, and (b) the distribution of torsional angles of CH_3 group in myoglobin.

Atoms H1, C1 and C2 and atoms R1, C2 and C1 form two planes and the dihedral angle between the two planes defines the conformation of the CH_3 group as being either staggered or eclipsed. In the myoglobin case, 92 CH_3 groups have been identified, their dihedral angles were calculated and their distribution is shown in Fig. 4(b). It is found that most of the CH_3 groups belong to the stable staggered conformations (angles close to 60°), but several % of them appear to have values close to eclipsed conformations (defined by angles close to 0°). The structural mechanism which allows H atoms to approach the higher-energy eclipsed conformations has been analyzed in the case of myoglobin, but these deviations may very well be due to steric interactions from neighboring atoms.

Hydrogen Bonds

Hydrogen bonds clearly play important roles in countless biological processes. The strengths of hydrogen bonds are intermediate between those of covalent bonds and van der Waals forces: they are directional and form different types of networks under various conditions. Along with other investigators, Baker and Hubbard have extensively discussed H-bonds in globular proteins, using hydrogen atom positions calculated using the atomic coordinates derived from high-resolution protein X-ray data[16]. It was pointed out that, while most of the H positions in a protein can be reliably predicted (for example, those of C-H bonds), H positions could not be uniquely defined for most O-H and S-H bonds, as well as some N-H bonds. In some cases (a) the orientation of the H atoms (e.g., alcoholic protons such as those of Ser, Thr and Tyr) may not be known, in other cases (e.g. Asp, Glu, His) the questions are (b) the extent of protonation (e.g., is the carboxylate group of an Asp residue protonated or not?), and (c) if protonated, which of the two competing sites (e.g., the two oxygens of Asp and Glu or the two nitrogens of His) is the protonated one? And these are often the most interesting protons as far as chemical reactivity is concerned. In their conclusion, Baker and Hubbard stressed the necessity of high-resolution neutron diffraction studies. In this article we will discuss several examples in which questions of this type have been answered-through single-crystal neutron work.

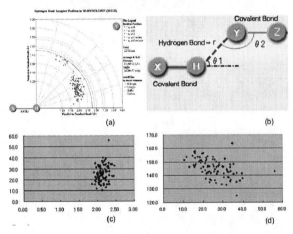

FIGURE 5. Hydrogen bonds configuration in α-helix of myoglobin. (a) Schematic statistics of hydrogen bonds N-H----O=C realize in a-helix of myoglobin provided by HHDB, (b) the geometrical configuration of the N-H----O=C hydrogen bonds, where the length between H and O is r and the angle NHO and COH are θ_1 and θ_2, respectively, (c) and (d) the correlation between r and θ_1, and between θ_1 and θ_2, respectively.

In the neutron analysis of the α-helices of myoglobin, the linearity of the X-H---Y hydrogen bonds has been analyzed. Fig. 5(a) shows the statistics of hydrogen bonds N–H----O=C of the α-helices of myoglobin as listed in the HHDB. Fig. 5(b) shows the geometrical configuration of the N–H----O=C

hydrogen bonds, where the length between H and O is r and the NHO and COH angles are θ_1 and θ_2, respectively. Fig. 5(c) and (d) show the correlation between r and θ_1, and between θ_1 and θ_2, respectively. We had already commented on the non-linearity of the actual N–H----O bond angle θ_1. The significance of Fig. 5(d), which shows several H----O=C angles θ_2, is that they deviate significantly from the value of 120° normally expected for the lone pair electrons of an sp^2-hydridized carbonyl oxygen. Once again, it seems that the steric constraints imposed by the packing of atoms in a protein molecule, in this case an α-helix, necessitates considerable deviation normally found in small-molecule structures.

Protonation states of certain amino acid residues

The charges of various amino acid side chains depend on the pH: for example, at a high pH (low acidity conditions) carboxylic acids tend to be negatively charged (deprotonated) and amines uncharged (unprotonated). At a low pH (high acidity), the opposite is true. The pH at which exactly half of any ionized amino acid is charged is called the pK_a of that amino acid. These pK_a values of such ionizable amino acid side chains are tabulated in standard textbooks. However, whether a certain amino acid side chain in a protein is charged or not cannot be estimated from standard pH values measured from isolated amino acids in solution because inside a protein it may vary significantly depending on the local environment surrounding the particular amino acid residue. The electrically charged states of the amino acid residues are very important in understanding the physiological function of the protein, the interaction between ligands and proteins, molecular recognition, structural stability and so on. Let us take one example: normally, the standard pK_a value of histidine is about 6.0. In insulin there are two histidine amino acid residues (both on chain B) and it is of interest to know whether at pD 6.6 they are protonated or deprotonated since pD 6.6 is clearly close to the borderline pK_a value. Moreover, it is also important to know which of the two nitrogen atoms (Nτ, Nπ) in the imidazole ring of histidine are ionized in order to discuss the physiological function of insulin.

Neutron diffraction experiments of porcine insulin (crystallized at pD 9.0 and 6.6) were performed at room temperature[17,18]. Fig. 6 (a) and (b) show the positive neutron density map of His B10 at pD 6.6 and pD 9.0, respectively and we have found that both Nπ and Nτ are protonated. On the other hand the pH dependence of protonation of His B5 is different from

His B10. Fig. 6(c) shows the positive neutron density map of the imidazole ring of His $B5$ at pD 6.6 in which both Nτ and Nπ are again found to be protonated.

However, at pD = 9.0 the Nτ nitrogen atom is not protonated (Figure 6(d)) and only the Nπ nitrogen atom is protonated. Thus, one can hypothesize that the environments of His $B10$ and His $B5$ are distinctly pH-dependent.

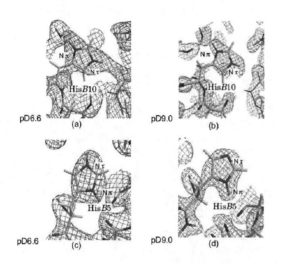

FIGURE 6. (a) $2|F_o|$-$|F_c|$ positive nuclear density map of His$B10$ at pD 6.6, (b) His B10 at pD 9.0, (c) His$B5$ at pD6.6 and (d) His$B5$ at pD 9.0.

Water molecules of hydration

We have categorized observed water molecules into the following classes based on their appearance in Fourier maps[19,20]: (1) triangular shape, (2) ellipsoidal stick shape, and (3) spherical shape, and in some cases molecules of the second category (ellipsoidal stick shapes) can be further sub-classified as either (2a) short and (2b) long. We found that this classification conveniently reflects the degree of disorder and/or dynamic behavior of a water molecule. Full details are given in the original publications[19] but the essential conclusions are summarized in Fig. 7. A triangular shape indicates a completely ordered water molecule, with all three atoms located (Fig. 7(a)), while a short ellipsoidal shape indicates that the O-D bond is visible while the second D atom is disordered (Fig. 7(b)). A long ellipsoidal contour (which is rarely observed) shows two ordered D atoms with the central O disordered (Fig. 7(c)), while the very common spherical shape (Fig. 7(d)) corresponds to a completely disordered water molecule. In an X-ray map, all four types of water molecules would appear simply as spherical peaks, and hence it is apparent that neutron maps are much more informative about hydration structure, especially concerning the water molecules

near the surface of a protein, which are much more likely to be ordered.

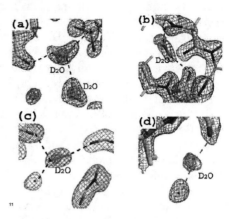

FIGURE 7. $2|Fo|$-$|Fc|$ neutron Fourier maps of water molecules of hydration for a rubredoxin mutant and and a myoglobin.examples shown are peaks having: (a) a triangular shape, (b) a short ellipsoidal shape, (c) a long ellipsoidal shapeand (d) a spherical shape. In these figures, the centralpeak (inner contours) embedded in the middle of eachwater contour corresponds to the oxygen atom positionderived from X-ray data. Note that, from the outercontours, in (a) all atoms of the central D_2O molecule arevisible, whereas in the other diagrams only some of the solvent atoms are visible in the neutron maps: O and D in (b); two D in (c); and only O in (d).

"Next generation" spallation neutron sources

An exciting new development currently under way is the construction of high-intensity, state-of-the-art pulsed neutron facilities. These are the Spallation Neutron Source (SNS) at Oak Ridge, Tennessee in the U.S.A. and the Japan Proton Accelerator Research Complex (J-PARC) in Ibaraki prefecture in Japan. Both SNS and J-PARC promise to deliver neutron intensities that are at least an order of magnitude (or more) higher than those at existing sources: traditional nuclear reactors such as those at ILL in Grenoble (France) and JRR-3 in JAEA (Japan), as well as existing pulsed (spallation) neutron sources such IPNS in Argonne (U.S.), ISIS at Didcot (England) and LANSCE at Los Alamos (U.S.).

At both new high-intensity spallation sources, diffractometers for macromolecular crystallography are either being built or planned. At J-PARC, construction of the so-called iBIX (ibaraki-prefectural BIological material X'tal diffractometer) instrument has already been in progress since 2004[21]; while at the SNS a corresponding instrument, to be called MaNDi (Macromolecular Neutron Diffractometer)[22] is most probably going to be built. These two diffractometers (iBIX and MaNDi) will differ somewhat in the type of moderator used (this is the material situated between the neutron target source and the sample crystal; its choice determines the intensities and wavelength distribution of the neutrons used in the diffraction experiment). iBIX will be using neutrons from a coupled moderator, which has a higher intensity but a wider wavelength range; while MaNDi is proposed to use a decoupled moderator, which is less intense but has a sharper distribution. One reason for this difference is that J-PARC has a lower pulsing frequency (25 Hz) than SNS (60 Hz): overlap between pulses that are separately more widely in time is therefore less serious of a problem at J-PARC than the closely-spaced pulses at SNS.

Regardless of these technical differences, the implication of the development of SNS and J-PARC on macromolecular research are obvious: if intensities can be increased by an order of magnitude or two, then in principle it will be possible to study crystals that are correspondingly smaller in volume. The consequences of this development will be enormous dramatic decrease in crystal size, which will in turn have an absolutely dramatic effect on the usefulness of single-crystal neutron diffraction in the field of structural biology. It is hoped that iBIX and MaNDi will both become operational around the period 2009-2010.

ACKNOWLEDGMENTS

Most of this research was supported in part by an "Organized Research Combination System (ORCS)" grant from the Ministry of Education, Culture, Sports, Science and Technology of Japan. We are grateful to the many scientists involved in that effort, including many who took part in several international workshops hosted in JAPAN and also to Profs. J. R. Helliwell and E. Westhof for their work on the ORCS Advisory Committee over a 5-year period. A part of this research was supported by a Grant-in-aid for Science Research from the Ministry of Education, Culture, Sports, Science and Technology of Japan.

REFERENCES

1. N. Niimura, Y. Karasawa, I. Tanaka, J. Miyahara, K. Takahashi, H. Saito, *Nucl. Instrum. Methods*, **A349**, 521-525 (1994)

2. N. Niimura, Y. Minezaki, T. Nonaka, J. C. Castagna, F. Ciprini, P. Hoghej, *Nat Struct Biol*, **4**, 909-914 (1997)

3. J. R. Helliwell, C.Wilkinson, "Neutron and Synchrotron Radiation for Condensed Matter Studies: Applications to Soft Condensed Matter and Biology", (ed. Baruchel *et al.*), Springer Verlag, 1994, Chapter XII, Vol. III.

4. I. Tanaka, K. Kurihara, T. Chatake, N. Niimura. *J. Appl. Cryst.* **35**, 34-40 (2002)

5. K. Kurihara, I. Tanaka, M. R. Muslih, A. Ostermann, N. Niimura, *J. Synchrotron Radiat.* **11**, 68-71 (2004)

6. B. P. Schoenborn, P. Langan *J. Synchrotron Rad.* **11**, 80-82 (2004)

7. A.Wlodawer, *Prog. Biophys. Mol. Biol.* **40**, 115(1982)

8. B. P. Schoenborn, *Methods Enzymol.* **114**, 510-529 (1985)

9. A. A. Kossiakoff. *Annu. Rev. Biochem.* **54**, 1195-1227 (1985)

10. J. R. Helliwell, *Nat. Struct. Biol.* **4**, 874-876 (1997)

11. N. Niimura, *Curr. Opin. Struct. Biol.* **9**, 602-608 (1999)

12. I. Tsyba, R. Bau, *Chemtracts*. **15**, 233-257 (2002)

13. N. Niimura, S. Arai, K. Kurihara, T. Chatake, I. Tanaka, R. Bau, *Cell. Mol. Life Sci.* **63**, 285-300 (2006)

14. I. Hazemann, M. T. Dauvergne, M. P. Blakeley, F. Meilleur, M. Haertlein, A. van Dorsselaer, A. Mitschler, D. A. A. Myles, A. Podjarny, *Acta Cryst.* **D61**, 1413-1417 (2005)

15. D.A.A. Myles, C. Bon, P. Langa, F. Cipriani, J.C. Castagna, M.S. Lehmann, *Phys.* **B241&243**, 1122-1130 (1998)

16. E.N. Baker, R.E. Hubbard, *Prog. Biophys. Molec. Biol.* **44**, 97-179 (1984)

17. T. Ishikawa, T. Chatake, Y. Ohnishi, I. Tanaka, K. Kurihara, R. Kuroki, N. Niimura, (2007) submitted.

18. M. Maeda, T. Chatake, I. Tanaka, A. Ostermann, N. Niimura, *J. Synchrotron Rad.* **11**, 41-44 (2004)

19. T. Chatake, A. Ostermann, K. Kurihara, F.G. Parak, N. Niimura, *Proteins* **50**, 516-523 (2003)

20. N. Niimura, T. Chatake, K. Kurihara. M. Maeda, *Cell Biochem Biophys*. **40**, 351-370 (2004)

21 I. Tanaka, N. Niimura, T.Ozeki, T. Ohhara, K.Kurihara, K. Kusaka, Y. Morii, K. Aizawa, M. Arai, T. Kasao, K. Ebata, Y. Takano, *ICANS-XVII Proceedings*, **LA-UR-3904 Vol. III**, 937-945 (2006),

22. A. J. Schultz, P. Thiyagarajan, J. P. Hodges, C. Rehm, D. A. A. Myles, P. Langan, A. D. Mesecar, *J. Appl. Cryst.* **38**, 964 (2005)

SANS and DLS Studies of Protein Unfolding in Presence of Urea and Surfactant

V. K. Aswal[1], S. N. Chodankar[1], J. Kohlbrecher[2], R. Vavrin[2] and A. G. Wagh[1]

[1]Solid State Physics Division, Bhabha Atomic Research Centre, Mumbai 400085, India
[2]Laboratory for Neutron Scattering, ETH Zurich & Paul Scherrer Institut, CH-5232 Villigen PSI, Switzerland

Abstract. Small-angle neutron scattering (SANS) and dynamic light scattering (DLS) have been used to study conformational changes in protein bovine serum albumin (BSA) during its unfolding in presence of protein denaturating agents urea and surfactant. On addition of urea, the BSA protein unfolds for urea concentrations greater than 4 M and acquires a random coil configuration with its radius of gyration increasing with urea concentration. The addition of surfactant unfolds the protein by the formation of micelle-like aggregates of surfactants along the unfolded polypeptide chains of the protein. The fractal dimension of such a protein-surfactant complex decreases and the overall size of the complex increases on increasing the surfactant concentration. The conformation of the unfolded protein in the complex has been determined directly using contrast variation SANS measurements by contrast matching the surfactant to the medium. Results of DLS measurements are found to be in good agreement with those obtained using SANS.

Keywords: Protein unfolding, small-angle neutron scattering, dynamic light scattering
PACS: 61.12.Ex, 87.14.Ee, 87.15.Nn

INTRODUCTION

The function of a protein depends absolutely on its three-dimensional folded structure. Protein unfolding process involves the disruption of H-bonds, disulphide bonds, salt bridges and hydrophobic interactions, leading to its successive alteration of quaternary, tertiary, and secondary structure. However, peptide bonds are not broken leaving the primary structure unaltered. Protein unfolding is one of the most widely studied topics in molecular biology due to its wide spread application in the industrial and scientific world [1]. The unfolding process can be brought about by various means and conditions. Each different route of unfolding has its own application and advantage in material processing and basic sciences. Urea is being used for long time to understand the fundamental process of protein folding/unfolding. On the other hand, the surfactant induced unfolding is known to play an important role in the pharmaceutical and cosmetic industry. Urea is known for its water breaking ability, causing an increase in the solubility of hydrophobic groups, thereby being able to solubilize both the hydrophobic and hydrophilic patches of the protein macromolecule leading to the unfolding of the whole protein. Protein unfolding in the case of amphiphilic molecules such as surfactant is caused due to binding of these molecules to the hydrophobic patches of the protein.

Earlier studies have proposed that the addition of urea unfolds a protein into a polypeptide chain which acquires a random coil conformation [2]. In the presence of surfactant, the binding of surfactant on protein results in micelle-like aggregates enclosing the hydrophobic patches on the protein backbone. This leads to acquisition of a necklace-bead structure of the protein surfactant complex [3]. Herein, we show the results of small-angle neutron scattering (SANS) and dynamic light scattering (DLS) to probe conformational changes during the protein unfolding induced by on addition of urea and surfactant. The experiments are performed on bovine serum albumin (BSA) protein, which is one of the commonly used proteins. Protein unfolding in presence of urea is studied over a wide concentration range of urea (0 - 8M). The systematic structural changes during protein unfolding on addition of varying concentration of anionic surfactant sodium dodecyl sulfate (SDS) have been examined. SANS with the possibility to vary the contrast is an ideal technique for studying hydrogenous systems such as protein solution. This technique provides information about the geometry and conformation of the scattering particles [4]. DLS is a complimentary technique to SANS, which gives size information by measuring the diffusion coefficient of the particle [5].

CP989, Neutron and X-ray Scattering in Materials Science and Biology, International Conference on Neutron and X-ray Scattering 2007, edited by A. Ikram, A. Purwanto, Sutiarso, A. Zulfia, S. Hendrana, and Z. Nurachman
© 2008 American Institute of Physics 978-0-7354-0508-0/08/$23.00

EXPERIMENTAL

BSA protein (catalogue no. 05480), SDS surfactant (catalogue no. 71727) and urea were purchased from Fluka. Samples for SANS experiments were prepared by dissolving known amount of BSA and other additives (surfactant or urea) in a buffer solution of D_2O. The use of D_2O as solvent instead of H_2O provides better contrast for hydrogenous components in neutron experiments. The interparticle interactions in these systems were minimized by preparing the samples in acetate buffer solution at pH 5.4, which is close to the isoelectric pH of BSA (4.9), and at high ionic strength of 0.5 M NaCl. Small-angle neutron scattering experiments were performed on the SANS-I instrument at the Swiss Spallation Neutron Source, SINQ, Paul Scherrer Institut, Switzerland [6]. The mean wavelength of the incident neutron beam was 6 Å with a wavelength resolution of approximately 10%. The scattered neutrons were detected using two-dimensional 96 cm × 96 cm detector. The experiments were performed at two sample-to-detector distances of 2 and 8 m, respectively to cover the data in the wave vector transfer Q range of 0.007 to 0.25 Å^{-1}. The measured SANS data were corrected and normalized to a cross-sectional unit using BerSANS-PC data processing software [7]. DLS measurements on above samples were carried out using a commercial ALV/LSE-5003 light scattering instrument featuring a multiple tau digital correlator. The light source was a helium neon laser operated at 6328 Å and the measurements were performed at a scattering angle of 90°. The temperature was kept fixed at 30°C for all the measurements.

DATA ANALYSIS

Small-angle neutron scattering

In small-angle neutron scattering one measures the coherent differential scattering cross-section per unit volume $[d\Sigma/d\Omega(Q)]$ as a function of Q. For a system of monodispersed interacting protein macromolecules, $d\Sigma/d\Omega(Q)$ can be expressed as [8].

$$\frac{d\Sigma}{d\Omega}(Q) = N_p V_p^2 (\rho_p - \rho_s)^2 \left[\left\langle F(Q)^2 \right\rangle + \left\langle F(Q) \right\rangle^2 \left(S_p(Q) - 1 \right) \right] + B \tag{1}$$

where N_p is the number density and V_p is the volume of the protein macromolecule. ρ_p and ρ_s are the scattering length density of the protein and the solvent, respectively. $F(Q)$ is the single particle form factor and $S_p(Q)$ is the interparticle structure factor. In general, charged colloidal systems such as protein solutions

show a correlation peak in the SANS distribution [9]. The peak arises because of the interparticle structure factor $S_p(Q)$ and indicates the presence of electrostatic interaction between the colloids. In the case of a solution with low protein concentration, having high salt concentration and pH close to isoelectric point of the protein, $S_p(Q)$ can be approximated to unity as the interparticle interactions are minimized so that

$$\frac{d\Sigma_n}{d\Omega}(Q) = N_p V_p^2 (\rho_p - \rho_s)^2 \left\langle F(Q)^2 \right\rangle + B \tag{2}$$

B is a constant term that represents the incoherent scattering background, which is mainly due to hydrogen in the sample. The single particle form factor of the protein macromolecules in their native conformation has been calculated by treating them as prolate ellipsoids. For such an ellipsoidal particle

$$\left\langle F^2(Q) \right\rangle = \int_0^1 \left[F(Q, \mu)^2 d\mu \right] \tag{3}$$

$$F(Q, \mu) = \frac{3(\sin x - x \cos x)}{x^3} \tag{4}$$

$$x = Q \left[a^2 \mu^2 + b^2 (1 - \mu^2) \right]^{1/2} \tag{5}$$

where a and b are, respectively, the semimajor and semiminor axes of the ellipsoidal protein macromolecules and μ is the cosine of the angle between the directions of a and the wave vector transfer Q.

In order to describe the unfolding of protein in presence of urea is modeled using a random coil Gaussian conformation

$$\frac{d\Sigma_g}{d\Omega}(Q) = I_0 \left[Q^2 R_g^2 - 1 + \exp(-Q^2 R_g^2) \right] / (Q R_g)^4 \tag{6}$$

where R_g is the radius of gyration of the unfolded protein polypeptide chain.

The unfolding of protein in presence of surfactant has been treated using the necklace model of protein-surfactant complexes that assumes micelle-like clusters of surfactant randomly distributed along the unfolded polypeptide chain. The cross-section for such a system can be written as [10]

$$\frac{d\Sigma_s}{d\Omega}(Q) = \frac{N_1^2}{N_p N} (b_m - V_m \rho_s)^2 P_m(Q) S_f(Q) + B \tag{7}$$

where N_1 is the number density of the total surfactant moecules in solution. V_m the volume of the micelle and N the number of such micelles attached to a polypeptide chain. b_m represents the scattering length of the surfactant molecule. $P_m(Q)$ denotes the normalized intraparticle structure factor of a single micelle-like cluster. $S_f(Q)$ for such a system is expressed as

$$S_f(Q) = 1 + \frac{1}{(QR)^D} \frac{D\Gamma(D-1)}{[1+(Q\xi)^{-2}]^{(D-1)/2}} \times \sin[(D-1)\tan^{-1}(Q\xi)]$$

(8)

where D is the fractal dimension of the micellar distribution in space and ξ the correlation length that is a measure of the extent of unfolding of the polypeptide chain.

Dynamic light scattering

The signal generated by the light scattering from diffusing particles can be analyzed by its intensity autocorrelation function $G^I(\tau)$ [5]

$$G^I(\tau) = \langle I(t)I(t+\tau) \rangle$$

(9)

where $I(t)$ is the scattered light intensity at time t and $I(t+\tau)$ the scattered light intensity at time t plus a log time τ. The normalized intensity autocorrelation function $g^I(\tau)$ is

$$g^I(\tau) = \frac{G^I(\tau)}{\langle I(t) \rangle^2}$$

(10)

The electric field autocorrelation function $g^E(\tau)$ is related to the normalized intensity autocorrelation function by the Siegert relation

$$g^I(\tau) = 1 + C[g^E(\tau)]^2$$

(11)

where C is an experimental parameter which mainly depends on the detection optics and alignment. For a mono-disperse system of particles, $g^E(\tau)$ follows a simple exponential decay with decay constant γ.

$$g^E(\tau) = \exp[-\gamma\tau]$$

(12)

The average decay rate (γ) of $g^E(\tau)$ has been estimated using a monomodal fit. The apparent diffusion coefficient (D_a) is obtained from the relation $\gamma = D_a Q^2$ and the corresponding effective hydrodynamic size (R_H) calculated using Stokes-Einstein relationship as given by

$$R_H = \frac{k_B T}{6\pi\eta D_a}$$

(13)

where k_B is the Boltzmann constant, T is the temperature and η is the viscosity of the solvent.

RESULTS AND DISCUSSION

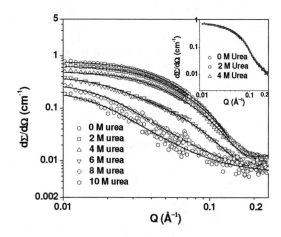

FIGURE 1. SANS data for 1 wt% BSA in presence of varying concentration of urea. Inset shows the scaling of data up to urea concentration of 4 M.

TABLE 1. Fitted parameters of SANS analysis for 1 wt% BSA in presence of varying urea concentrations. The protein has a prolate ellipsoidal shape up to 4 M urea beyond this concentration it unfolds into a random coil conformation.

[Urea] (M)	Semi-minor axis a (Å)	Semi-major axis $b = c$ (Å)	R_g (Å)
0	22.2 ± 0.8	71.0 ± 5.1	-
2	22.2 ± 0.8	71.0 ± 5.1	-
4	22.2 ± 0.8	71.0 ± 5.1	-
6	-	-	55.0 ± 2.9
8	-	-	84.0 ± 4.1
10	-	-	93.5 ± 6.4

SANS data for 1 wt% BSA in presence of a varying concentration of urea are shown in figure 1. The scattering cross-section decreases with increasing urea concentration. It is observed that up to 4 M concentration of urea, there is a continuous decrease in the scattering cross-section, however the functionality of the scattering pattern does not change. The inset of the figure 1 shows the scaling of the data suggesting the same functionality of the scattering profiles for urea concentrations in the range 0 to 4 M. The decrease in scattering cross-section can be explained in terms of decrease in contrast $(\rho_p - \rho_s)^2$ as the scattering length density of deuterated solvent (ρ_s) decreases on addition of hydrogenous urea to protein solution [11]. There is a change in the functionality of the scattering profile beyond 4 M urea and it is interpreted in terms of unfolding of the protein. It is believed that the solvation of hydrophobic portions of the protein at high urea concentrations leads to the unfolding of a protein. The unfolded protein is fitted as random Gaussian coil using equation (6). It is found that radius of gyration (R_g) of the unfolded protein increases with increasing urea concentration. The value of R_g increases from 55 to 93.5 Å as the urea concentration is increased from 6 M to 10 M (Table 1).

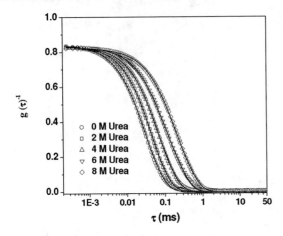

FIGURE 2. DLS data for 1 wt% BSA in presence of varying concentration of urea.

TABLE 2. Fitted parameters of DLS analysis for 1 wt% BSA in presence of varying urea concentrations.

[Urea] (M)	Diffusion coefficient D_a (10^{-8} cm²/s)	Solvent viscosity η (mPa s)	Hydrodynamic radius R_H (Å)
0	64.3	1.03	33.5
2	63.6	1.12	34.1
4	61.2	1.25	34.5
6	43.6	1.44	68.5
8	18.3	1.71	136.5

DLS data on addition of urea to protein solution are shown in Figure 2. DLS measures the time-dependent fluctuations in the intensity of scattered light [5]. These fluctuations happen as a result of the Brownian motion. Small particles diffuse rapidly and yield fast fluctuations, whereas large particles and aggregates generate relatively slow fluctuations. The rate of the fluctuations is determined through the autocorrelation analysis technique. The calculated autocorrelation function enables the determination of the diffusion coefficient [Equation (11)], which then can be converted to a size using the Stokes-Einstein relationship [Equation (13)]. There is a decrease in the average decay constant indicating a slowing down of the diffusion of the protein macromolecules on increasing urea concentration. The calculated hydrodynamic size from the DLS data is shown in Table 2. The hydrodynamic size remains similar up to 4 M urea concentration and increases beyond this

concentration as the protein unfolds, which is consistent with the SANS results (Table 1).

Figure 3 shows SANS data for 1 wt% BSA in presence of varying SDS concentration. Based on the features of the scattering profiles, the data can be grouped in two different sets as the surfactant concentration is increased. The first data set corresponds to proteins at low surfactant concentrations (0 to 10mM), where the scattering data show similar behavior to that of pure protein solution. In this data set, the overall scattering cross-section increases with increase in surfactant concentration. It can be explained in terms of Equation (1) if the individual surfactant molecules bind to protein and the volume of the scattering particle increases. The features of the scattering data in the second data set at higher surfactant concentrations (>10mM) are very different to those of the first data set. One of the interesting features is the linearity of the scattering profiles on log-log scale in the intermediate Q range with a Q range of linearity increasing with surfactant concentration. This is an indication of formation of fractal structure by the protein-surfactant complex [12]. The build up of scattering cross-section in the higher cut-off of the linearity of scattering data suggests the formation of surfactant aggregates and the lower cut-off corresponds to the overall size of the protein-surfactant complex. It is observed that the position of high Q cut-off remains almost same while the position of low Q cut-off shifts to smaller Q values with increasing surfactant concentration. The calculated structural parameters in this system are given in Table 3.

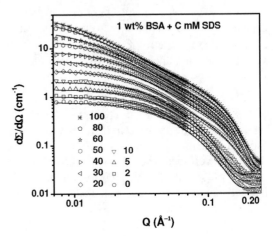

FIGURE 3. SANS data for 1 wt% BSA in presence of varying concentration of surfactant.

At low surfactant concentrations, Table 3 shows changes in the dimensions of the protein on increasing binding of surfactant molecules as a function of surfactant concentration. The semiminor axis remains almost same while the semimajor axis increases with increasing surfactant concentration. The size of the protein elongates from 70 to 88 Å along the semimajor axis on the addition of 10 mM of surfactant. Similar results of the elongation of the BSA protein have also been observed with cationic surfactant Azobenzene trimethyl ammonium bromide. It is believed that the six protein sub-domains forming BSA remain intact but separates from each other, leading to an elongation of the protein on addition of surfactant [13]. The fractal structure of the protein-surfactant complex at higher surfactant concentrations is modeled on the basis of the necklace model which considers micelle-like clusters of the surfactant formed along the unfolded polypeptide chain of the protein. It is found that the fractal dimension decreases and the overall size of the complex increases on increasing surfactant concentration. The size of micelle-like clusters (R) does not change while the number of such micelle-like clusters (N) in protein-surfactant complex increases with the surfactant concentration (Table 3).

TABLE 3. Fitted parameters of SANS analysis of protein-surfactant complex for

(a) 1 wt% BSA with low SDS concentrations 0 to 10 mM

[SDS] (mM)	Semi-major axis a (Å)	Semi-minor axis $b = c$ (Å)
0	70.2 ± 5.1	22.2 ± 0.8
2	77.3 ± 5.8	22.2 ± 0.8
5	80.0 ± 6.1	22.2 ± 0.8
10	88.0 ± 6.4	23.0 ± 0.9

(b) 1 wt% BSA with high SDS concentrations 20 to 100 mM

[SDS] (mM)	Fractal dimension D	Correlation length ξ (Å)	Micelle radius R (Å)	No. of micelles N
20	2.27 ± 0.15	38.0 ± 1.9	18.0	2
30	2.15 ± 0.14	43.0 ± 2.7	18.0	3
40	2.05 ± 0.13	54.0 ± 3.8	18.0	4
50	1.95 ± 0.10	67.8 ± 4.9	18.0	6
60	1.88 ± 0.09	87.9 ± 5.4	18.0	8
80	1.79 ± 0.06	117.9 ± 6.9	18.0	10
100	1.71 ± 0.04	144.3 ± 7.5	18.0	13

The calculated aggregation number of micelle-like clusters in the complex decreases from 51 to 42 on increasing surfactant concentration from 20 to 100mM. It is interesting to note that these values of aggregation numbers are much smaller than as one would have found about 70 in pure surfactant solution for the similar size of micelles [14]. This indicates the participation of the hydrophobic portions of the unfolded protein chain in the micellar formation. The

participation of the unfolded protein in the formation of micelle-like clusters is enhanced with the increase in unfolding and this result in decreasing aggregation number of micelle-like clusters. Also all the surfactant molecules probably participate in the micelle-like clusters to avoid the exposure of hydrophobic portions of the protein on its unfolding with increase in surfactant concentration.

The conformational changes of protein in protein-surfactant complex have been examined by contrast variation SANS by contrast matching the surfactant. The surfactant is contrast matched using deuterated SDS (d-SDS) and the sample prepared in D_2O. Figure 4 shows the SANS data for 1 wt% BSA in presence of varying d-SDS concentration. Unlike in Figure 3 where the scattering cross-section increases on increasing SDS concentration, Figure 4 shows decrease in scattering cross-section with increase in d-SDS concentration. This can be understood in terms of equation (1) that the contrast of protein decreases as the size of complex increases on addition of surfactant (Tables 3).

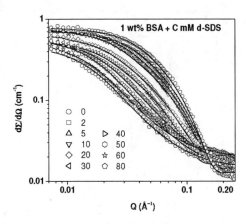

FIGURE 4. SANS data for 1 wt% BSA in presence of varying concentration of deuterated surfactant.

Similar to Figure 3, Figure 4 also shows two sets of data as obtained the first set at low and the second set at high surfactant concentrations. The first data set corresponds to the binding of individual surfactant molecules and hence the size of the complex increases as more and more surfactant is added (Table 4). This data set is fitted with Equation (1). The analysis of the data (Table 4) gives similar dimensions of the complex as obtained in Table 3. For the data of second set micelle-like clusters of surfactant molecules are formed in the complex (Table 3). The unfolded protein in the complex is fitted by the scattering function obeying random coil Gaussian conformation. It may be mentioned that at d-SDS concentration of 20 and 30mM, where the complex already consists of micelle-like clusters the data are still best fitted with a prolate ellipsoidal shape of the protein. Unfolding in these systems is limited due to formation of small number of micelles in the complex (Table 3).

TABLE 4. Fitted parameters of SANS analysis of protein-surfactant complex for 1 wt% BSA with varying d-SDS concentration.

[d-SDS] (mM)	Folded structure		Unfolded Structure
	Semi-major axis a (Å)	Semi-minor axis $b = c$ (Å)	Radius of gyration R_g (Å)
0	71.0 ± 5.1	22.2 ± 0.8	-
2	77.3 ± 5.8	22.2 ± 0.8	
5	82.0 ± 6.3	22.2 ± 0.8	-
10	88.0 ± 6.4	23.0 ± 0.9	-
20	94.0 ± 6.7	25.8 ± 1.1	-
30	99.0 ± 7.1	27.1 ± 1.3	-
40	-	-	60.1 ± 1.6
50	-	-	70.3 ± 1.8
60	-	-	85.5 ± 2.4
80	-	-	102.3 ± 4.6

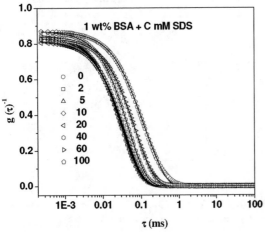

FIGURE 5. DLS data for 1 wt% BSA in presence of varying concentration of surfactant.

DLS data for 1 wt% BSA on addition of surfactant are shown in Figure 5. These data show the slowing down of intensity auto correlation function with increasing surfactant concentration. The analysis suggests that irrespective of the surfactant concentration, all the data fit to the single diffusion coefficient for the structure of the protein-surfactant complex. The fitted values of the apparent diffusion coefficient (D_a) and the corresponding hydrodynamic radius (R_h) are given in table V. There is decrease in the value of D_a and increase in R_h with increase in surfactant concentration. The diffusion coefficient decreases as the amount of surfactant in the complex increases either through the binding of surfactant

molecules as individuals or through the formation of micelle-like clusters with the protein. The increase in overall size of complex is significant at higher surfactant concentrations as related to the unfolding of protein. These observations are in agreement with the results obtained using SANS. The larger values of the sizes of the complex obtained using DLS than SANS (Table 5) are expected because DLS measures structure along with its hydration.

TABLE 5. Fitted parameters as obtained by DLS for the protein-surfactant complex with 1 wt% BSA and varying SDS concentrations. SANS results are compared with effective size of protein-surfactant complex calculated from table III as $(ab^2)^{1/3}$ for the folded structure and as ξ for the unfolded structure.

[SDS] (mM)	Diffusion coefficient D_a (10^{-8} cm^2/sec)	Hydrodynamic radius R_h (Å)	Effective size R_e (Å)
0	64.3	33.5	32.6
2	64.3	33.5	33.5
5	59.0	36.5	34.1
10	55.2	39.0	36.0
20	51.9	41.5	38.0
40	35.7	60.3	54.0
60	23.6	91.2	87.9
100	14.3	150.5	144.3

REFERENCES

1. C.M. Dobson, *Nature* **426**, 884 (2003).
2. R.S. Tu and V. Breedveld, *Phys. Rev.* **E72**, 041914 (2005).
3. S.H. Chen and J. Teixeira, *Phys. Rev. Lett.* **57**, 2583 (1986).
4. D.I. Svergun and M.H.J. Koch, *Rep. Prog. Phys.* **66**, 1735 (2002).
5. R. Pecora, *Dynamic Light Scattering*, New York, Plenum, 1985.
6. J. Kohlbrecher and W. Wagner, *J. Appl. Cryst.* **33**, 804 (2000).
7. U. Keiderling, *Appl. Phys.* **A 74**, S1455 (2002).
8. J.S. Pedersen, *Adv. Colloid Interface Sci.* **70**, 171 (1997).
9. S. Chodankar and V.K. Aswal, *Phys. Rev.* **E72**, 041931 (2005).
10. X.H. Guo, N.M. Zhao, S.H. Chen and J. Teixeira, *Biopolymers* **29**, 335 (1990).
11. V.K. Aswal, J. Kohlbrecher, P.S. Goyal, H. Amenitsch and S. Bernstorff, *J. Phys.: Condens. Matter* **18**, 11399 (2006).
12. S. Chodankar, V.K. Aswal, P.A. Hassan and A.G. Wagh, *Phys.* **B398**, 112 (2007).
13. T.C. Lee, K.A. Smith, T.A. Hatton, *Biochemistry* **44**, 524 (2005).
14. V.K. Aswal and P.S. Goyal, *Phys. Rev.* **E67**, 051401 (2003).

Synthesis and X-ray Structural Study on the Complexes of Silver(I) Halide with Tricyclohexylephosphine, Diphenyl-(2,4,6-trimethoxy)phenylphosphine, Phenyl-2,4,6-trimethoxyphenyl phosphine, and Tris(2,4,6-trimethoxy)phenylphosphine

Effendy[1] and A. H. White[2]

[1] *Jurusan Kimia, FMIPA Universitas Negeri Malang, Jl. Surabaya 6 Malang 65145, Indonesia*
[2] *Chemistry, School of Biomedical, Biomolecular, and Chemical Sciences, The University of Western Australia, 35 Stirling Highway, Crawley WA 6009, Australia*

Abstract. A diverse array of structures for the complexes of silver(I) halide with triphenylphosphine (PPh_3) has been studied. The complexes may be described as being of the type $[AgX(PPh_3)_n]$ (X = Cl, Br or I). The value of n varies in the range of 1-3. This also indicates that the stoichiometry of the complexes is in the range of 1-3. The complex with stoichiometry 1:1 is a tetramer. There are two structural types of tetramer reported, termed cubane and step or chair. The cubane structure has been reported for $[AgX(PPh_3)]_4$ (X = Cl, Br or I), while the step structure has only been reported for $[AgI(PPh_3)]_4$. The complex with stoichiometry 1:2 may be a monomer or a dimer. The monomer has a quasi trigonal planar structural type and has only been reported for $[AgBr(PPh_3)_2]$. The dimer has been reported for $[(PPh_3)_2Ag(\mu-X)_2Ag(PPh_3)_2]$ (X = Cl or Br) with silver atom in the distorted tetrahedral environment. The complex with stoichiometry 1:3 has a distorted tetrahedral structural type and has been reported for $[AgX(PPh_3)_3]$ (X = Cl, Br or I). Changing PPh_3 with more hindered ligand such as tricyclohexylephosphine (Pcy_3) or derivative of PPh_3 such as diphenyl-2,4,6-trimethoxy(phenyl)phosphine (dpmp), phenyl-bis{2,4,6-trimethoxy(phenyl)} phosphine (pdmp), or tris{2,4,6-trimethoxy(phenyl)}phosphine (tmpp) may give complexes with various structural types but with lower range of stoichiometry. Synthesis and X-ray structural study of these complexes has been done with the results summarized below. Silver(I) halide and PCy_3 give complexes with stoichiometry 1:1 and 1:2. The complex with stoichiometry 1:1 is a dimer or cubane. The dimer is observed for $[(Pcy_3)Ag(\mu-X)_2Ag(Pcy_3)]$ (X = Cl or Br). The unusual dimer is observed for $[(Pcy_3)Ag(\mu-I)_2(\mu-py)Ag(Pcy_3)]$ where the pyridine ligand is bonded to two silver atoms. The cubane is observed for $[AgI(Pcy_3)]_4$. The complex with stoichiometry 1:2 has a quasi trigonal planar structural type and has been observed for $[AgX(Pcy_3)_2]$ (X = Cl, Br, I). Silver(I) halide and dpmp give complexes with stoichiometry 1:1 and 1:2. The complex with stoichiometry 1:1 is a dimer and has been observed for $[(dpmp)Ag(\mu-X)_2Ag(dpmp)]$ (X = Cl, Br or I). The complex with stoichiometry 1:2 has a quasi trigonal planar structural type and has been observed for $[AgX(dpmp)_2]$ (X = Cl, Br, I). Silver(I) halide and pdmp also give complexes with stoichiometry 1:1 and 1:2. The complex with stoichiometry 1:1 is a dimer and has been observed for $[(pdmp)Ag(\mu-X)_2Ag(pdmp)]$ (X = Cl, Br or I). The complex with stoichiometry 1:2 has a quasi trigonal planar structural type and has been observed for $[AgX(pdmp)_2]$ (X = Cl, Br, I). Silver(I) halide and tmmp only give complexes with stoichiometry 1:1. This complex is a monomer and has been observed for $[AgX(tmpp)]$ (X = Cl or Br). In this complex the silver atom is in a quasi linear environment. Based on the bond lengths between silver and phosphorous atoms in the complexes obtained, it can be concluded that bulky ligands tend to give complexes with lower range of stoichiometry. In addition, the bulkier the ligand the longer the bond length between the silver and phosphorous atoms.

CP989, *Neutron and X-ray Scattering in Materials Science and Biology, International Conference on Neutron and X-ray Scattering 2007*, edited by A. Ikram, A. Purwanto, Sutiarso, A. Zulfia, S. Hendrana, and Z. Nurachman
© 2008 American Institute of Physics 978-0-7354-0508-0/08/$23.00

Powder Diffraction Studies of Phase Transitions in Manganese Perovskites

B. J. Kennedy

School of Chemistry, The University of Sydney, Sydney, NSW 2006 Australia

Abstract. The results of recent structural studies of some Manganese perovskites are presented, in particular oxides in the system $Ca_{1-x}Sr_xMnO_3$ and $SrRu_{0.5}Mn_{0.5}O_3$. In the first series we firstly show the power of synchrotron X-ray powder diffraction to refine accurate and precise structures for oxides containing first row transition metals and then show the presence of a direct orthorhombic *Pbnm* to tetragonal *I4/mcm* transition associated with the tilting of the MnO_6 octahedra. The inclusion of a heavier second row transition metal reduces the precision of the structure, however the details of the tetragonal to cubic phase transition in $SrRu_{0.5}Mn_{0.5}O_3$ are still established.

Keywords: Perovskite, Phase Transition, Synchrotron X-ray Powder Diffraction, Jahn-Teller Distortion
PACS: 61.05.Cp; 61.50.Ks; 61.66.Fn

INTRODUCTION

The, often unique, electronic and magnetic properties of perovskite-type oxides continue to attract substantial interest, both from a fundamental science point-of-view, but increasingly due to the potential and actual technological importance of devices incorporating perovskite oxides [1]. Perovskite oxides with ABO_3 stoichiometry have a structure based on a three-dimensional framework of corner-sharing BO_6 octahedra, with the A-type cation occupying the resulting high coordination sites. Very few perovskites have the archetypal cubic structure, rather a number of lower symmetry variants exist as a result of the ability of the corner-sharing octahedral framework to undergo cooperative tilting distortions. Octahedral tilting occurs as a consequence of the mismatch in size between the *A*-site cation and the corner sharing octahedral network

Mixed valence manganites such as those based on Sr doped $LaMnO_3$ can exhibit colossal magnetoresistance near the ferromagnetic transition temperature that can be tuned to near room temperature [2,3]. The vastly different crystal chemistry requirements of Mn^{3+} and Mn^{4+} undoubtedly contribute to the structural complexity of these materials and this in turn influences their important properties.

The Mn^{III} cation has the electron configuration $(t_{2g})^3(e_g)^1$ and is susceptible to a Jahn-Teller (JT) induced distortion. In molecular species this is most commonly observed as a large tetragonal elongation of the MX_6 octahedra, although in metal oxides other effects can inhibit this [4,5]. The overlap of the t_{2g} orbitals in perovskites is very sensitive to the tilting of the MO_6 octahedra, and orbital ordering associated with the JT effect can play an important role in influencing this. In the present paper I describe the use of synchrotron X-ray powder diffraction studies of some perovskites ranging from a simple Mn^{IV} oxide $CaMnO_3$ including variable temperature structural studies of a Mn^{III} containing oxide $SrRu_{0.5}Mn_{0.5}O_3$.

EXPERIMENTAL

Synchrotron X-ray diffraction data were collected on the Debye Scherrer diffractometer at the Australian National Beamline Facility, Beamline 20B at the second generation synchrotron Photon Factory, Tsukuba, Japan [6]. Each sample was finely ground and loaded into a 0.3-mm quartz capillary that was rotated during the measurements. All measurements were performed under vacuum to minimize air scatter. Data were recorded using two Fuji image plates. Each image plate is 20 x 40 cm and covers 40° in 2θ. A thin strip *ca.* 0.5 cm wide is used to record each diffraction pattern so that up to 30 patterns could be recorded before reading the image plates. The wavelength of the incident X-rays was selected using a Si (111) monochromator and calibrated using NIST Si 640.

CP989, Neutron and X-ray Scattering in Materials Science and Biology, International Conference on Neutron and X-ray Scattering 2007, edited by A. Ikram, A. Purwanto, Sutiarso, A. Zulfia, S. Hendrana, and Z. Nurachman

The peak-width resolution of the diffractometer gives a minimum full width at half height of the peaks of 0.03°.

Data were recorded in the angular range 5 < 2θ < 85°, step size 0.01°. Data collection times were 10 minutes each. Variable temperature data were collected, using a custom built furnace, at temperatures of up to 800 °C. The collection of 30 diffraction patterns over this temperature range can be completed in under 12 hours of beam time. The structures were refined using the program RIETICA [7]. The samples were prepared using conventional solid-state methods and characterized by electron microscopy, including EDAX prior to the diffraction studies.

RESULTS AND DISCUSSION

Perovskites exhibit an almost pathological tendency to high pseudo-symmetry and determining the correct space group of perovskites from powder diffraction data can be far from trivial [8,9]. The protocol we have developed over the last several years for determining the correct symmetry and space groups for perovskite-type oxides involves a combination of careful examination of the splitting of the strongest Bragg reflections, to determine the cell metric, and establishing the systematic presence and/or absence of the weaker superlattice reflections resulting from cation ordering or displacement, octahedral tilting, or a combination of these. The space group is then assigned to the highest possible symmetry that accounts for all the observed reflections and is in-keeping with the group theoretical analysis of Howard and Stokes [10]. The success of this approach is reliant on high quality diffraction data.

The synchrotron X-ray diffraction pattern for $CaMnO_3$ at room temperature, shown in Figure 1, illustrates the high quality of the diffraction data, in terms of both signal-to-noise and resolution, obtained in our studies. The exceptional signal-to-noise ratio of the data, achieved at our synchrotron diffractometer, allows the weak superlattice reflections associated with the tilting of the MnO_6 octahedra to be easily observed and ultimately fitted in the Rietveld refinements. The excellent peak-shape resolution allows high quality data to be collected over a wide angular range (7-85° 2θ or 6.5 to 0.59 Å^{-1} in d for λ = 0.82625 Å).

Based on the peak splitting, the diffraction pattern of $CaMnO_3$ was indexed to an orthorhombic cell. The pattern contains a number of superlattice reflections, associated with different instabilities, from R-point distortions due to MnO_6 octahedra undergoing out-of-phase or "minus" rotations, M-point distortions due to in-phase or "plus" rotations and X-point reflections

arising from a coupling of the R- and M-point distortions [10]. These reflections show the correct space group is *Pbnm* that has a Glazer tilt system $a^+b^-b^-$ [11]. The structure was refined in this space group. The combination of high resolution and excellent signal-to-noise together with the absence of any "very" heavy elements in the sample allows the positions of the oxygen anions to be determined with a precision normally associated with neutron diffraction measurements, Table 1.

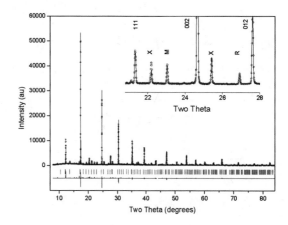

FIGURE 1. The observed, calculated and difference synchrotron x-ray diffraction pattern of $CaMnO_3$ at room temperature. The inserts show the detail of the fit to some of the superlattice reflections arising from the tilting of the MnO_6 octahedra. The small discontinuity near 2θ = 45° is the gap between the two image plate detectors.

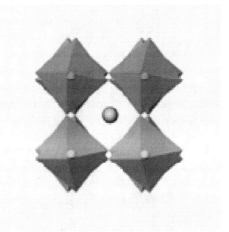

FIGURE 2. View of the orthorhombic perovskite structure of $CaMnO_3$, looking down the 110 axis to illustrate the out-of-phase tilting of the MnO_6 octahedra. The Mn^{IV} cations are at the centre of the octahedra and the Ca^{2+} cations occupy the 12-coordinate sites.

The structure of $CaMnO_3$ is illustrated in Figure 2 and we find the absence of any appreciable distortion of the MnO_6 octahedra with the refined Mn-O(1) distance being 1.897(2) Å and the two Mn-O(2) bond distances 1.898(2) and 1.907(2) Å, for an average Mn-O distance of 1.900 Å [12]. The Bond Valence Sum (BVS) of the Mn is 4.0. In general we find the BVS to be a better indicator of validity of the structural model than the average metal oxygen bond distances.

The tilting of the MnO_6 octahedra occurs as a consequence of the mismatch in size between the A-site cation and the corner sharing octahedral network. In $CaMnO_3$ the Ca^{2+} cation (ionic radii 1.34 Å [13]) is too small for the cuboctahdral site and, to optimize its bonding, cooperative tilting follows. This tilting can be reduced by progressively increasing the size of the A-type cation through the formation of solid solutions such as $Ca_{1-x}Sr_xMnO_3$ [14]; the ionic radius of Sr^{2+} is 1.44 Å. Examination of the synchrotron diffraction patterns for the samples with x ≤ 0.35 showed that these oxides all adopted the same orthorhombic structure, in *Pbnm*, seen for undoped $CaMnO_3$.

It should be stressed that for the x = 0.5 sample the structure actually appears metrically cubic, in that no splitting of the Bragg reflections indication of orthorhombic symmetry is readily available. Examination of the profile, however reveals differences in peak widths as well as the presence of weak superlattice reflections that demonstrate the material remains orthorhombic.

TABLE 1. Results of the Rietveld refinements for $CaMnO_3$ and $SrRuO_3$ using synchrotron X-ray diffraction data

Atom	Site	X	y	z	B(iso)
\multicolumn{6}{c}{$CaMnO_3$ a = 5.26746(5) b = 5.28287(5) c = 7.45790(7) Å R_p 4.81 % R_{wp} 4.67 %}					
Ca	4c	0.0056(3)	0.5330(1)	0.25	0.68(1)
Mn	4a	0	0	0	0.42(1)
O(1)	4c	-0.0657(8)	-0.0092(5)	0.25	0.79(5)
O(2)	8d	0.2116(4)	0.2870(4)	0.0332(4)	0.59(3)
\multicolumn{6}{c}{$SrRuO_3$ a = 5.57039(4) b = 5.53168(4) c = 7.84748(5) Å R_p 1.43 % R_{wp} 2.02 %}					
Sr	4c	0.0005(10)	0.5181(3)	0.25	0.18(2)
Ru	4a	0	0	0	0.09(2)
O(1)	4c	-0.0524(16)	0.0002(26)	0.25	0.10(16)
O(2)	8d	0.2187(21)	0.2784(22)	0.0262(16)	1.10(22)

The diffraction pattern for the x = 0.4 sample, $Ca_{0.4}Sr_{0.6}MnO_3$, appeared very different to that of the samples with less Sr and the cell metric is tetragonal, Figure 3. However examination of the diffraction pattern showed the presence of both R and M point reflections that are inconsistent with any of the tetragonal space groups identified by Howard and Stokes [10]. Accordingly we initially attempted to fit data using a model in *Pbnm* however the resulting R-factors were unacceptable, R_p ~ 17.8 and R_{wp}~ 23.1 %. Indeed the pattern could be better fitted in *I4/mcm*, R_p ~ 9.8 and R_{wp} 12.7%, as suggested by Chmaissem and co-workers [15] although this model failed to fit the weak M-point reflections (these are forbidden in *I4/mcm*). Ultimately the pattern was fitted using a mixture of orthorhombic *Pbnm* and tetragonal *I4/mcm*, the final R-factors being 4.57 and 6.39%. This two-phase model accounted for both the R- and M-point reflections as well as providing an excellent fit to the main Bragg reflections. The co-existence of the *Pbnm* and *I4/mcm* phases is indicative of a first order transition between these and in consistent with Howard and Stokes [10] group theoretical analysis that demonstrated that any transition between these two space groups for ABO_3 perovskites must be first order. Heating the sample 300 °C resulted a single phase (*I4/mcm*) structure confirming that the co-existence of the two phases at room temperature is a consequence of an incomplete transition.

A second approach to tune the structure of these materials is substitution onto the B site of the perovskite. To this end we have examined the structure of the mixed Ru-Mn perovskite $SrRu_{0.5}Mn_{0.5}O_3$. The room temperature synchrotron diffraction pattern of $SrRuO_3$ shows this to be orthorhombic in *Pbnm* [16]. Here the increased X-ray scattering power of the heavier Ru cations lowers the precision of the structure refinement, Table 1. This is most evident in the larger errors associated with the oxygen positional parameters. Crudely the superlattice reflections contribute proportionally less to the pattern, yet these are critical in establishing the precise positions of the anions. Using an even higher resolution synchrotron x-ray powder diffractometer can yield more precise values, however that is outside the scope of the present work. Despite the weakness of the superlattice reflections associated with the tilting of the RuO_6 octahedra the refined structure is chemically sensible.

The RuO$_6$ octahedra shows a small distortion with the refined Ru-O(1) distance being 1.984(2) Å and the two Ru-O(2) bond distances being 1.957(9) and 2.014(9) Å. The average Ru-O distance is 1.985 Å which is noticeably smaller than the average Mn-O distance in CaMnO$_3$ 1.900 Å. The bond valence sum for the Ru is slightly lower than expected, being 3.4, and this is thought to reflect the accuracy of the structural refinement, rather than partial reduction of the Ru cation.

The room temperature synchrotron diffraction pattern for SrRu$_{0.5}$Mn$_{0.5}$O$_3$ shows this has a tetragonal structure and whilst a number of R-point reflections indicative of out-of-phase tilts are observed there was no evidence for any M-point reflections in the diffraction pattern [17]. The appropriate space group is *I4/mcm* with the tilt system $a^0\ a^0\ c^-$. In this structure the Ru and Mn are disordered over a single crystallographic site and the tetragonal cell related to

the parent structure $a \approx \sqrt{2}a_p$ and $c \approx 2a_p$ where a_p is the ideal primitive cubic perovskite lattice parameter. The tilt angle, derived from the Rietveld refinement is 6.2(6)°. In the tetragonal structure the two M-O(2) bond distances become equivalent and we observe these to be shorter than the M-O(1) distance 1.9797(2) compared to 1.9418(2) Å; the average M-O distance is 1.954 Å. The octahedra are best described as elongated. This arrangement is thought to reflect a formal valence arrangement with Ru^{+5} and Mn^{+3}, as previously identified using X-ray spectroscopy [18]. The Mn^{3+} ion has the Jahn-Teller active electron configuration $(t_{2g})^3(e_g)^1$ where the single e_g electron is in the d_{z^2} orbital, consequently lengthening two of the bonds. In brief the *I4/mcm* structure is characterized by cooperative tilting of the (RuMn)O$_6$ octahedra that also exhibit a large JT type elongation.

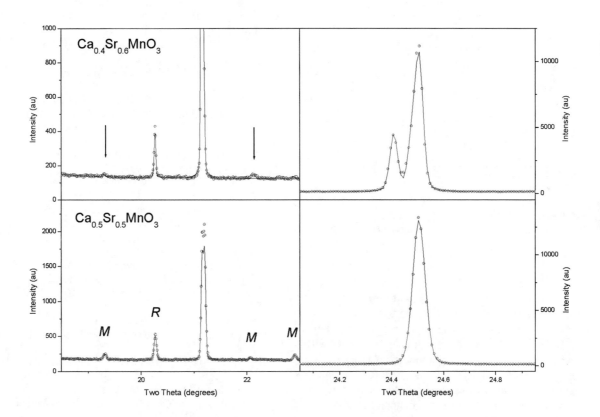

FIGURE 3. Portions of the synchrotron powder diffraction pattern for Ca0.4Sr0.6MnO3 and Ca0.4Sr0.6MnO3 illustrating the change in the cell metric and highlighting the presence of the M-point reflections in both samples. The Rietveld refinement for Ca0.4Sr0.6MnO3 shows the sample contains 30(1)% orthorhombic Pbnm phase in addition to the I4/mcm phase.

Numerous studies of perovskites have demonstrated that the in-phase tilting characteristic of

the *I4/mcm* perovskites can be removed by heating the sample [8,9,14,16]. Likewise heating is capable of melting the orbital ordering associated with the JT distortion of the Mn^{3+} ions [19]. An interesting question, and one we sought to answer, is are these two processes linked? Heating reduces the magnitude of the tetragonal distortion in $SrRu_{0.5}Mn_{0.5}O_3$, however this appears to occur in two steps. The first involves a rapid reduction in the magnitude of the tetragonal distortion of both the unit cell and of the MO_6 octahedra. Above 275 °C this the rate change of both these features is noticeably less and the structure finally becomes cubic near 400 °C. The first transition does not involve a change in symmetry and is associated with orbital melting and the removal of the cooperative JT distortion of the $(RuMn)O_6$ octahedra. The second transition involves a transition to the cubic structure and is due to the removal of the cooperative tilting of the octahedra. The observation of the two successive transitions in $SrRu_{0.5}Mn_{0.5}O_3$ answers the question regarding the coupling of the two instabilities in this material.

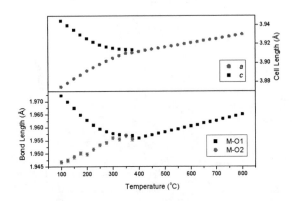

FIGURE 4. Temperature dependence of the tetragonal lattice parameters and M-O bond distances deduced for SrRu0.5Mn0.5O3. Where not apparent the error bars are smaller than the symbols.

In summary I have attempted to illustrate the power of high resolution powder diffraction methods for the study of metal oxides. The high intensity available with synchrotron based instruments coupled with efficient detectors allows parametric studies to be conduced in reasonable timeframes. More importantly the high resolution available when using a synchrotron X-ray source enables us to obtain precise structures and so to probe subtle structural effects.

ACKNOWLEDGMENTS

The Australian Research Council supported this work. The synchrotron measurements at the Australian National Beamline Facility were supported by the Australian Synchrotron Research Program, which is funded by the Commonwealth of Australia under the Major National Research Facilities program. The efforts of Drs James Hester and Garry Foran in meeting our needs at the ANBF over the past decade in gratefully acknowledged. The Author has enjoyed a long collaboration with Dr. Chris Howard from ANSTO and thanks him and other co-workers named in the references for their contributions.

REFERENCES

1. R.H. Mitchell, Perovskites: Modern and Ancient, Almaz Press, Ontario Canada, 2002
2. A.P. Ramirez, *J. Phys.: Condens. Matter* **9**, 8171 (1997)
3. C.N.R. Rao, *J. Phys. Chem.* B **104**, 5877 (2000)
4. A.R. Oki A R and D.J. Hodgson D J, *Inorg. Chim. Acta* **170**, 65 (1990)
5. I. Loa, P. Adler, A. Grzechnik A, K. Syassen, U. Schwarz, M. Hanfland M, G.K. Rozenberg, P. Gorodetsky and M.P. Pasternak M P, *Phys. Rev. Lett*, **87**, 125501 (2001)
6. T. M. Sabine, B. J. Kennedy, R. F. Garrett, G. J. Foran, and D. J. Cookson, *J. Appl. Cryst.* **28**, 513 (1995)
7. C.J. Howard and B.A. Hunter, A Computer Program for Rietveld Analysis of X-Ray and Neutron Powder Diffraction Patterns, NSW, Australia: Lucas Heights Research Laboratories, 1998, 1-27
8. B.J. Kennedy, C.J. Howard and B.C. Chakoumakos, *J. Phys.C Conden. Matter*, **11**, 1479 (1999)
9. C.J. Howard, K.S. Knight, B.J. Kennedy and E.H. Kisi, *J. Phys.C Conden. Matter*, **12**, L677 (2000)
10. C.J. Howard and H.T. Stokes, *Acta Crystallogr.* B **54**, 782 (1998)
11. A.M. Glazer, *Acta Crystallogr., Sect B: Struct. Crystallogr. Cryst. Chem*, B **28**, 3384 (1972)
12. Q. Zhou and B. J. Kennedy *J. Phys.Chem. Solids* **67**, 1595 (2006)
13. R.D. Shannon, *Acta Crystallogr.* **A32**, 751 (1976)
14. Q. Zhou and B. J. Kennedy *J. Solid State Chem.* **179**, 3568 (2006).
15. O. Chmaissem, B. Dabrowski, S. Kolesnik, J. Mais, D.E. Brown, R.Kruk, P. Prior, B. Pyles and J.D. Jorgensen, *Phys. Rev.* B **64**, 134412 (2001)
16. B.J. Kennedy, B.A. Hunter and J.R. Hester, *Phys. Rev.* B **65**, 224103 (2002)
17. Q. Zhou and B. J. Kennedy, To Be Published
18. R.K. Sahu, Z. Hu, M.L. Rao, S.S. Manoharan, T. Schmidt, B. Richter, M. Knupfer, M. Golden, J. Fink and C.M. Schneider, *Phys. Rev.* B **66**, 144415 (2002)
19. T. Chatterji, F. Fauth, B. Ouladdiaf, P. Mandal, and B. Ghosh, *Phys. Rev.* B **68**, 052406 (2003)

SANS and SAXS Study of Block and Graft Copolymers Containing Natural and Synthetic Rubbers

H. Hasegawa

Department of Polymer Chemistry, Graduate School of Engineering, Kyoto University, Kyoto 615-8510, Japan

Abstract. Small-angle neutron scattering (SANS) and small-angle X-ray scattering (SAXS) are excellent techniques to study nano-scale concentration fluctuations in the two-component polymer systems such as block and graft copolymers and polymer blends. The miscibility, phase transitions, microphase-separated structures and interface thicknesses were investigated by SANS and SAXS for the block and graft copolymers, which at least contain natural or synthetic rubber as one component.

Keywords: Block copolymer, graft copolymer, SAXS, SANS
PACS: 61.05.cf, 61.05.fg, 61.41.+e

INTRODUCTION

Well-defined block copolymer samples prepared by living anionic polymerization can provide excellent data for SANS and SAXS analyses because of their highly regular nanostructures. The information such as Flory-Huggins interaction parameter between two constituent polymers, order-order and order-disorder transitions, microphase-separated structures and interface thicknesses can be obtained by such analyses. On the other hand, because of less regular structures less information is available for natural rubber thermoplastic elastomers (NR-TPE) such as natural rubber grafted with poly(methyl methacrylate) (PMMA). In this paper, we report the SANS studies on the miscibility of polyisoprene-*block*-poly(d$_6$-butadiene) and the interface thickness of PMMA-grafted natural rubber, and the SAXS study on the microdomain structures and order-order transitions of polystyrene-block-polyisoprene.

EXPERIMENTAL METHOD

Polyisoprene-*block*-poly(d$_6$-butadiene) sample (HPI-DPB: M_n= 4.5×10^5 g/mol, M_w/M_n = 1.24, DPB mole fraction 0.31) and polystyrene-*block*-polyisoprene sample (S-I: M_n = 2.64×10^4 g/mol, M_w/M_n = 1.02, I volume fraction 0.638) were prepared by sequential living anionic polymerization. The NR-TPE samples (MG40 and MG25) were prepared and supplied by Mr. Le Hai in Nuclear Research Institute in Dalat, Vietnam by γ-ray radiation polymerization of MMA monomer in natural rubber latex solution. SANS measurements were performed using SANS-U and SANS-J at JRR-3M in JAEA, Tokai, Japan, and SAXS measurements were performed using BL-15A in KEK and BL45XU in SPring-8, Japan.

RESULTS AND DISCUSSION

Order-Order Transition of S-I

It has been known that the complex microdomain morphologies such as Gyroid bicontinuous network structure and hexagonally perforated layer (HPL) phase appear in a narrow composition region called "complex phase window" between lamellar and cylindrical morphologies.[1-2] In this composition range recent theoretical studies[3-4] predict a new bicontinuous morphology, which has an orthorhombic unit cell with the symmetry of *Fddd* space group. This gives us a strong motivation to investigate the complex phase window thoroughly.[5] S-I is the sample having the suitable molecular weight and composition for this study.

Figure 1 shows the temperature dependence of the SAXS profiles of S-I obtained from the 2D SAXS patterns by azimuthal averaging. The scattered intensity in arbitrary units is plotted as a function of the wavenumber q (q = (4π/λ) sinθ); 2θ is the scattering angle and λ is the wavelength) for each profile.

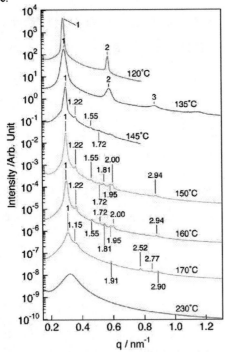

FIGURE 1. SAXS profiles for S-I obtained at designated temperatures. (120, 145 °C measured with BL45XU and the others with BL-15A)

At 230 °C, a single broad peak suggests that S-I is in the disordered state ($T_{ODT} \sim$ 190 °C). Below 170 °C, several distinct peaks appear in the profiles, indicating that the S-I is in its ordered state. The ratios of q/q_m (q_m: q at the first-order peak) indicated in the figure suggest the microdomain morphology of lamellar phase at 120 and 135 °C and Gyroid phase at 170 °C. The peaks in the SAXS profiles at 150 and 160 °C can be indexed with the orthorhombic unit cell (a:b:c = 1:2.00:3.51) having the *Fddd* space group symmetry as 111 (q/q_m = 1), 113 (1.22), 131 (1.55), 133 (1.72), 202 (1.81), 220 (1.94), 222 (2.00), 242 (2.49), 062 (2.65), 313 (2.75), and 315 (2.93). The *Fddd* structure appears irrespective of the thermal history in S-I. Moreover, the Gyroid-*Fddd*-lamellae transition sequence with decreasing temperature observed in our experiment also agrees with the sequence obtained by Tyler et al. [5]

Structure Analysis of NR-TPE

Grafting PMMA onto natural rubber improves the mechanical property of natural rubber because the glassy PMMA microdomains work as the crosslinks to form thermoplastic elastomers. MMA monomer and lauryl sulfate (emulsifier) was added to the natural rubber latex suspension after removing proteins and γ-ray (10 kGy) was applied to initiate radical polymerization of MMA. The PMMA radicals are grafted onto natural rubber molecules to form the graft copolymer (MG). The residual PMMA homopolymer was removed from the suspension with toluene and the products were cast into films. MMA and natural rubber content in the reaction mixture was 25/75 and 40/60 wt%/wt% for MG-25 and MG-40, respectively. Figure 2 shows the SANS profiles for MG-25 and MG-40 after corrections for background and incoherent scattering. A single broad peak observed for MG-25 suggests a microphase-separated structure with the domain spacing of ca. 130 nm in agreement with TEM observation. The linear dependency of the scattering intensity in the double logarithmic plot with slope -4 suggests a narrow interface.

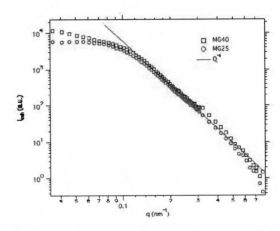

FIGURE 2. SANS profiles for MG-25 and MG-40 obtained for the as-cast films. (measured with SANS-J)

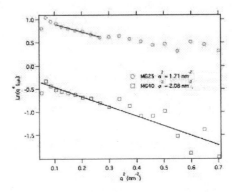

FIGURE 3. Porod analysis for MG-25 and MG-40.

Porod analysis was performed on both samples:

$$\ln\left(Iq^4\right) = \ln\left(2\pi\Delta\rho^2\Sigma\right) - \sigma^2 q^2 \qquad (1)$$

where $\Delta\rho^2$ is the contrast factor for the two phases, Σ is the interfacial area, σ is related to the interface thickness t by $t = \sqrt{2\pi}\sigma$. From the slopes of the linear regions in Figure 3 the interface thickness t was estimated to be 3.3 and 3.5 nm for MG-25 and MG-40, respectively.

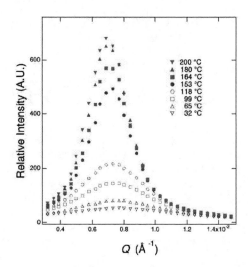

FIGURE 4. Temperature dependence of SANS profiles for HPI-DPB. (measured with SANS-U)

CONCLUSIONS

SANS and SAXS are powerful tools to investigate block and graft copolymers.

ACKNOWLEDGMENT

This research is partially supported by Japan Society for the Promotion of Science, Grant-in-Aid for Scientific Scientific Research(S), 17105004. The author thanks Prof. M. Takenaka and Kyoto University group, and Mr. Le Hai and NRI Dalat group for their collaborations.

REFERENCES

1. H. Hasegawa, H. Tanaka, K. Yamasaki And T. Hashimoto, *Macromolecules* **20**, 1651 (1987).
2. A. K. Khandpur, S. Foerster, F. S. Bates, I. W. Hamley, A, J. Ryan, W. Bras, K. Almdal and K. Mortensen, *Macromolecules* **28**, 8796 (1995).
3. C. A. Tyler And D. C. Morse, *Phys. Rev. Lett.* **94**, 208302 (2005).
4. K. Yamada, M. Nonomura And T. Ohta, *J. Phys. Condens. Matter* **18**, L421 (2006).
5. M. Takenaka, T. Wakada, S. Akasaka, S. Nishitsuji, K. Saijo, H. Shimizu, M. I. Kim And H. Hasegawa, *Macromolecules* **40**, 4399 (2007).
6. H. Hasegawa, S. Sakurai, M. Takenaka, T. Hashimoto And C. C. Han, *Macromolecules* **24**, 1813 (1991).
7. H. Hasegawa, N. Sakamoto, H. Takeno, H. Jinnai, T. Hashimoto, D. Schwahn, H. Frielinghaus, S. Janssen, M. Imai And K. Mortensen, *J. Phys. Chem. Solids* **60**, 1307 (1999).

Small Angle X-ray and Neutron Scattering in the Study of Polymers and Supramolecular Systems

X. B. Zeng[1], F. Liu[1], F. Xie[1], G. Ungar[1], C. Tschierske[2], J. E. Macdonald[3]

[1] Department of Engineering Materials, Sheffield University, Sheffield S1 3JD, United Kingdom
[2] Institute of Organic Chemistry, University Halle, Kurt-Mothes-Strasse 2, D-06120 Halle, Germany
[3] Department of Physics and Astronomy, Cardiff University, Cardiff, CF24 3AA, United Kingdom

Abstract. Some recent work carried out in our research group on complex structures found in polymers and supramolecular systems, using Small Angle X-ray and Neutron Scattering (SAXS and SANS) methods, are reviewed. These include, Combined SAXS and SANS study of superlattice structures in pure and mixed model polymers; Real-time SANS study of transient phases during polymer crystallization; Columnar phases with polygonal cross-sections in T-shaped polyphilic compounds;Complex 3-d phases formed by packing spherical objects (e.g. micelles self-assembled from tree-like molecules), including the recently discovered liquid quasi-crystals which possess 12-fold rotational symmetry. Examples of powder, fibre or surface oriented, and single-domain diffractions will be given. Reconstruction of electron density maps as well as computer modelling are also applied to help solving various complex structures.

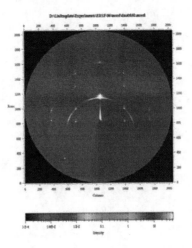

Figure 1. A Grazing-Incidence SAXS pattern of a thin film of T-shaped polyphilic liquid crystal forming compound on silicon surface, showing a less oriented phase on the top surface and another well-oriented phase deeper into the film.

Keywords: SAXS, SANS, GI-SAXS, supramolecules, liquid crystal

CP989, Neutron and X-ray Scattering in Materials Science and Biology, International Conference on Neutron and X-ray Scattering 2007, edited by A. Ikram, A. Purwanto, Sutiarso, A. Zulfia, S. Hendrana, and Z. Nurachman
© 2008 American Institute of Physics 978-0-7354-0508-0/08/$23.00

PARTICIPANT CONTRIBUTIONS

ALLOYS

Uniform Corrosion of Zirconium Alloy 4 Under Isothermal Oxidation at High Temperature

D. H. Prajitno[1]

[1]Metalurgy Laboratory,Nuclear Technology Center for Materials and Radiometry, BATAN
Jl. Tamansari 71 Bandung 40132 Indonesia

Abstract. Many studies have suggested that oxygen at the metal/oxide interface could play a detrimental role on the gas-side oxidation rate of zirconium alloys. However, mechanism of the role of oxygen is not clearly understood yet. Recently it has been shown that oxygen may lower the stress built up in the metal/oxide interface during the oxidation and thus promote the oxide phase transformation, leading to the oxidation enhancement. In this paper, isothermal oxidation of zirconium alloy 4 or zircalloy-4 at high temperature has been studied. High temperature oxidation carried out at tube furnace in air atmosphere at 400, 500 and 600°C. Oxidized sample was characterized by X-ray diffraction and optical microscope. In this study, kinetic oxidation of zirconium-4 at high temperature has also been studied. X-ray diffraction examination show that the oxide scale formed during oxidation of zirconium alloys is ZrO_2. Characterization by optical microscope showed that microstructure of zirconium alloys 4 relatively unchanged after oxidation. Kinetic curves of oxidation of zirconium alloy 4 showed that increasing oxidation temperature will increase oxidation rate.

Keywords: Oxidation, zircalloy-4, high temperature, kinetic.
PACS: 81.65.Kn

INTRODUCTION

Zirconium is a commercially available refractory metal with excellent corrosion resistance, good mechanical properties, very low thermal neutron cross section, and can be manufactured using standard fabrication techniques. The unique properties of zirconium made ideal cladding material for the U.S. Navy nuclear propulsion program in the 1950's[1]. The initial commercial nuclear power reactors used stainless steel to clad the uranium dioxide fuel due to cost. But by mid-1960 zirconium alloys were the principle cladding material due to the superior neutron economy and corrosion resistance. These same zirconium alloys are available to designers of high level nuclear waste disposal containers as internal components or external cladding. Additional advantages of zirconium alloys for long term nuclear waste disposal include excellent radiation stability and 100% compatibility with existing zirconium alloys or zircalloy fuel cladding to alleviate any concerns of galvanic corrosion. The various zirconium alloy grades used in water-cooled nuclear reactors are also available for nuclear waste disposal components. Reactor grade designates that the material has low hafnium content suitable for nuclear service. The hafnium is typically 0.010% maximum. The American Society for Testing and Materials (ASTM) offers widely recognized grades of zirconium alloys. Zircaloy-2 (Grade R60802) is composed of Zr-1.5%Sn-0.15%Fe-0.1%Cr-0.05%Ni and has been predominantly used as fuel cladding in Boiling Water Reactors (BWR) and as calandria tubing in CANadian Deuterium Uranium (CANDU) reactor types. Zircaloy-4 (Grade R60804) has removed the nickel and increased the iron content for less hydrogen uptake in certain reactor conditions. The alloy is typically used as fuel cladding in Pressurized Water Reactors (PWR) and CANDU reactors with the nominal Zircaloy-4 composition is Zr-1.5%Sn-0.2%Fe-0.1%Cr. Refinements in the ingot homogeneity have allowed tighter control of the alloy elements within the ASTM specification. Controlled Composition Zircaloy offers optimized in-reactor corrosion resistance by adjusting the alloy aim point within the ASTM specification ranges. Controlled Composition Zircaloy-4 has lower tin (1.3%) and higher iron (0.22%) than the standard grade. Zr-2.5Nb (Grade R60904) is a binary alloy with niobium to increase the strength. The alloy has been utilized for pressure tubes in CANDU reactors[2,3].

The high temperature oxidation behavior of zirconium alloys as cladding materials for nuclear power reactor has been discussed for many years. A number of papers has been published on this subject. It

CP989, Neutron and X-ray Scattering in Materials Science and Biology, International Conference on Neutron
and X-ray Scattering 2007, edited by A. Ikram, A. Purwanto, Sutiarso, A. Zulfia, S. Hendrana, and Z. Nurachman
© 2008 American Institute of Physics 978-0-7354-0508-0/08/$23.00

has been established that thermal oxidation kinetics of zirconium and zircalloy at atmospheric pressure obeys a parabolic or cubic law preceding the linear one. In the pre transition step, the oxide film growth mechanism is the migration of oxygen through the oxide scale film with the formation of a new oxide at the metal-oxide interface[4,5]. This mechanism has been modelled and compared to experimental data. Let us remind that the zirconia crystallographic structure presents three polymorphs of ZrO_2. At atmospheric pressure zirconia is monoclinic below 1300 K. Around 1400 K, a phase transition occurs and the crystal takes a tetragonal structure. Near 2600 K a second phase transition leads to the cubic structure. In the 500-1100 K temperature range, the crystalline structure of the oxide, formed in the pre-transition state during thermal oxidation, is monoclinic with a proportion of tetragonal phase stabilized by local stress field[6,7].

Early studies show that the corrosion kinetic curves possess two type divided by what is known as a breaking point The first type is described by an almost parabolic equation[8] :

$$x = K_p \, t^{1/2} \tag{1}$$

where x is the thickens of the oxide film, K_p is a constant of the parabolic equation, and t is the duration of the test or the service time. The second type oxidation kinetic rate is commonly describable by an equation of a straight line

$$x = K_1 t + C \tag{2}$$

where K_1 is a constant (rate) of linear oxidation (corrosion) and C is a constant of the linear equation.

The aim of the present work is to studies the high temperature oxidation kinetics zircalloy-4 at level temperature and different time.

METHODOLOGY

The samples used were 1 cm diameter rod of Zircalloy-4. The chemical composition of the zircaalloy-4 was (weight percent): Fe, 0.25; Cr, 0.13; Sn, 1.74; O, 0.13; H, 0.0013; Zr, balance. Oxidation samples were prepared by cutting rod of thick 4 mm by slow speed diamond cutting machine, and grinding their surfaces to a 1200-grit finish and ultrasonically cleaning with alcohol. Isothermal oxidation tests were carried out at 400, 500 and 600°C in still air furnace. Weight change measurements of the samples were made by using an analytical balance. X-ray analysis for identification of various phases present in external oxidation products of the specimen after isothermal oxidation zircalloy-4 alloys have been carried out by

X-ray diffraction (XRD) technique using Cu Ka radiation. Microstructure examination was done by using standard metallographic techniques. Samples were grinding with abrasive paper from 500 to 4000-grit finish and ultrasonic cleaning by alcohol. After that the samples were characterization by optical microscope.

RESULTS AND DISCUSION

Figure 1 showed macrostructure of the zircalloy-4 before and after oxidation at high temperature. From this figure showed that uniform corrosion take place to the entire of the surface of metals and the colour of oxide change from black to redish and diameter of the sample was change due to swelling during isothermal oxidation. The swelling take place because on the surface of the zircalloy-4 during oxidation formed and growth of zirconium oxide. The external oxide layer zirconium oxide is dense and adherent to the surface of the zirconium alloys. The oxide layer is also free of pore as seen in figure 1.

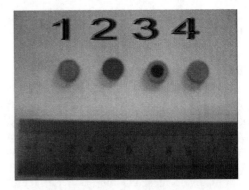

FIGURE 1. Macro structure of alloy Zr – 4 before and after oxidation. (1) Before oxidation (2) After oxidation at 400°C (3) 500°C and (4) 600°C

Figure 2a and b depicted X-ray diffraction pattern of the zircalloy-4 before and after oxidation. From this figure showed that the phase present in the alloy before oxidation dominated by α-zirconium phase. While after oxidation phase present in the oxide are monoclinic zirconium oxide ZrO_2 with deference crystal structure orientation.

Figure 3 shows kinetic oxidation of zircalloy-4 at elevated temperature from 400 to 600°C with different time. Generally, the graphic showed that increasing temperature oxidation will increased weight gain of the specimen. The graphic showed that increasing time is also increasing weight gain. Oxidation of zirconium alloys at elevated temperatures is a natural process. Under certain conditions the presence of impurities and intermetallic compounds in the zirconium alloy can lead to accelerated local corrosion. The kinetics of the process oxidation in the initial stages is sharp increases of weight gain (linier) because the large volume fraction changes accompanying the formation of zirconium dioxide.

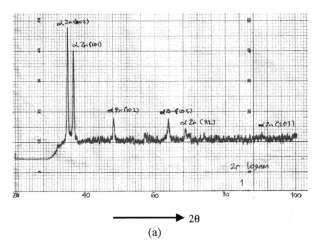

(a)

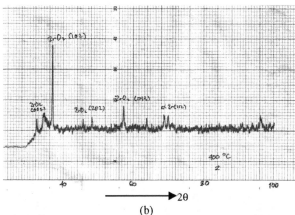

(b)

FIGURE 2. X-ray diffraction pattern of alloy zircalloy-4 (a) before and (b) after oxidation at 600°C

X-ray diffraction confirm that the external oxide layer formed during high temperature oxidation is dominated by monoclinic zirconium oxide After that the oxidation kinetics turn from linier to parabolic rate by gradual decrease weight gain at a relatively constant rate. Gradual decrease oxidation rate take because the external oxide layer zirconium oxide formed during oxidation is dense and adherent to the surface of the zirconium alloys and also the zirconium oxide layer free of pore.

Cross section characterization of the sample by optical microscope is seen in Figure 4. From figure 4 it can be seen that uniform oxide layer appeared in the surface of zircalloy. The microstructure zircalloy-4 after oxidation relatively unchanged. The micro-structure of zircalloy-4 is absolutely contains of equaxed shape structure.

Oxidation kinetics

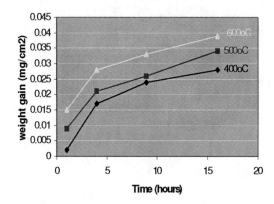

FIGURE 3. Isothermal oxidation kinetics of alloy Zr – 4 at high temperature

FIGURE 4. Cross Section of alloy Zr – 4 after Isothermal oxidation at high temperature.

CONCLUSION

In correspondence with isothermal oxidation of zircalloy-4 at different temperature it can be concluded that the external oxide layer zirconium oxide formed during oxidation is dense and adherent and free of pore. The main phase present in the zircalloy is α-zirconium. The phase present examination by X-ray diffraction examination of oxide scale zircalloy-4 formed during oxidation is ZrO_2. Kinetic curves of oxidation of zirconium alloy 4 showed that increasing oxidation temperature and time will increase oxidation rate. The microstructure of zirconium alloys 4 after oxidation at high temperature relatively unchanged.

REFERENCES

1. C. Wah, "Reactor Grade Zirconium Alloys for Nuclear", Technical Data Sheet, Allegheni Technology, 2003.
2. C. Lemaignan, A.T. Motta; "Zirconium alloys in Nuclear Applications" in *Materials Science and Technology*, R.W. Cahn, P. Haasen, E.J. Kramer (Ed), Vol. 10B, 1994.

3. B. Cox, "Oxidation of zirconium and its alloys in Advance" in *Corrosion Science and Technology*, F. Staeble (Ed), Vol. 5, 1976, 173.

4. V. I. Perekhozhev, V. N. Konev, M. G. Golovachev, et al., "Effect of reactor radiation on oxidation of zirconium-niobium alloys in gas media with complex composition," *VANT, Ser. FRP RM*, Issue 5(33), 54 – 59 (1984).

5. A. V. Matveev, V. I. Perekhozhev, L. P. Sinel'nikov, et al., "A study of nodule corrosion of process channels of the first generating unit of the Kursk NPP", *VANT, Ser. FRPRM*, Issues 1(58) and 2(59), 112 – 115 (1992).

6. A. V. Matveev, V. I. Perekhozhev, A. G. Surnin, et al., "Accelerated oxidation of alloy Zr – 1% Nb in contact with stainless steel", *VANT, Ser. Materialoved. Nov. Mater.*, Issue 1(32), 18 – 22 (1990).

7 V. I. Perekhozhev, A. G. Surnin, A. G. Mizyukanov, "Mechanism and mathematical model of nodule corrosion of zirconium alloys", in: *Proc. Int. Conf. of Radiation Mater., Alushta, May 22 – 28, 1990, Vol. 8* [in Russian], Kharkov (1991), pp. 144 – 150.

8. A. I. Stukalov, V. M. Gritsina, T. P. Chernyaeva, D. A. Baturevich, *Mater. Sci.* **36(5),** 2000

Degradation of Aluminide Coatings in Fe-Al-Cr Alloy on the Isothermal Oxidation

L. Juwita[1], D. H. Prajitno[2], J. W. Soedarsono[3], A. Manaf[4]

[1]Materials Science Department, Indonesia University, Jl. Salemba Raya 4, Jakarta 1043, Indonesia.
[2]Metallurgy Laboratory, PTNBR-BATAN, Jl .Taman Sari 71, Bandung, Indonesia
[3]Metallurgy Engineering Department, Indonesia University, Kampus UI Depok, Jawa Barat 16424, Indonesia
[4]Materials Science Department, Indonesia University, Jl. Salemba Raya 4, Jakarta 10430, Indonesia

Abstract. Fe base superalloy has a good mechanical strength to be used as component operating at high temperature with oxidative environment. Although, the oxidation rate can not be tolerated as it will be oxidized and form oxide scale of un-protective FeO. Coating is a proper solution that this alloy can be used at high temperature. In this research, pack aluminizing on sample was conducted with temperatures of 900°C, 1000°C and 1100°C for 10 hours in inert (argon) environment and then an oxidation test was carried out at temperature of 650°C by an isothermal method for 10 hours in air environment. It was carried out an analysis for characteristics of coating and oxide scale formed in Fe-Al-Cr super alloy resulted from pack aluminizing. From this experiment, it was indicated by XRD analysis that the coating formed on substrate was a layer of $FeAl_2$ compound, other than coating it was found a diffused zone, where in this area it occurred movement of Fe and Cr atoms from substrate toward coating, while Al atoms moved from coating to substrate. The increase of temperature of pack aluminizing process will affect settling rate of Al and coating growth.

Keywords: Superalloy, pack aluminizing, isothermal oxidation.
PACS: 81.65.Mq

INTRODUCTION

Metals/alloys for equipment components operating at high temperature such as in equipments for gas turbine, blade, compressor disk and gas turbine cooling fan of airplane, exhaust gas pipe of airplane and gas turbine in chemical industries, heat treatment, oil removing process and others will usually be oxidized in oxidative environment and at high temperature. Such equipment components must have adequate mechanical characteristic including surface resistance and oxidation resistance depending on oxide scale formation. Oxide scale regarded as having adequate protection at present are Al_2O_3, Cr_2O_3 and SiO_2 with each advantage and disadvantage. The most protective and stable one at higher temperature is Al_2O_3. Beside this Al can also be a good protector for sulfur compound. While SiO_2 and Cr_2O_3 tend to form oxide phase easily vaporize, such as SiO in reductive environment and CrO_3 in oxidative environment at temperature of more than 1000°C. Surface coating is one of the most versatile approaches to enhance the lifetime and to improve the performance of components of interest [1]. Diffusion coating aluminum is a kind of coating frequently applied for protecting high temperature material, one of them by pack cementation (pack aluminizing). Aluminide coatings are of interest for many high temperature applications because of the possibility of improving the oxidation resistance of structural alloys by forming a protective external alumina scale [2-6]. The objectives of this research work include : (1) to understand the mechanism of aluminide layers formation, (2) to study the effect of various temperatures to the characteristics of coating formation (3) to investigate the resistance of aluminide coating formation to isothermal oxidation of 650 °C in air condition for 10 hours.

APPROACH AND METHODS

Three main work in this research are sample preparation of Fe-Al-Cr alloy substrate, diffusion process (pack aluminizing), and oxidation resistance examination at high temperature. The test alloy was produced by first melting iron (99.9%Fe), aluminium (99.9%Al) and chrom (99.9%Cr) in a vacuum arc melting furnace which employed a water-cooled copper hearth and a nonconsumable tungsten electrode. High purity argon was used as an inert gas, and

CP989, Neutron and X-ray Scattering in Materials Science and Biology, International Conference on Neutron and X-ray Scattering 2007, edited by A. Ikram, A. Purwanto, Sutiarso, A. Zulfia, S. Hendrana, and Z. Nurachman

titanium getters were melted to minimize interstitial gases including oxygen and nitrogen from the furnace atmosphere before melting. Homogenization of Fe-Al-Cr alloy substrate using horizontal tube furnace.

Aluminide layer were produced by using pack aluminizing process of Fe-Al-Cr alloy substrate. Pack aluminizing is a diffusion process of aluminum which will form aluminide layer on material surface. In application, it occurs an oxidation forming protective oxide scale of Al_2O_3 which is best for increasing the corrosion resistance of material. In this research, it will be conducted pack aluminizing to Fe-Al-Cr material. After deposition, the coatings were examined by energy-dispersive X-ray spectroscopy (EDX).

Oxidation resistance examination at high temperature using thermal gravimetric equipment Magnetic Suspension Balance (TGA/MSB), to study coating resistance against isothermal oxidation for 10 hours in air environment. Measurement of formed layer thickness carried out by micro-hardness test. Hardness test using micro vicker equipped with diamond indentor with an angle of 136° and applied load of 100 grams with loading time of 15 seconds. Micro structural analysis using an optical microscope. Analysis of phases formed on specimen surface using XRD method (X-Ray Diffraction). Analysis of coating chemical composition using SEM-EDX method (Scanning Electron Microscope-Energy Dispersive Analysis of X-Ray).condition for 10 hours.

RESULT AND DISCUSSION

Material and Coating Characterization

Melting using single arc furnace with target of the composition of Fe-16Al-4Cr (% weight), XRD analysis showed that in alloy of melting result (as cast) and homogenization result (as homogenized) homogenized for 10 hours at a temperature of 1100^oC in inert environment (argon gas), phase formed was (α-Fe,Cr).

In this research aspect covering micro structure observation, hardness, XRD and SEM-EDAX for pack aluminizing process and isothermal oxidation at temperature of 650^oC for 10 hours in air to substrate of Fe-16Al-4Cr alloy (% of weight) will be discussed. It was carried out micro structure taking of pack aluminizing result and high temperature isothermal result after specimen have been tested, with a purpose of clearly observing coating, zone of inter-diffusion, substrate and formed oxide. A Method which is frequently applied to determine phases existing on coating surface is X-Ray Diffraction (XRD) on specimen and substrate surfaces. From this analysis it

is determined that pack aluminizing process produces coating i.e. $FeAl_2$ phase. Figure 1 exhibits formed coating micro structure of pack aluminizing at a temperature of 1000^oC for 10 hours with pack composition (20% of Al + 2% of NH_4Cl + 78% of Al_2O_3). The formed coating consists of a layer with thickness of ~ 34 µm below coating layer being an inter-diffusion zone with thickness of ~ 210 µm.

Micro structure observation of pack aluminizing at temperature of 900^oC, 1000^oC and 1100^oC for 10 hours in inert environment (argon gas).

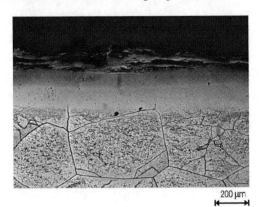

FIGURE 1. Microstructure of Pack Aluminizing at 1000° C for 10 hours.

From Figure 1 it can be compared each sample with different coating temperature, it was observed the formed inter-diffusion zone becomes thicker for longer coating temperature. With various temperatures variation (of 900^oC, 1000^oC, 1100^oC), the result of pack aluminizing with composition (20% of weight of Al powder as layer source of coating, 2% of weight of NH_4Cl as activator, and the rest is 78% of Al_2O_3 inert filler) and the same period (of 10 hours) caused the sample weight to vary. This collected data indicate that the longer the temperature period will result in the higher weight of settling Al.

High Temperature Isothermal Oxidation

Oxidation tests were done to the coated sample substrate at a temperature of 900^oC, 1000^oC and 1100^oC for 10 hours in inert environment (argon gas) by isothermal method at temperature of 650^oC for 10 hours in air environment. From a series of these isothermal tests it was found the variation of specimen weight. The data of weight variation will be used to determine the isothermal oxidation rate of specimens at the temperatures. Graphically the relationship between weight variation per area unit versus oxidation time is demonstrated in Figure 2. These are weight variation values resulted from specimen weighing before and after heating during isothermal oxidation on substrates of homogenizing result (as homogenized) without coating and with coating by pack aluminizing process with the same percentage of coating source and activator. It appears that coated specimen of pack aluminizing result at a temperature of 1100°C has the best corrosion resistance (lower oxidation rate) compared with at temperature of 900°C and 1000°C, and as homogenized. After oxidation for 10 hours, specimen which have best corrosion resistance successively are at 1100°C, 1000°C, 900°C and the last being homogenizing result (as homogenized) without coating.

It was found that order of high temperature oxidation rate i.e. first is the specimen of homogenizing result (as homogenized) without coating, then specimen coated at a temperature of 900°C, 1000°C and the last being at 1100°C. Specimen of homogenizing result (as homogenized) without coating, has a highest oxidation rate, this is caused that the sample was uncoated, consequently from the beginning until the end it has been oxidized fast.

From the curve of high temperature isothermal oxidation, it may be concluded that substrate of homogenizing result (as homogenized) without coating and coated substrate show different oxidation behavior i.e. in substrate without coating, from the beginning until the end, the oxidation rate was high or oxidized faster than coated substrate. In coated substrate, higher the temperature of pack aluminizing process, the oxidation rate is lower or the resistance to oxidation is better.

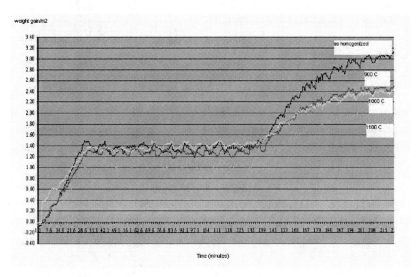

FIGURE 2. Kinetics of isothermal oxidation of the Fe-Al-Cr alloy as homogenized and pack aluminizing at 900°C, 1000°C, 1100°C.

In this oxidation kinetics of Fe-16Al-4Cr alloy (% of weight) of coating result by pack aluminizing process, the weight variation occurring is not so high, decrease of weight as well as the increase of weight. The weight variation is caused by the formation of chromium oxide (Cr_2O_3), the weight variation will decrease for longer period, but this increase does not improve the oxidation cycle. The oxide scales that formed on Fe_3Al-Cr alloys consisted primarily of α-Al_2O_3 containing a small percentage of Fe and less than 1% of Cr. They were non-adherent and fragile. The doping effects of Fe and Cr increased the growth rate of α-Al_2O_3 differently. The selective oxidation of Al to form the oxide scale led to the formation of an Al-free, Fe-enriched zone underneath. Around this zone, a Cr-free region was formed. This implies that the outward diffusion of Al and Cr in the alloy to form the oxide scale occurred[7].

High temperature oxidation of Fe-16Al-4Cr alloy (% weight) is always followed by formation of protective layer α-Al_2O_3. Good oxidation resistance needs a slow oxidation growth speed and the ability of layer to not cracking or detaching at high temperature.

In cyclic oxidation process of Fe-Al-Cr sample with 2-3 mm of thickness, the resistance of oxide not to detach (spelling) is an important factor which has a role in oxidation resistance of the alloy. Initial process and releasing speed is not affected only by the binding power of oxide to metal substrate but also affected by other following factors:

1. Growth speed of aluminum layer
2. Expansion strength of material
3. Component or specimen geometry, thin specimen layer is more resistant than thick specimen.[8]

SEM–EDAX analysis shows that in all samples which have undergone isothermal oxidation for 10 hours in air environment. It was found Al_2O_3 compound (corundum), Cr_2O_3 and FeO (wustite). SEM-EDAX chemical composition shows that iron content is always higher than Cr in Fe-16Al-4Cr (% weight), this is caused by the Fe concentration is higher than the base alloy. High temperature isothermal oxidation as homogenized on without coating alloy of pack aluminizing result produces a different composition.

For alloy with coating, after spot EDAX is carried out on each point, it was found composition values which are nearly similar or uniform, while for alloy with coating of pack aluminizing result as spot EDAX is carried out on each point, it was found varying composition values where significant difference exists. After SEM-EDAX was carried out, there was observed chemical composition close to target designed in calculation of element percentage weight before melting although still shows the existence of an increase and decrease of designed element percentage weight. This is caused by several things, first is that feed material, when melting was carried out, is spilled and the melted material is spattered, adheres on protection glass and furnace material, or feed material existing at crucible angle is not melted, because of that oxidation occurred when smelting, as melting was not carried out in vacuum furnace, while argon gas used as protector of process did not fully eliminate the environment effect.

CONCLUSIONS

XRD test result on smelting using single arc furnace with target of Fe-16Al-4Cr(% of weight) composition, shows that in alloy of melting result (as cast) and homogenization result (as homogenized) which has been homogenized for 10 hours at temperature of $1100^{\circ}C$ in inert environment (argon gas) are found (α-Fe, Cr) phase. XRD analysis on

coating of pack aluminizing, phase formed in substrate is $FeAl_2$ compound layer.

With several temperature variation ($900^{\circ}C$, $1000^{\circ}C$ and $1100^{\circ}C$), result of pack aluminizing process with the same composition and period results in varying of sample weight variation , this indicates that the longer the temperature period will result in more increase of settled Al weight.

$FeAl_2$ becomes thicker as the higher temperature of pack aluminizing process. Other than coating resulted, it was also obtained inter-diffusion zone. Formed oxide scale of isothermal oxidation at temperature of $650^{\circ}C$ for 10 hours in air environment is FeO_2, Al_2O_3 and Cr_2O_3.

ACKNOWLEDGMENT

LJ thanks to the staff members in the Metallurgy Laboratory of PTNBR-BATAN, Bandung, Indonesia in supporting this work.

REFERENCES

1. J.W. Lee, J.G. Duh, S.Y. Tsai, *Surface and Coatings Technology* **153**, 59–66, (2000).
2. P. F. Tortorelli, K. Natesan, *Mater. Sci. Eng.*, **A258** 115 (1998)
3. P. F. Tortorelli, J. H. DeVan, G. M. Goodwin, M. Howell in: *Elevated Temperature Coatings: Science and Technology I*, Edited by N. B. Dahorte, J. M. Hampikian, J. J. Stiglich, TMS, Warrendale, PA, 1995, 203.
4. F. D. Geib and R. A. Rapp, *Oxid. Met.* **40**, 213 (1993)
5. M. Zheng and R. A. Rapp, *Oxid. Met.* **49**, 19 (1998)
6. B. A. Pint, Y. Zhang, P. F. Tortorelli, J. A. Haynes and I. G. Wright, *Mater. High Temp.* **18** (2001)
7. D.B. Lee, G.Y. Kim, J.G. Kim, *Mater. Sci. Eng.* **A 339**, 109-114 (2003)
8. B. A. Pint, Y. Zhang, J. A. Haynes and I. G. Wright, "High Temperature Oxidation Performance of Aluminide Coatings", *Metals and Ceramics Division Oak Ridge National Laboratory*, Oak Ridge, TN 37831-6156.

Oxidation of Ni and Ni-5%W

Z. Lockman[1], M. H. Jamaluddin[1], R. Nast[2]

[1] School of Materials and Mineral Resources Engineering, Universiti Sains Malaysia,
14300 Nibong Tebal,Penang, Malaysia,
[2] Forschungszentrum Karlsruhe, Institut für Technische Physik, Karlsruhe, Germany

Abstract. Oxidation of cube textured (100) <001> Ni and Ni-5%W foils were studied in order to verify if NiO surface layer could be formed on Ni-W similar to that Ni. The principle aim of this study is to prescribe the basic conditions for growing a compact, adherent, smooth, and cube texture NiO suitable to be used as a buffer layer in coated conductor, high temperature superconductor architecture. It was found that for Ni 30μm thick (002) NiO were formed at oxidation temperature of 1250 ± 10°C in air for 60 min. Under the same oxidation condition, (002) NiO with duplex-type morphology was formed on Ni-5%W which reduces the mechanical integrity of the sample. Furthermore, due to oxygen diffusion during the oxidation process, spherical $NiWO_4$ formed inside the Ni-W substrate.

Keywords: Oxidation, nickel, nickel oxide, nickel-tungsten, rolling assisted biaxially textured substrate
PACS: 74.72.Bk Y, 81.65.Mq

INTRODUCTION

The critical step towards the development and commercialisation of high temperature superconductors, based on second generation superconductors in power industry relies greatly on the development of low-cost substrate-buffer system. One way in achieving this is via a so called surface oxidation epitaxy (SOE) process [1] whereby Ni substrates are oxidised to produce NiO as an alternative buffer layer oxide on the surface of Ni. NiO can be used as a buffer layer since the lattice parameter (4.17Å) of this material is close to that of $YBa_2Cu_3O_{7-x}$ (YBCO). The formation of the NiO must however be controlled since only (001) oriented NiO is preferred. Recently, rolling assisted biaxially textured (RABiT) Ni-W foils have been used as substrate for YBCO deposition. The use of Ni-W is advantageous in many ways for the formation of long length coated conductor. However, to produce NiO on Ni-alloys by SOE is not exactly as straight forward as when Ni is used as the substrate. This is due to the formation of secondary and/or ternary internal oxides which often causes delamination of the surface oxides. For example for the case of SOE Ni-Cr [2], the formation of thin Cr_2O_3 layer underneath fast growing NiO layer has effectively stopped Ni^{2+} and O^{2-} diffusions. As a result, porous inner layer formed which degrade the mechanical properties of the substrate. Furthermore, internal oxides precipitations along the grain boundaries induce rough surface oxide [2], which must first be treated before it can be used as a template for subsequent buffer layer or cap layer deposition.

In this work, Ni-5%W was oxidised at different temperature regimes in order to asses if SOE can be adapted for the formation of NiO buffer layer. The aim was to find a window which thin, smooth and single phased (001) NiO can be formed. Oxidation of pure Ni was also conducted for comparison. The mechanism of oxides formation on Ni-5%W is considered and discussed.

EXPERIMENT

RABiT Ni-5%W (Ni-W) and pure Ni substrates were produced by cold rolling followed by heat treatment in inert atmosphere to induce cube-textured grains ((100)<001>) formation [3]. The foils were cleaned with deionised water and ultrasonically washed in methanol for 30 min. Oxidation of the cleaned foils was carried out in a high temperature horizontal furnace at oxidation temperatures ranging from 500-1300°C at heating and cooling rate of 10°C/min. Oxidation was conducted in laboratory air for 60 min. Varieties of standard techniques were used to analyse the oxidised foils. Texture measurements were done by using X-ray diffraction method. Microstructures of the surface morphology and cross section of the as-oxidised sample were viewed by using field emission scanning electron microscopy (FE-SEM). Electron dispersive X-ray (EDX) was used to identify phases formed in the oxide.

RESULTS AND DISCUSSION

Randomly oriented polycrystalline YBCO has low critical current density where the J_C is less than 500Acm^{-2}. Values of J_C were shown to decrease significantly as the misorientation angle of two YBCO grains increases [4]. To eliminate this so-called weak link behaviour biaxially textured YBCO is required. Epitaxial growth of YBCO can be achieved by utilising a so-called coated conductor architecture where YBCO is coated onto textured buffer-substrate system. In a coated conductor, several layers of oxides like Y:ZrO_2 or CeO_2 are deposited on a Ni-based substrates (RABiTS). Since we require a registry of texture between the RABiTS and the buffer layers, epitaxial growth of the buffer layer on the metal template is preferred. When NiO is used as a buffer layer, texturing of NiO is obviously essential. The main requirement of the NiO is therefore; a high degree of in-plane and out-of-pane textures (biaxial texture) as illustrated in Fig 1. To achieve this, oxidation must be carefully controlled.

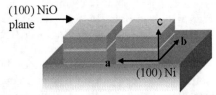

FIGURE 1. Cube oriented NiO: (100) <001> NiO on RABiTS substrate

In this work, oxidation was conducted at temperatures ranging from 500 to 1350°C in order to determine the exact temperature where (001) textured NiO could be produced (which is indexed as (002) in the X-ray diffractogram). Oxidised foils made at temperatures < 1000°C were found to consist of a mixture of (111) NiO and (002) NiO phases. At temperature between >1000-1260°C, oxidised foils consisted of mostly (002) NiO. Temperature > 1260°C were too high which resulted in too thick of a NiO layer. For Ni-W, at 1350°C, the foil was broken, indicating that the whole foil was completely oxidised at this temperature.

Fig. 2 shows X-ray diffraction patterns of oxidised Ni-W (a) and Ni (b) at 1250°C for 60 min in air. For both samples, strong (002) NiO peaks are observed. Peak at $2\theta = 52°$ is associated with the underlying substrate: (002) Ni-W or Ni. Peak at $2\theta = 37°$ is (111)

NiO peak. The formation of (111) NiO cannot be avoided as (111) is the more kinetically stable plane for cubic oxide like NiO [5]. Nonetheless, as shown in the diffractogram, (002) is the main grain orientation, confirming the registry between the oxide and the underlying metals: (002) substrate|(002)oxide. (002) NiO is the phase required if NiO were to be used as a buffer layer in coated conductor architecture. 1250°C is therefore an optimum temperature for the formation of single phased (002) NiO formation.

Whilst it is known that the oxidation of pure Ni would result in the formation of pure NiO layer on the surface of the oxide, oxidation of Ni-W is often associated with the formation of both external and internal oxides. Furthermore according to Wagner-Hauff theory the existence of higher valance W^{6+} in NiO, could increase the oxidation rate and hence thickens the oxide rapidly.

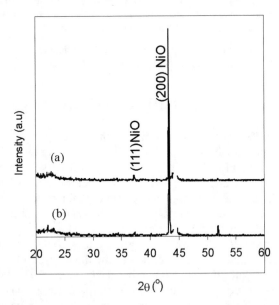

FIGURE 2. XRD pattern for oxidised (a) Ni-W and (b) Ni at 1250°C, air, 60 min

Surface and cross sectional morphology of the oxides were investigated by FESEM and EDX in SEM. These were done to investigate the effect of W addition to the oxidation of Ni on the morphology of the oxides hence a growth mechanism can be deduced. Fig.3 shows the cross sectional morphology of oxidised Ni (a) and Ni-W (b) foils at 1250°C for 60 min in air. From these figures, it is obvious that the oxide formed on pure Ni is much thinner compared to Ni-W which indicates the dissolution of tungsten oxides in NiO layer increases the rate of oxide formation. The substitution of W^{6+} with Ni^{2+} in NiO possible because the ionic radius of Ni^{2+} and W^{6+} is reasonable close [6]. Similarly, oxidation of system containing higher valance cations like Cr^{3+} would result in higher oxidation of rate as seen in the oxidation of Ni-Cr [2]. Substitution of W^{6+} in the NiO layer leads to a creation of more Ni vacancies (V_{Ni}'') to balance electroneutrality of the system. This will increase the conductivity of V_{Ni}'' which result in the rise of the oxidation rate.

A layer of approximately 30μm is present on both side of the Ni foil (Fig 3 (a)). A single columnar layer of Ni can be observed in the micrograph indicating that fast nickel cation diffusion of the bulk of the oxides. In contrast, for Ni-W, the formation of internal oxides can be clearly seen (Fig. 3 (b)). The oxide layers can be roughly divided into three distinct regions (as marked in the microstructure):

A) Surface (external oxide) consisting of compact, columnar oxide layer with thickness of ~ 30μm
B) Internal oxide layer consisting of equiaxed oxides
C) Underneath this equiaxed oxides, a region comprising of a large number of bright round precipitates within the remaining of the substrate can be seen

To investigate the nature of each of this layer EDX analyses were performed in several spots as seen in Fig. 4.

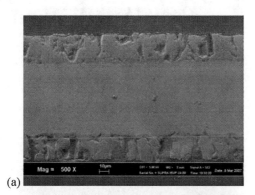

(a)

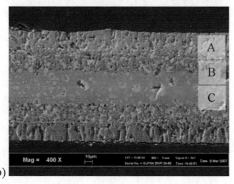

(b)

FIGURE 3. Cross section of Oxidised (a) Ni and (b) Ni-W foils at 1250°C, in air, for 60 min

Point A is taken at the external columnar layer. The point contains and 2.32at% of W. This region is indentified as NiO (43.52at% O and 54.17at%Ni) containing a small amount of W (2.32at)%. This is also consistent with the XRD result in Fig. 3 which shows strong diffraction line from (002) NiO.

The thickness of the internal layer is around 30μm. Two spots were analysed by EDX, a grey spot which has similar colour that that of point A (i.e. Point B as seen in Fig. 4) and point C which is a brighter coloured grains). EDX analysis performed on point B has similar results to that of point A. This region is identified as NiO equiaxed inner layer. Point C consists of 55.73at%O, 22.40at%Ni and 21.87at%W. This point is likely to correspond to a stable ternary phase of Ni-W-oxide alloy of NiWO$_4$

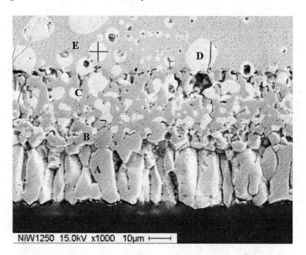

FIGURE 4. Cross section back-scattered image of Ni-W foil after oxidation at 1250oC, in air, for 60 min to show the EDX spot analyses conducted in this specimen

NiO is the fast growing oxide on Ni-W. A duplex-type morphology comprising columnar NiO and equiaxed internal oxide consisting of NiO and NiWO$_4$ is formed. There are various different models that have been proposed to account for this duplex-type morphology. The most plausible explanation has been proposed by Atkinson et al. [6] which stated that during oxidation, the inner equiaxed grains form due to the movement of oxygen inwards through oxide lattice or grain boundaries which then oxidise the substrate at the metal/oxide interface. However, when the diffusivity of oxygen (D_O) was measured by tracer studies they found that the diffusivity of oxygen through NiO lattice (or even grain boundaries) was in order of magnitude less than diffusion of Ni D_{Ni} i.e. the movement of oxygen is not fast enough. Then they proposed a model to accommodate the fast diffusion of oxygen inwards by stating that oxygen was ingressed through fissures that may be transient in nature i.e. the fissure formed, admit gas, and then reseal during isothermal oxidation.

There is a temperature region which such morphology could be produced in pure nickel. Normally it is around 1000-1200°C, whereas above 1200°C, single columnar growth dominates due to high magnitude of D_{Ni}. This is indeed what has been observed in Figure 3 (a) where only a single columnar NiO can be seen. For the case of oxidation of Ni with dilute W content, the existence of W in the NiO must have allowed the ingress of oxygen through the NiO columnar layer to form the equaixed NiO and $NiWO_4$. This is in agreement with the oxidation of Ni-W coating on steel as reported by Lee et al [7].

Point D in Fig. 4 corresponds to a region which consisted of 57.47at%O, 20.78at%Ni and 21.75at%W which could also correspond to the $NiWO_4$ phase. The rest of the area in region (point E) consist of 91.2at% of Ni, 3.5at% of W and 5.3wt% O since the amount of oxygen is very small, we could conclude that this region is the remaining of the substrate i.e. the unoxidised Ni-W foil. As expected the substrate is significantly depleted in W from the starting value of ~ 5 at%. This is because the W has segregated to form the $NiWO_4$ phase.

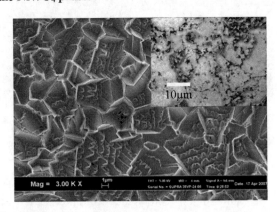

FIGURE 5. Surface Morphology of the Ni-W foil after oxidation foils at 1250°C, in air, for 60 min. Inset is the surface morphology of oxidised pure Ni foil

The surface morphology of the oxide is shown in Fig. 5. Inset in Fig. 5 is a representative surface morphology of pure Ni oxidised under the same condition. As can be seen, upon the addition of W in the substrate, not only that the rate of the oxidation increases, the surface morphology has also changed. Terrace-like surface oxide could be seen on Ni-W foil. Such changes is attributable to the different mechanism of growth when W is present in the NiO layer. The surface oxide formed on pure Ni consisted of flat regions with large square-like grains. This could be the (002) NiO grains which grow abnormally on the surface of the oxide.

Bending test performed on oxidised Ni-W foil resulted in severe crack upon a small angle of bending. This is because the oxide on Ni-W is thicker as well as the existence of the inner equiaxed layer in the foil has significantly reduced the mechanical integrity of this sample.

CONCLUSION

Cube textured Ni and Ni-5 at% W foils were oxidised in flowing oxygen at temperature ranges from 500-1300°C. Similar to Ni, at 1250°C (60 min in air) (002) NiO peak is observed for the Ni-W samples indicating the possibility of the use of this foil as a buffer layer for coated conductors. Nonetheless, even though the surface oxide was found to consist of mostly (002) NiO, the existence of duplex-type morphology in the oxidised Ni-W foil reduces the robustness of the foil since surface delamination often occurs. Furthermore, the rate of oxidation is significantly increased resulting in > 60μm thick oxide. This oxide can be divided into two different regions: NiO external columnar layer, NiO + $NiWO_4$ internal equiaxed layer. $NiWO_4$ also forms in the remaining of the substrate as round, white precipitates. The surface morphology of the oxide consisted of terrace-like morphology.

ACKNOWLEDGMENTS

The author would like to thank Dr. Rainer Nast, Forschungszentrum Karlsruhe, Institut für Technische Physik, Karlsruhe, Germany for the Ni-W foils.

REFERENCES

1. T. K. Matsumoto, I. Hirabayashi, K. Osamura, *Phys.* **C 378** (2), 922-926 (2002)
2. Z. Lockman, X. Qi, A. Berenov, W. Goldacker, R. Nast, B. deBoer, B. Holzapfel and J. L. MacManus-Driscoll, *Phys.* **C 383** (1-2), 127-139 (2002)
3. Rainer Nast, Forschungszentrum Karlsruhe, Institut für Technische Physik, Karlsruhe, Germany, *private communication*
4. P. Dimos, P. Chaudhuri, J. Mannhart, F. Le-Goues, *Phys. Rev. Latt.* **61**, 219-222 (1988)
5. Z. Lockman, X. Qi, A. Berenov, W. Goldacker, R. Nast, B. deBoer, B. Holzapfel and J. L. MacManus-Driscoll, *Phys.* **C 351** (1), 34-37 (2001)
6. Atkinson, D. W. Smart, *J. Electrochem. Soc.* **135** (11), 2886-2893 (1988)
7. D. B. Lee, J. H. Ko, S. C. Kwon, *Mater. Sci. Eng.* **A 380**, 73 – 78A (2004)

Texture and Structure Analysis of Aluminum A-1050 using Neutron Diffraction Technique

T. H. Priyanto, N. Suparno, Setiawan, M. R. Muslih

Neutron Scattering Laboratory, Center of Technology for Nuclear Industrial Materials
National Nuclear Energy Agency of Indonesia - BATAN, Kawasan Puspiptek Serpong, Tangerang 15314, Indonesia

Abstract. Analysis of bulk aluminum A-1050 has been performed using FCD/TD-BATAN. From neutron diffraction pattern of bulk sample, two Bragg peaks, (220) and (200) are observed while three Bragg peaks of (111), (311) and (222) are over dumped. Due to preferred orientation, crystallites are highly oriented to <110> direction. Texture analysis for pole figure reconstruction is performed for (200) and (220). From analysis result using MAUD program, a goodness of fitting ($\sigma = 1.259$), crystallographic weighted error ($R_w = 32.81$ %) and texture error ($R_p = 19.03$ %) are obtained. Analysis of powder type of A-1050 is also carried out. From neutron diffraction measurement, Bragg peaks of (111), (200), (220), (311) and (222) are observed with highest intensity at (111). From analysis result $\sigma = 1.008$, and $R_w = 16.68$ % are obtained.

Keywords: Pole figure, texture, Rietveld texture analysis (RTA), E-WIMV
PACS: 61.05.fg

INTRODUCTION

Preferred orientation or texture is increasingly measure texture recognized as an important feature in many polycrystalline materials. A key problem in materials science is to relate anisotropic physical properties of polycrystalline materials to the preferred alignment of its crystallite constituent in certain sample directions, which is often linked to the way these materials are made or processed. In the industrial use of metals as well as polymers, texture control is essential to improve strength and lifetime of component and structural materials. In many cases texture needs also to be accounted for in crystal structure solution and refinement from powder diffraction patterns. In an effort to characterize textures a large facility such as neutron source has become important [1].

In this paper, aluminum is used as a sample. Aluminum with group 1xxx is chosen because this group includes super-purity aluminum (99.9%) and various grades of commercial-purity (CP) aluminum containing up to 1% of impurities or minor addition [2]. Aluminum is a soft metal and it very useful in technology application because it has corrosion resistance, high electric and thermal conductivity. Structure and texture analysis is carried out using neutron diffraction technique. This technique is appropriate for characterization bulk materials due to low absorption. [3].

The Rietveld Texture Analysis (RTA) procedure

Starting from a collection of several spectra, measured at different sample orientation ϕ-χ to cover a contiguous part of a pole figure, we have adopted an iterative fitting technique based on Rietveld method for profile analysis and direct Wiliam-Imholf-Matties-Vinel (WIMV) algorithm for texture determination, to simultaneously obtain the crystal structure of the sample and its orientation distribution with well-defined distribution.

The intensities of the diffraction peaks depend on the crystal structure and on the sample texture. If crystallites are randomly oriented, we can use the Rietveld method to fit the spectra and refine the crystal structure. If the sample under consideration shows preferred orientation (or texture) there are deviation from the theoretical intensities computed by the Rietveld procedure that depend on sample orientation. Using an appropriate algorithm such as the Le Bail method, it is possible to extract the ratio between the experimental intensity of a peak and the theoretical random one. From all the deviation of a (hkl) reflection we can build the corresponding (hkl) pole figure. If the

CP989, *Neutron and X-ray Scattering in Materials Science and Biology, International Conference on Neutron and X-ray Scattering 2007*, edited by A. Ikram, A. Purwanto, Sutiarso, A. Zulfia, S. Hendrana, and Z. Nurachman
© 2008 American Institute of Physics 978-0-7354-0508-0/08/$23.00

(*hkl*) peak is well separated from others, the construction of a pole figure is straightforward. However, if there are partial or complete overlaps, as in the case of trigonal and lower symmetries, the Le Bail algorithm does not provide a unique solution. The strategy adopted here uses the first inaccurate pole figure data from the Le Bail algorithm to obtain an approximation of the orientation distribution function (ODF). The recalculated pole figures from this first trial provide a better starting point for the subsequent refining iteration.

Texture and WIMV

The objectives of texture analysis are to determine orientation characteristic of a polycrystalline sample from diffraction experiments. A commonly used description of the texture is by three dimensional orientation density function *f(g)* (ODF), e.g. Euler angles. For the commonly used angular grid of 5° the orientation space for triclinic-triclinic symmetries contains about 186.624 (72x36x72). Depending on crystal and sample symmetry, the number is reduced but it still large. The ODF analysis requires more experimental data than just a single diffraction profile that is sufficient for a crystallographic Rietveld analysis of a powder sample and complex

There are standard algorithms for reconstructing the ODF from pole figures. They have been divided into two categories: the first solves the problem in Fourier space expressing pole figures and the orientation distribution with spherical harmonic functions (harmonic expansion method). The second approach is to solve the problem directly (direct methods) by discretization of the ODF into cells.

In this paper a direct method, WIMV, is applied for the texture analysis. This method was chosen for its flexibility and its ability to extract the ODF in cases of very sharp textures.

For the following discussion it is necessary to introduce some texture terminology: a pole figure $p(\phi, \chi)$ can be viewed as a two-dimensional, *hkl* specifics projection of three-dimensional orientation space *f(g)* (for brevity, *hkl* is omitted). If Euler angles $\alpha, \beta,$ and γ are used, *f(g)* is conveniently viewed as a cylindrical whose circular base is a pole figure in spherical projection with pole distance $(\beta \to 90^\circ-\chi)$ and azimuth $(\alpha \to \phi)$. Pole figures are projections of the cylinder space along different projection path. In the case of an (001) pole figure the path are straight line parallel to the cylinder axis (γ), for other pole figures paths are complicated curves. In the sense of projection of a given object *f(g)* all pole figures are compatible.

If pole figures are measured by diffraction experiments, $p(\phi, \chi)$ values are no longer ideally compatible due to experimental error and counting statistics. The WIMV ODF, calculated from experimental pole figures, is the closest approximation. The overall error can be assessed with an R_p value.

$$R_p = 100\% \sum \frac{p(\phi, \chi) - h(\phi, \chi)}{p(\phi, \chi)} \qquad (1)$$

where $h(\phi, \chi)$ are pole figure values recalculated from the ODF. The sum is only performed if $p(\phi, \chi) > \varepsilon$, where ε is usually 0.005 to avoid emphasizing low pole figure value where deviations are mainly due to counting statistics. Using a 5° x 5° angular grid and ideal data, the best R_p value s is in the order of 1%. Excellent experimental data have R_p value of 2% – 5%, but R_p also depends on the texture. R_p will be used as a discriminating parameter to determine the quality of the texture refinement procedure. Another index is the R_{p1} value that has the same definition of R_p except that it uses an ε value of 1, thus emphasizing high orientation densities.

MATERIALS AND PROCEDURES

A bulk sample of aluminum A-1050 with cubic shape with dimension of 20mm x 20mm x 20mm is used for this experiment. A-1050 aluminum type has impurity 0.5% [4].

Traditionally texture analysis has relied on pole figure measurement [5]. Reflecting method is applied for data collection of neutron diffraction. Monochromator Si(311) is used to determine monochromatic neutron beam. A neutron wavelength of 1.2706Å is used for neutron diffraction and texture data collection. Schematic diagram of reflecting method is shown in Figure 1

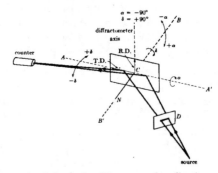

Figure 1. Schematic Diagram of reflection method for texture measurement [6]

An Euler cradle (with ϕ-χ rotations of sample corresponding to spherical coordinates on the pole figure) was placed in the diffractometer ω so that the normal to the plane of the χ-circle bisected the angle between the incident beam and the detector bank at 2θ. The pole figure was covered from $0°$ to $360°$ in ϕ (azimuthal angle) and from $0°$ to $90°$ in χ (polar angle) with an angular resolution of $5°$. Counting system with preset count mode of 2000 neutron/second was used for this experiment. The first beam narrower with dimension of 20mm x 20mm is placed just after monitor detector and the second beam narrower of 30 mm x 30 mm is placed between the sample table and the main detector.

RESULTS AND DISCUSSIONS

To show whether there is preferred orientation (texture) in aluminum A-1050, both neutron diffraction of powder sample and bulk sample are taken to be observed. These experiments were carried out before texture experiment. Aluminum has FCC structure with space group of *Fm-3m*. Offset angle $2\theta = -0.2° \pm 0.03°$. All neutron data are refined using MAUD program based on Rietveld fitting. Rietveld, Marquardt Least Square is used in model refinement. [7]

Neutron Diffraction of Powder Aluminum A-1050

By the used of $\lambda = 1.2706$ Å, a neutron diffraction pattern from powder sample is shown in Figure 2. All Bragg peaks of (222), (311), (220), (200) and (111) are observed in the range of scattering angle $2\theta = 20° - 70°$. From calculation of Rietveld fitting it is obtained that lattice parameter $a = 4.0105$Å ± 0.0026 Å, reliability $R_w = 16.68\%$ and goodness of fitting $\sigma = 1.008$

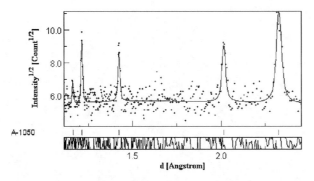

Figure 2. Neutron diffraction pattern (intensities$^{1/2}$ *vs* lattice spacing *d*) for powder sample of aluminum A-1050 measured using FCD/TD. Data was collected from scattering angle $2\theta = 20° - 70°$ ($d = 1.1$ Å - 2.45 Å). Several Bragg peaks of (222), (311), (220), (200) and (111) are shown in the figure.

Neutron Diffraction of Bulk Aluminum A-1050

A bulk aluminum with cube shape of 20 mm x 20 mm x 20 mm is used as a sample to observe neutron diffraction and texture. In bulk type, several Bragg peaks other than (220) and (200) are dumped due to crystallite orientation. To analyze neutron diffraction of the bulk sample, some Bragg peaks with intensities lower than 2% compare to maximum intensity are excluded. So that Bragg peaks in the range of $2\theta = 20 - 34.9°$ and $60 - 70°$ which are related to (222), (311) and (111) Bragg peaks are not refined. From data refinement, lattice parameter $a = 4.0060 \pm 0.0026$ Å is obtained and lattice spacing for (220) plane is $d = 1.418$ Å ± 0.001 Å. Compare with the same Bragg peak of powder sample for (220) plane, $(\Delta d/d)_{<220>} = 0.141$. For (200) plane $d_{bulk} = 2.0030$ Å ± 0.001 Å, $d_{powder} = 2.0052$ Å ± 0.001 Å, and $(\Delta d/d)_{<200>} = 0.109\%$. Therefore $(\Delta d/d)_{<220>}/(\Delta d/d)_{<200>} = 1.294$. This value indicate that crystallite has stronger oriented to <110> direction than <100> direction. To reduce texture error, R_p, analysis is carried out just in the range of $d = 1.4 - 2.15$ Å. In the range of its value, $R_w = 32.81\%$, $\sigma = 1.259$ and $R_p = 19.02\%$ are obtained. Neutron diffraction pattern in the range of $d = 1.4 - 2.15$ Å is shown in Figure 3.

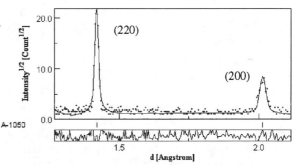

Figure 3. Neutron diffraction pattern of bulk aluminum A-1050. Refinement is carried out just in the range of $d = 1.4 - 2.15$ Å. It is done by eliminated over dumped peaks (222), (311) and (111) to obtain better quality of texture error.

Texture analysis of bulk Aluminum A-1050

Pole figure indicate pole distribution to lattice plane. In this experiment two Bragg peaks (220) and (200) in neutron diffraction are chosen as a representative of pole figures. For WIMV calculation, pole figures are weighted according to diffraction intensities. Bragg peaks with intensities lower than 2% compare to maximum intensity are negligible, so that pole figures of (222), (311) and (111) are not investigated. By using 15% ODF resolution, two reconstruction pole figures, (200) and (220) are shown in Figure 4. Some results of RTA refinement using E-WIMV is shown in Table 1.

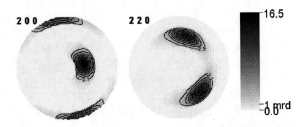

Figure 4. (200) and (220) pole figures calculated using E-WIMV method.

CONCLUSIONS

The RTA methods have been applied successfully for analysis of aluminum A-1050. From the results of data refinement, it is observed that the bulk sample is highly crystallite oriented to <110> direction. Due to preferred oriented, several Bragg peaks; (222), (311) and (111), are over dumped and intensities are reduce less then 2% compare to maximum intensity; (220), For the A-1050 powder sample R_w =16.68%. Large value of R_w > 15% is mainly because of statistical error of neutron data. σ value for both sample types are lower than 2%. For texture analysis, in the range of d = 1.4 – 2 Å, R_p = 19.02%.

TABLE 1. Crystallography parameters of the RTA refinement for aluminum A-1050, a is lattice parameter, R_w is crystallographic weighted error, R_p is texture error, and σ is goodness of fitting.

A-1050	d range (Å)	a (Å)	R_w (%)	R_p (%)	σ	Number of peaks to be refined	Pole figure to be used
powder	1.1 – 2.45	4.011	16.68	13.22	1.008	5	-
Bulk	1.4 – 2.0	4.006	32.81	19.02	1.259	2	2

ACKNOWLEDGMENTS

We would like to thank Mr. Nobuaki Minakawa from JAEA, Japan in providing the sample for the texture experiment.

REFERENCES

1. H.R. Wenk and S.Grigull, *J. Appl. Cryst.* **36**, 1040-1049 (2003).
2. J. Polmear, "Wrought aluminum alloys", in *Light Alloys Metallurgy of the Light Metals*, Metallurgy and Materials Science series, Edward Arnold, 1981, 70-72.
3. H.R Wenk, *J. Appl. Cryst.* **24**, 920-927 (1991).
4. http://mdmetric.com. Maryland Metrics. Comparison Chart Aluminum/Aluminum: USA popular grades vs several overseas grades.
5. L. Lutterotti, S. Matthies, H.R. Wenk, A.S. Shultz, J.W. Richardson, Jr. *J. Appl. Phys.* **81** (2), 594-600 (1997).
6. M. Matherley, W.B. Hutchinson, "An Introduction to texture in Metals", The Institution of Metallurgist, 1979, 20-37.
7. http://www.ing.unitn.it/luttero/maud/tutorial. Luca Lutterotti, MAUD tutorial – Computing ODF from traditional Pole Figure Using, WIMV, December 2000.

The Effect of Sintering Soaking Time on Microstructural and Properties of $CaCu_3Mn_4O_{12}$ System

A.R. Mohd Warikh[1], A.Z. Ahmad Zahirani[2], S.D. Hutagalung[2], A. Zainal Ahmad[2]

[1]Department of Engineering Material, Faculty of Manufacturing Engineering,
Universiti Teknikal Malaysia Melaka, Locked Backed 1200, Ayer Keroh, 75450 Melak, Malaysia.
[2]School of Materials and Mineral Resources Engineering, Engineering Campus,
University Sains Malaysia, 14300 Nibong Tebal, Penang, Malaysia

Abstract. The synthesis of $CaCu_3Mn_4O_{12}$ (CCMO) has been accomplished via solid-state reaction. The mixture was calcined at 850 ^0C for 12 hours. Temperature of 1090 ^oC has been chosen with 6 differ soaking time ranging from 1 hour to 24 hours. The CCMO formation was confirmed using X-ray diffraction (XRD). The microstructure analysis was carried out using field emission scanning electron microscopy (FESEM), while electrical properties have been studied using AutoLab PGSTAT 30 Frequency Analyser. Results shows that differ soaking time for sintering introduce unique properties for CCMO. Microstructural analysis reveal that soaking time more than 12 hours produce grains with almost uniform shape. The physical densification of the sintered pellets promotes the creation of new conduction channels and increase of intergrain effective area for transport current under conductivity properties. Results for bulk conductivity vary from 0.2148×10^{-4} to 0.5825×10^{-4} S/cm.

Keywords: $CaCu_3Mn_4O_{12}$, solid-state reaction, bulk conductivity
PACS: 81.05.-t

INTRODUCTION

Recently, the complex perovskite $CaCu_3Mn_4O_{12}$ [1] has attracted the attention of researches. The crystal structure of $CaCu_3Mn_4O_{12}$ [2] has the rare feature of containing Cu^{2+} (or other Jahn-Teller transition metal cations, such as Mn^{3+}) at the A positions of the ABO_3 perovskite. This Jahn-Teller cation and Ca^{2+} are 1:3 ordered in a $2a_0$x $2a_0$x $2a_0$ cubic cell of *Im3* symmetry (a_0: unit cell of the aristotype).

Perovskite-type (ABO_3) doped manganese oxide has generated a considerable interest because of their many electronic, magnetic and structural properties and potential applications. Extensive theoretical and experimental efforts have been made to understand their complicated mechanism [3].

In this paper, the effect of $CaCu_3Mn_4O_{12}$ at different sintering soaking time was reported. The characterization of microstructure observation and also electrical properties are also given.

EXPERIMENTAL WORK

CCMO was prepared by mixing stoichiometric proportions of $CaCO_3$ (99 + % Aldrich), CuO (99 + % Aldrich) and MnO_2 (90-95 %, Merck). The mixed powders were ball milled for 1 hour and then calcined at 850 ^0C with soaking time 12 hours. The phase of the calcined powders was examined by XRD using a Siemen D5000 Diffractometer. After calcinations, the sample were uniaxially pressed into pellets (13 mm in diameter and 2 mm in thickness) and sintered at 1090 ^oC for 1, 3, 6, 12, 18 and 24 hours, respectively.

The morphology of the sintered samples was investigated by field emission scanning electron microscope (FESEM) using Leo Supra 35VP system. Electrical conductivity measurements were performed using Autolab PGSTAT 30 Frequency Analyser (Eco Chemie B.V.) in a frequency range between 1 Hz to 1 MHz. The measurements were carried out at room temperature.

CP989, *Neutron and X-ray Scattering in Materials Science and Biology, International Conference on Neutron and X-ray Scattering 2007*, edited by A. Ikram, A. Purwanto, Sutiarso, A. Zulfia, S. Hendrana, and Z. Nurachman

RESULTS AND DISCUSSION

X-Ray Analysis

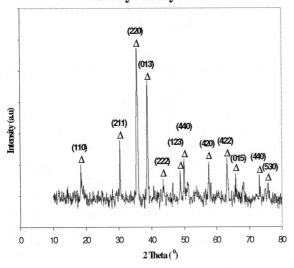

FIGURE 1. XRD pattern on CaCu$_3$Mn$_4$O$_{12}$ at 850 ^{0}C.

The diffractograms of CCMO was obtained as shown in Fig. 1. The mixture is fully reacted to form CCMO at 850 0C. The diffractograms shows the presence of CCMO (ICSD 01-072-0401) with the compound crystallizing in a body-centered cubic perovskite-related structure (groups: Im3).

Microstructural Analysis

SEM of fractured and thermally etched surfaces of the CaCu$_3$Mn$_{4-x}$Mn$_4$O$_{12}$ sample are shown in Fig. 2. From the micrograph, it is shown that microstructure at longer soaking time will have tendency to form grains with 6 corner (hexagonal). Furthermore, the micrograph for longer soaking time also introduce microstructure with near-uniform shape compared to shorter soaking time. In addition, the presence of porosity can also be noted at early soaking time and slowly began to disappear with increasing soaking time. Porosity is an empty that reduces the amount of active material present and replaces it with air (or other trapped gas) - any device is therefore running below its optimum performance.

Electrical Properties

Based on fig. 3, the value of conductivity increases with longer soaking time. The straight lines represent the value of conductivity meanwhile the long dotted line represent density value of each soaking time. The line for density was included in fig. 3 to correlate density with conductivity. As a result, it is proven for CCMO system, higher density will introduces higher value of conductivity where as the current were given more space to pass for a period of time.

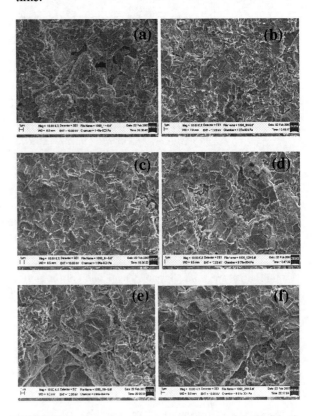

FIGURE 2. Scanning electron micrograph of CaCu$_3$Mn$_4$O$_{12}$ at sintered of 1090 ^{0}C at different soaking time. (a) 1h, (b) 3h, (c) 6h, (d) 12h, (e) 18h and (f) 24h.

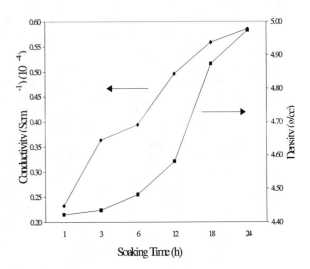

FIGURE 3. Relationship between conductivity and density versus soaking time sintering.

Furthermore, with longer soaking time, the presence of porosity was also greatly reduced. This situation acts as a stepping stone for conductivity to increase its magnitude where the barriers for current to pass were reduced also. Thus, the increase of sintering temperature promotes the microstructural densification. It benefits the conduction of electric carriers and is thus responsible for the improvement of the conductivity [4].

With the increasing of soaking time for sintering, the grain size and mechanical connection between grains are expected to play an important role in the electronic conduction. It is known that the conductivity is mostly influenced by the presence of grain boundary which acts as region of enhanced scattering for the conduction electrons.

CONCLUSION

Single phase $CaCu_3Mn_4O_{12}$ (CCMO) with a perovskite-related structure was synthesized by solid-state reaction at calcinations temperature of 850 ^{0}C. Furthermore, based on the results, it is shown that soaking characteristic has an important role to determining the properties of CCMO. Microstructural observation using FESEM reveal that for sintering at 12 hours and longer soaking time will produce an near-uniform grain close to hexagonal shape. As a result, the properties of conductivity will also increase because the closed arrangement of grains. The result of conductivity greatly proves that soaking time for sintering process possesses a vital role on determining CCMO properties.

ACKNOWLEDGMENTS

A gratefully acknowledge to Universiti Sains Malaysia and also for the support provided by Universiti Teknikal Malaysia Melaka.

REFERENCES

1. Z. Zeng, M. Greenblatt, J.E. Sunstrom, *J. Solid State Chem.* **147**, 185-198 (1999).
2. J. Chenavas, J. C. Joubert, M. Marezio, and B. Bochu. *J. Solid State Chem.* **13**, 162 (1975).
2. M. P. Brown and K. Austin, *Appl. Phys. Lett.* **85**, 503-2504 (2004).
3. J. Sanchez-Benitez, J.A Alonso, A. de Andreas, M.J. Martinez-Lope, M.T. Casais, J.L. Martinez, *J. Magn. Magn. Mater.* **272-276**, suppl 1, E1407-E1409 (2004).
4. C.K. Suman, K. Prasad, R.N.P. Choudhary, *Mater. Chem. Phys.* **97**, 425-430 (2006)
5. R. Przenioslo, M. Regulski, I. Sosnowska, R. Schneider, *J. Phys: Conden. Matter* **14**, 1061-1065 (2001)
6. Maria de F.V. de Moura, Jivaldo do R. Matos, Robson F. de Farias, *Thermochim. Acta* **414**, 159-166 (2004)
7. A. Kumar, B.P. Singh, R.N.P. Choudhary, A.W. Thakur, *Mater. Chem. Phys.* **99**, 150-159 (2006)
8. J. Sanchez-Benitez, A. de Andreas, M. Garcia-Hernandez, J.A. Alonso, M.J. Martinez-Lope, *Mater. Sci. Eng.* **B126**, 262-266 (2006)

Internal Stress Distribution Measurement of TIG Welded SUS304 Samples Using Neutron Diffraction Technique

M. Refai Muslih[1], I. Sumirat[1], Sairun[1], Purwanta[2]

[1]Neutron Scattering Laboratory, Center of Technology for Nuclear Industrial Materials - BATAN
Gedung 40 Kawasan Puspiptek Serpong, Tangerang 15313, Indonesia
[2]Center of Technology for Nuclear Fuel Element – BATAN
Kawasan Puspiptek Serpong, Tangerang 15313, Indonesia

Abstract. The distribution of residual stress of SUS304 samples that were undergone TIG welding process with four different electric currents has been measured. The welding has been done in the middle part of the samples that was previously grooved by milling machine. Before they were welded the samples were annealed at 650 degree Celsius for one hour. The annealing process was done to eliminate residual stress generated by grooving process so that the residual stress within the samples was merely produced from welding process. The calculation of distribution of residual stress was carried out by measuring the strains within crystal planes of Fe(220) SUS304. Strain, Young modulus, and Poisson ratio of Fe(220) SUS304 were measured using DN1-M neutron diffractometer. Young modulus and Poisson ratio of Fe(220) SUS304 sample were measured in-situ. The result of calculations showed that distribution of residual stress of SUS304 in the vicinity of welded area is influenced both by treatments given at the samples-making process and by the electric current used during welding process.

Keywords: Residual stress, strain, SUS304, annealing, welding.
PACS: 61.05.fm

INTRODUCTION

Welding is one of the most significant causes of residual stresses and typically produces large tensile stresses whose maximum value is approximately equal to the yield strength of the materials being joined, balanced by lower compressive residual stresses elsewhere in the component. Welding methods that involve the melting of metal at the site of the joint necessarily are prone to shrinkage as the heated metal cools. Shrinkage, in turn, can introduce residual stresses and both longitudinal and rotational distortion. Heat from welding may cause localized expansion, which is taken up during welding by either the molten metal or the placement of parts being welded. When the finished weldment cools, some areas cool and contract more than others, leaving residual stresses. Welding has significant effect to the mechanical properties of materials[1]. Residual stress within materials can be a detrimental factor to the quality of materials and has strong relation with fatigue behavior of materials[2,3]. Generally, two broad methods are widely used for assessment of residual stress. They are: destructive test such as drilling[4], and non destructive test such as X-ray and neutron diffraction[5], and numerical analytic such as finite element method[6,7,8]. Several researchers have also measured residual stress of materials caused by welding process[9,10]. In this research we have measured the residual stress caused by welding process within SUS304 materials using neutron diffraction technique. The principle of residual stress measurement using neutron diffraction technique can be shortly described as follows. First, the lattice strain of material under investigated is determined using relation:

$$\varepsilon = \frac{d - d_0}{d_0} \tag{1}$$

where: ε is lattice strain of a material,
d_0 is lattice spacing of stress-free material,
d is lattice spacing of stressed material

Then, the d_0 and d of the under investigation materials were evaluated from the peak position of each diffraction curve and were calculated using Bragg's equation:

$$\lambda = 2 d_0 \sin \theta_0 \tag{2}$$

$$\lambda = 2 d \sin \theta \tag{3}$$

where:

CP989, Neutron and X-ray Scattering in Materials Science and Biology, International Conference on Neutron and X-ray Scattering 2007, edited by A. Ikram, A. Purwanto, Sutiarso, A. Zulfia, S. Hendrana, and Z. Nurachman
© 2008 American Institute of Physics 978-0-7354-0508-0/08/$23.00

λ is neutron wavelength, d_0 is lattice spacing of material of stress-free material, d is lattice spacing of stressed material, θ_0 is peak position of diffraction curve of stress-free material, and θ is peak position of diffraction curve of stressed material.

Finally, the residual stresses for the principal axes were calculated using Hooke's equation:

$$\sigma_{1,2,3} = \frac{E}{(1+\upsilon)}\left\{\varepsilon_{1,2,3} + \frac{\upsilon}{(1-2\upsilon)}(\varepsilon_1 + \varepsilon_2 + \varepsilon_3)\right\} \quad (4)$$

where:
  σ is residual stress,
  ε is lattice strain of a material,
  υ is Poisson ratio of material,
  E is Young modulus of material,
  1, 2, 3 are the index of principal axis

EXPERIMENTS

Samples Preparation

The materials used in this research were stainless steel type SUS304 plate with initial size of 20 x 20 x 2 cm³. The plate then was cut into several samples with the dimension as shown in Figure 1. Each sample was

scrabbled so its final thickness is 10 mm. At the middle part of the sample we made a V-shape grooved with 50 mm in length, 8 mm width, and 4 mm in depth. The final dimension of the sample is 100 x 60 x 10 mm³. The cross section of the sample used in this experiment is shown in Figure 1.

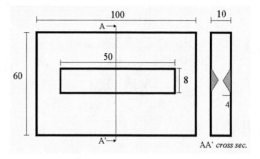

FIGURE 1. Sample dimension.

All samples were annealed for one hour at a temperature of 650 °C. Then, the samples were welded using Tungsten Inert Gas (TIG) with four passes welding method and A316L-Si used as filler. Four values of electric current welding were used, they were 33 A, 43 A, 53 A, and 63 A. Samples code and treatments of is shown in table 1.

Table 1. Samples code and treatment type.

Sample Code	Treatment
A	As machining (directly measured without welded first)
B	A + Anneal: 650 °C, 1 hour + Weld: *33 Ampere*, 1 pass then turn back (x 4 passes)
C	A + Anneal: 650 °C, 1 hour + Weld: *43 Ampere*, 4 passes then turns back
D	A + Anneal: 650 °C, 1 hour + Weld: *43 Ampere*, 1 pass then turn back (x 4 passes)
2A	As machining
2B	2A + Anneal: 650 °C, 1 hour + Weld: *53 Ampere*, 1 pass then turn back (x 4 passes)
2C	2A + Anneal: 650 °C, 1 hour+ Weld: *63 Ampere*, 1 pass then turn back (x 4 passes)
2D	2A + Anneal: 650 °C, 1 hour

Experimental method

For measuring the strain on lattice crystal we have to calibrate the wavelength of neutron came out from the reactor. We also have to measure in-situ the Young modulus and Poison ratio for the (220) plane of SUS304 samples. Silicon powder NBS 640b has been used for neutron wavelength calibration in present measurement. From the calibration we have that the wavelength of neutron was 0.1833753 nm. From in-situ measurement of the SUS304 Fe(220) mechanical constants we have 133.86 GPa and 0.32 for Young's Modulus and Poisson ratio, respectively.

Strain measurements were made on the DN1-M neutron diffractometer at National Nuclear Energy

Agency – Indonesia. The measurement was carried out for (220) lattice plane. Two directions of strain measurements were done. The first is what we called transmission direction (Tr) and the other direction is what we called reflection direction (Ref). Strain measurements were made at the centre of the plate and at 4, 6, 8, 12, 14, 18, and 25 mm from the centre of the groove.

RESULTS AND DISCUSSION

The stress distributions of the samples for transmission and reflection directions are shown respectively on Figure 2 and 3. From the Figures we can see that both distributions show similar tendency. The curve for sample 2D shows that annealing treatment was effectively reduced the residual stress of materials that generated during machining. This can be compared with the sample 2A which was measured directly after machining/grooving process without given any heat treatment.

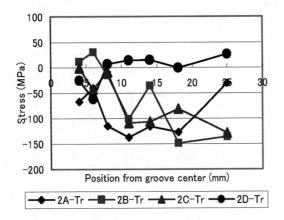

FIGURE 2. Residual stress of (220) plane of the samples for transmission (Tr) direction

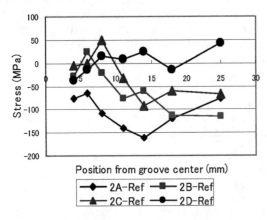

FIGURE 3. Residual stress of (220) plane of the samples for reflection (Ref) direction.

Both for measurement on transmission and reflection directions of the samples which were welded with welding electric current of 53 A (sample 2B) and 63 A (sample 2C) shows almost the same pattern for their residual stress distribution. So, it can be assumed that for samples with similar treatment during making-process, the electric current used during welding process do not have significance effect on their residual stress.

Figure 4 and 5 respectively show the stress distribution of the samples for transmission and reflection directions for four different of electric currents welding. For measurement in transmission direction (Fig. 4) samples that were welded using 43 A produced tensile residual stresses. Samples that were welded with 33, 53, and 63 A produced compressive residual stress for almost all position greater than 8

mm from the center of groove. From measurement in transmission direction it is shown that the higher the electric current used the bigger the residual stress produced. This is because the amount of heat energy received by samples is higher when the electric current is high.

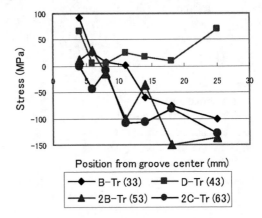

FIGURE 4. Residual stress of (220) plane of the samples for transmission (Tr) direction with four different of electric currents welding.

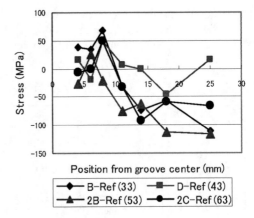

FIGURE 5. Residual stress of (220) plane of the samples for reflection (Ref) direction with four different of electric currents welding.

Residual stress distribution for measurement in reflection direction (Fig. 5) shows similar pattern with the measurement in transmission direction. Except for sample D in reflection direction which posses both tensile and compressive residual stress. Also, in reflection direction residual stress for both tensile and compressive stresses are lower compare to the transmission direction result.

CONCLUSION

The neutron diffraction method has been used to measure the distribution of residual stress of SUS304 type stainless steel samples that were undergone welding process. For all samples, about 200 MPa of compressive residual stress were produced after the samples undergone machining process. With particular value of the electric current of welding process we could reduced the compressive residual stress in an order of hundreds MPa. From this experiment, it was found that for the samples which undergone different treatment of their making-process their residual stress distribution were determined by the electric current of welding process.

ACKNOWLEDGMENTS

The MRM thanks to Dr. Abarrul Ikram for his enlighten discussion on neutron diffraction technique.

REFERENCES

1. D. Dye, O. Hunziker, S. M. Roberts, and R. C. Reed, *Metall. Mater. Transactions A*, **32A**, 1713-1725 (2001).
2. M. E. Fitzpatrick and L. Edwards, *J. Mater. Eng. Performance* **7,** 190-198 (1998).
3. M. N. James, D. J. Hughes, Z. Chen, H. Lombard, D. G. Hattingh, D. Asquith, J. R. Yates, and P. J. Webster, *Eng. Failure Analysis* **14,** 384-395 (2007).
4. ASTM E 837-89, *Standard Test Method for Determining Residual Stress by the Hole-Drilling Strain Gage Method*, American Society for Testing and Materials, 1995.
5. A. J. Allen, M. T. Hutchings and C. Windsor, *Adv. Phys.* **34**, 445-473 (1985).
6. M. Gugliano, *J. Mater. Eng. Performance* **7**, 183-189 (1998).
7. L. Yajiang, W. Juan, C. Maoai and S. Xiaoqin, *Bull. Mater. Sci.* **27**, 127-132 (2004).
8. Z. Feng, X. L. Wang, C. R. Spooner, G. M. Goodwin, P. Z. Mezlasz, C. R. Hubbard, and T. Zacharia, "A Finite Element Model for Residual Stress in Repair Welds", ASME Conference, Montreal Canada, July 21-26, 1996.
9. S. Okido, M. Hayashi, Y. Mori, and N. Minakawa, *J. Phys. Soc. Jpn.* **70,** Suppl. A.526-527 (2001)
10. S. Spooner, S. A. David, J. H. Root, T. M. Holden, M. A. M. Bourke, and J. Goldstone, *Residual Stress and Strain Measurements in an Austenitic Steel Plate Containing a Multipass Weld*, Presented at the 3rd International Conference on Trends in Welding Research, Gatlinburg, TN, 1 June 1992.

Residual Stress Estimation of Ti Casting Alloy by X-ray Single Crystal Measurement Method

A. Shiro[1], M. Nishida[2], T. Jing[3]

[1]Advanced Course of Mechanical System engineering, Kobe City College of Technology, Kobe 651-2194, Japan
[2]Department of Mechanical engineering, Kobe City College of Technology, Kobe 651-2194, Japan
[3]School of Materials Science and Engineering, Harbin Institute of Technology, Harbin 150001, China

Abstract. Recently, titanium casting technology attracts attention in the industrial fields. These casting metals are including a various residual stresses due to the heat shrinkage and inclusion particles, et al. In order to apply the casting technology, the accurate estimation of residual stresses is desired in many cases. In this study, it aims at the nondestructive stress evaluation of titanium casting material by the X-ray stress measurement technique. At first in this study, the $\sin^2\psi$ method which used in the usual X-ray stress measurement was tried to measure the residual stresses. However, it was unsuitable for measurement of titanium casting material because of including coarse grains. Therefore, another X-ray method for single crystal system was employed to coarse crystal grains, in order to investigate the possibilities of residual stress estimating. Four-axes sample table which can set the both of tilt angle and rotate one on the sample surface was prepared. The stereographic diagrams and the theory of elasticity were used to measure the single crystal stresses on the sample surface.

Keywords: Titanium (Ti), X-ray, residual stress measurement, casting alloy
PACS: 61.10.-i

INTRODUCTION

Titanium has the several advantageous properties, such as high strength, light-weight, resistance to heat and corrosion and so on. From these characters, the demand of titanium is expected to rise on the industrial fields in future. Casting processing technology is one of the effective forming methods for titanium. In this process, residual stresses are generated from nonuniform temperature distribution in cooling stage. These residual stresses are influence to the mechanical properties in directly. Therefore, it is necessary to measure the residual stresses accurately in order to apply the casting technology in the industrial fields.

In this study, it aims at the nondestructive stress evaluation of titanium casting material used the X-ray stress measurement technique. We must note, that most examples of X-ray stress measurement of titanium are not seen in the past, because titanium absorbs X-ray easily. Moreover, many peaks appear in X-ray diffraction profile, and the intensity of each peak is very weak. These phenomena are caused from HCP crystal system of titanium, and these characteristics make it difficult to measure the residual stresses. At first in this study, we tried the $\sin^2\psi$ method which used in the usual X-ray stress measurement. However, the large crystal grains and very week peak intensity disturbed the reliable measurement. Therefore, another X-ray stress measurement method for the single crystal system was employed, and residual stresses were calculated from the theory of elasticity with axial transformations. The stereographic projection was used to determine the crystal orientation. The four-axes sample table was prepared for the flexible measurement of X-ray diffraction.

Principle of Single Crystal Stress Measurement by X-ray Diffraction

The three types of coordinate system are defined as shown in Figure 1. One is sample coordinate P_i, two is laboratory coordinate L_i, and another is crystal coordinate C_i. Where i=1, 2, 3 respectively. Especially P_3 direction is coincident with the surface normal of the sample, L_3 direction defines the normal line of the diffraction plane, C_i defines the crystal system and diffraction direction. The surface normal strain ε_{33}^L in L_3 direction are measured by X-ray diffraction in this theory.

CP989, Neutron and X-ray Scattering in Materials Science and Biology, International Conference on Neutron and X-ray Scattering 2007, edited by A. Ikram, A. Purwanto, Sutiarso, A. Zulfia, S. Hendrana, and Z. Nurachman
© 2008 American Institute of Physics 978-0-7354-0508-0/08/$23.00

The relationships of the three types coordinate system are shown in Figure 2. The symbols of π, γ and ω mean the directional cosines between each coordinate axes.

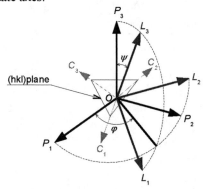

FIGURE 1. Relation among three coordinate system.

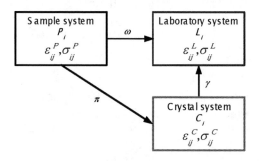

FIGURE 2. Relation of the transformation matrices.

The normal strain $\varepsilon_{33}{}^L$ in L_3 direction is transformed to the $\varepsilon_{ij}{}^C$ in crystal coordinate C_i by the following tensor equation.

$$\begin{aligned}
\varepsilon_{33}^L &= \gamma_{3i}\gamma_{3i}\varepsilon_{ij}^C \\
&= \gamma_{31}^2\varepsilon_{11}^C + \gamma_{32}^2\varepsilon_{22}^C + \gamma_{33}^2\varepsilon_{33}^C + 2(\gamma_{32}\gamma_{33}\varepsilon_{23}^C \\
&\quad + \gamma_{33}\gamma_{31}\varepsilon_{31}^C + \gamma_{31}\gamma_{32}\varepsilon_{12}^C).
\end{aligned} \quad (1)$$

In HCP system of titanium, stress-strain relation is descried as following equations using the elastic compliance S_{ij}.

$$\left.\begin{aligned}
\varepsilon_{11}^C &= S_{11}\sigma_{11}^C + S_{12}\sigma_{22}^C + S_{12}\sigma_{33}^C \\
\varepsilon_{22}^C &= S_{12}\sigma_{11}^C + S_{11}\sigma_{22}^C + S_{13}\sigma_{33}^C \\
\varepsilon_{33}^C &= S_{13}\sigma_{11}^C + S_{13}\sigma_{22}^C + S_{33}\sigma_{33}^C \\
2\varepsilon_{23}^C &= S_{44}\sigma_{23}^C \\
2\varepsilon_{31}^C &= S_{44}\sigma_{31}^C \\
2\varepsilon_{12}^C &= 2(S_{11}-S_{12})\sigma_{12}^C
\end{aligned}\right\}. \quad (2)$$

As the main purpose is description of the relation between normal strain $\varepsilon_{33}{}^L$ with stress $\sigma_{ij}{}^P$ in sample coordinate system, the new relationship between $c_{ij}{}^C$ and $\sigma_{ij}{}^P$ is defines as following equation with matrix π in Figure 2.

$$\sigma_{ij}^C = \pi_{ik}\pi_{jl}\sigma_{kl}^P. \quad (3)$$

Substituting eq. (2), (3) to (1), $\varepsilon_{33}{}^L$ is shown with plane stress condition of $\sigma_{33}=\sigma_{13}=\sigma_{23}=0$ as follow.

$$\begin{aligned}
\varepsilon_{33}^L =\ & (A_1\pi_{11}^2 + A_2\pi_{12}^2 + A_3\pi_{13}^2 + A_4\pi_{12}\pi_{13} \\
& + A_5\pi_{13}\pi_{11} + A_6\pi_{11}\pi_{12})\sigma_{11}^P \\
& + (A_1\pi_{21}^2 + A_2\pi_{22}^2 + A_3\pi_{23}^2 + A_4\pi_{22}\pi_{23} \\
& + A_5\pi_{23}\pi_{21} + A_6\pi_{21}\pi_{22})\sigma_{22}^P \\
& + \{2A_1\pi_{11}\pi_{21} + 2A_2\pi_{12}\pi_{22} + 2A_3\pi_{13}\pi_{23} \\
& + A_4(\pi_{12}\pi_{23} + \pi_{22}\pi_{13}) \\
& + A_5(\pi_{13}\pi_{21} + \pi_{23}\pi_{11}) \\
& + A_6(\pi_{11}\pi_{22} + \pi_{21}\pi_{12})\}\sigma_{12}^P,
\end{aligned}$$
(4)

Where

$$\begin{aligned}
A_1 &= \gamma_{31}^2 S_{11} + \gamma_{32}^2 S_{12} + \gamma_{33}^2 S_{13}, & A_2 &= \gamma_{31}^2 S_{12} + \gamma_{32}^2 S_{11} + \gamma_{33}^2 S_{13} \\
A_3 &= \gamma_{31}^2 S_{13} + \gamma_{32}^2 S_{13} + \gamma_{33}^2 S_{33}, & A_4 &= \gamma_{32}\gamma_{33}S_{44} \\
A_5 &= \gamma_{33}\gamma_{31}S_{44}, & A_6 &= 2\{\gamma_{31}\gamma_{32}(S_{11}-S_{12})\}.
\end{aligned}$$

From eq. (4), if three number of $\varepsilon_{33}{}^L$ values are measured, the stress components $\sigma_{11}{}^P$, $\sigma_{22}{}^P$ and $\sigma_{12}{}^P$ can be calculated and decided.

Furthermore, as shown in Figure 3, HCP coordinate system (a_1, a_2, c) is different from Cartesian coordinate system (x_1, x_2, x_3). In this study, the coordinate transformation and stress strain relation based on the Cartesian coordinate system. Therefore the HCP coordinate system should be change to Cartesian coordinate system. Eq. (5) shows transformation from Miller indices (hkl) in HCP system to Cartesian coordinate system (HKL). By the same token, eq. (6) is direction indices case. Where $\lambda=$ c/a is lattice parameter ratio.

$$\begin{pmatrix} H & K & L \end{pmatrix} = \begin{pmatrix} h & \dfrac{h+2k}{\sqrt{3}} & \dfrac{l}{\lambda} \end{pmatrix} : \text{plane indices.} \quad (5)$$

$$\begin{bmatrix} U & V & W \end{bmatrix} = \begin{bmatrix} u & \dfrac{u+2v}{\sqrt{3}} & \lambda w \end{bmatrix} : \text{direction indices.} \quad (6)$$

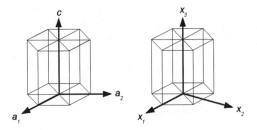

FIGURE 3. Convention of the coordinate system.

The other hand, direction cosine of the coordinate transformation between L_i and C_i is defined by the γ_{3k}. When diffraction plane is shown by HKL, γ_{3k} is described following equations,

$$\gamma_{31} = \frac{H}{\sqrt{H^2 + K^2 + L^2}}, \quad \gamma_{32} = \frac{K}{\sqrt{H^2 + K^2 + L^2}},$$

$$\gamma_{33} = \frac{L}{\sqrt{H^2 + K^2 + L^2}}. \tag{7}$$

Furthermore, directional cosine matrix π which is used to transform between P_i and C_i is as follows,

$$\pi_{ij} = \begin{pmatrix} \cos\phi\cos\psi & \sin\phi\cos\psi & -\sin\psi \\ -\sin\phi & \cos\phi & 0 \\ \cos\phi\sin\psi & \sin\phi\sin\psi & \cos\psi \end{pmatrix}. \tag{8}$$

SAMPLE PREPARATION

The measuring sample is Ti-6Al-4V manufactured by vacuum casting method. After buff polishing and chemical corrosion of sample surface, the coarse crystal grains are observed on sample surface, and it could be look by the naked eye very easily. The components of corrosive liquid are shown in Table 1. The photograph of this result is shown in Figure 4.

$\phi\,38\text{mm}$

FIGURE 4. Surface of titanium casting material.

TABLE 1. Components of corrosive liquid.

Escharotics	Amount
Distilled water	20 mL
Hydrogen peroxide (30%)	5 mL
Potassium hydrate (40%)	10 mL

EXPERIMENTAL PROCEDURE

X-ray stress measurement is using the X-ray diffraction based on the Bragg law (eq. (9)). By using it, we can measure interplanar spacing precisely.

$$\lambda = 2d\sin\theta \tag{9}$$

were λ: X-ray wavelength, d: interplanar spacing, θ: diffraction angle.

The strains are calculated from following simple equation;

$$\varepsilon = \frac{d - d_o}{d_o} \tag{10}$$

Where d_0 is stress free lattice spacing, d is measurement values by X-ray diffraction.

In this study, Schulz reflection method is used to make stereographic diagram. It requires a three-axes sample table which allows rotation of the sample in surface normal axis and horizontal axis; these axes are shown as ψ axis, χ axis and θ-2θ axis in Figure 5.

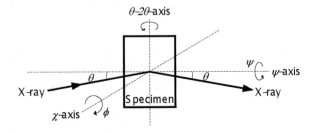

FIGURE 5. Schematic diagram of four axes ψ, χ and θ-2θ relation.

The three axes ψ, χ and θ-2θ are cross each other in the same position on the sample surface. This cross point also coincidents with the irradiation position of X-ray beam.

Conditions of X-ray stress measurement in this study are shown in Table 2, and elastic compliances of titanium are shown in Table 3.

TABLE 2. Conditions of X-ray stress measurement

Characteristic X-rays	CuKα
X-ray optics	Collimator & Parallel beam slit
Tube voltage	40 kV
Tube current	20 mA
hkl plane	004 2θ= 82.5°
Diffraction angle	213 2θ= 141.5°
Filter	Nickel
Irradiated area	ϕ 1 mm

TABLE 3. Elastic compliances of titanium.

S_{11}	S_{12}	S_{13}	S_{33}	S_{44}
0.958	-0.462	-0.189	0.698	2.141

EXPERIMENTAL RESULT AND DISCUSSIONS

In this study, the residual stresses are calculated form normal strain $\varepsilon_{33}{}^{L}$. These strains are based on relation of stress free lattice d_0 and measured spacing d (see eq. (10)). Previous to the stress measurement, the stress free lattice d_0 should be investigated in detail.

The titanium powder was grinded from the part of measurement sample by draw-filing. X-ray diffraction 103 and 102 was employed to measure the stress free lattice d_0 for powder sample in order to obtain high intensity and clear peak shape. The lattice parameters a and c were calculated from these diffraction peaks, because 004 and 213 diffraction planes of stress measurement were different from the d_0 of 103 and 102 plane. Therefore, the stress free lattice d_0 in 004 and 213 plane were calculated from these parameters a and c. Table 4 shows the results of d_0 measurement.

Figure 6 shows the results of crystal situation in the three measurement position. From these stereographic diagrams, it confirms that the behaviors of coarse crystal grain are almost similar to a single crystal. The several poles appeared very clearly in diagrams, and these poles are related by [004] axis and [213] circle.

TABLE 4. Results of lattice parameters and stress free lattice d_0.

Parameter(Å)	a = 2.9261
	c = 4.69681
Lattice spacing (Å)	d_0(004) = 1.1742
	d_0(213) = 0.81704

Figure 7 shows the result of residual stress alterations from edge side position to center position on sample surface. Measurement positions are indicated in figure, and interval distance is 2.5mm. From this result, residual stresses in center and edge

position are tensile state. The other hand, residual stress in between center and edge position is compression stress. Under normal circumstances, it must be balance between tensile residual stresses with compressive one. However, this balance is not equal in this result. That is, compressive residual stresses much larger than tensile stresses. It is seems that the reason of this phenomenon is effect of buff polishing. In addition, tensile stress in radial direction is smaller than it in hoop direction.

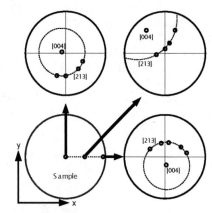

FIGURE 6. Schematic diagrams of stereographic and crystal direction in measurement points. [004] axis and [213] circle are appeared in each positions clearly.

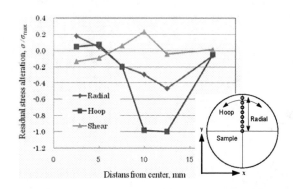

FIGURE 7. Residual stress alterations from edge side to center position on sample surface. Radial and Hoop directions are shown in this figure.

Ratio of the measurement stress and the maximum stress, σ/σ_{max} applied to vertical value in this figure. This tendency is coincide with the several results of neutron stress measurement in other reports. Then, the qualitative tendency is estimated in correct. However, in this measurement, maximum stress value is 5000MPa. It is ten times larger than usual value. The reason of this phenomena is assumed that the effect of inaccurate value of d_0. The d_0 calculation used lattice parameter a and c which measured from different planes. If we would like to be high precision measurement, it is the best in both of same *hkl* plane between d_0 calculation and stress measurement. However, the peak profile from 004 and 213 planes disappeared in this powder sample because of the characteristic in HCP crystal system of titanium. It is necessary to be more high precision measurement for d_0, it is also improvement problem in future.

CONCLUSIONS

The X-ray stress measurement was tried to titanium casting materials using the technique for single crystal materials. The conclusion is as follows.

The sample of titanium casting material includes large crystal grains. Therefore the usual $\sin^2\psi$ method in X-ray stress measurement can not apply to this sample.

X-ray stress measurement method for single crystal system is possible to measure the residual stresses in this sample. The coarse crystal grains can be taken as almost single crystals. The clear and high intensity profile is obtained by this method.

The compressive residual stresses existed on the sample surface. However, these values were very large and near the yield stress of titanium. It is consider that these results were influence to the precision of the stress free lattice spacing d_0.

REFERENCES

1. T. Hanabusa and H. Fujiwara, *The Soc. of Mat. Sci., Japan.* **33**, 372-377 (1984)
2. B. D. Cullity: Elements of X-Ray Diffraction, 1978, 308-309.
3. The Japan Inst. Metals: Data Book of Metals, 1996, 32.

Residual Stress Measurement of Titanium Casting Alloy by Neutron Diffraction

M. Nishida[1,*], T. Jing[2], M. R. Muslih[3] and T. Hanabusa[4]

[1]Department of Mechanical Engineering, Kobe City College of Technology, Kobe, Japan
[2]School of Materials Science and Engineering, Harbin institute of Technology, Harbin, China
[3]Neutron Scattering Laboratory, BATAN, Kawasan Puspiptek Serpong , Indonesia
[4]Dept. of Mechanical Engineering, Institute of Technology and Science, Tokushima University, Tokushima, Japan

Abstract. Neutron stress measurement can detect strain and stress information in deep region because of large penetration ability of neutron beams. The present paper describes procedure and results in the residual stress measurement of titanium casting alloy by neutron diffraction. In this study, the three axial method using Hooke's equation was employed for neutron stress measurement. This method was applied to the cylindrical shape sample of titanium casting alloy (Ti-6Al-4V). Form the results of this study, this sample has large crystal grain in the inside whole position, it is assumed this large grain was grown up during casting manufacture process. Furthermore, the peak profile used to the stress measurement appears in very weak because of the HCP crystal system of titanium character and effect of large crystal grain. These conditions usually make difficult to measure the accuracy values of residual stresses. Therefore, it had to spend a long time to measure the satisfied data from titanium sample. Regarding to the results of stress measurement, the stress values in the cylindrical sample of three directions is almost same tendency, and residual stresses change from the compressive state in the outer part to the tensile state in the inner part gradually.

Keywords: Titanium, Residual stress, Neutron diffraction, Nondestructive estimation.
PACS: 61.05.fm

INTRODUCTION

Residual stresses are generated in almost of all engineering components and constructions, and will have either beneficial or detrimental effects depending on their magnitude and signs. Unexpected failure may occur if residual stress does not be understood. Therefore, the assessment of residual stress is very important to ensure material properties.

Casting is a common engineering technique that fabricates mechanical parts having complex figures. Although a casting technique has a long history in manufacturing fields, many important techniques of casting are depended on the experiences and skilled job. Therefore, it is important work to estimate these traditional techniques by newest science technology. Shape change after casting is a big problem in cast products. This phenomenon such as a dimensional instability is deeply related with residual stresses.

In this study, residual stresses of titanium casting alloy is measured by neutron diffraction measurement. Although several trials were produced to estimate residual stresses in cast products, there is few report of the internal stress measurement on experimental non-destructive technique. As first trial of neutron stress measurement on casting sample, this study reports the result of residual stress distribution in internal position.

NEUTRON STRESS MEASUREMENT

Principle of neutron stress measurement

In the triaxial stress measurement, stress free lattice spacing, d_0 must be determined by following Bragg's equation in order to calculated strains of several directions,

$$\lambda = 2d \sin \theta \tag{1}$$

After that, strains are calculated following simple equation

$$\varepsilon = \frac{d - d_o}{d_o} \tag{2}$$

In the neutron stress measurement, stress values were calculated from the strains in the three principal

CP989, *Neutron and X-ray Scattering in Materials Science and Biology, International Conference on Neutron and X-ray Scattering 2007*, edited by A. Ikram, A. Purwanto, Sutiarso, A. Zulfia, S. Hendrana, and Z. Nurachman
© 2008 American Institute of Physics 978-0-7354-0508-0/08/$23.00

directions with Hooke's law given by the following equation:

$$\sigma_1 = \frac{E}{(1+\nu)(1-2\nu)}\{(1-\nu)\varepsilon_1 + \nu(\varepsilon_2 + \varepsilon_3)\}$$

$$\sigma_2 = \frac{E}{(1+\nu)(1-2\nu)}\{(1-\nu)\varepsilon_2 + \nu(\varepsilon_1 + \varepsilon_3)\} \qquad (3)$$

$$\sigma_3 = \frac{E}{(1+\nu)(1-2\nu)}\{(1-\nu)\varepsilon_3 + \nu(\varepsilon_1 + \varepsilon_2)\}$$

If it is measured in the cylindrical sample, the measurement directions are defined by Axial, Hoop and Radial. The equation is possible to change as follows:

$$\sigma_A = \frac{E}{(1+\nu)(1-2\nu)}\{(1-\nu)\varepsilon_A + \nu(\varepsilon_R + \varepsilon_H)\}$$

$$\sigma_H = \frac{E}{(1+\nu)(1-2\nu)}\{(1-\nu)\varepsilon_H + \nu(\varepsilon_A + \varepsilon_R)\} \qquad (4)$$

$$\sigma_R = \frac{E}{(1+\nu)(1-2\nu)}\{(1-\nu)\varepsilon_R + \nu(\varepsilon_A + \varepsilon_H)\}$$

where suffixes show each direction of axial, hoop and radial.

In this study, two kind of the neutron diffractometer were employed because of the beam time limit in each operation schedules. One is the diffractometer for "REsidual Stress Analysis (RESA)" designed and manufactured by Japan Atomic Energy Agency (JAEA). Another one is the "Neutron Diffractometer No.1 (DN1-M)" by the National Nuclear Energy Agency of Indonesia (Badan Tenaga Nuklir Nasional, BATAN) [1-3]. In this work, preceding the stress measurement, the neutron wavelength λ was defined by the peaks of the standard silicon sample (NBS 640B). Five types of silicon diffraction planes, 111, 200, 311, 400 and 331 were used in this measurement. The neutron wavelength is defined in this study, and the value is λ = 2.072336 Å.

After this, the wavelength is used to the stress and strain calculation.

SAMPLE AND MEASUREMENT CONDITIONS

Measurement of crystal condition in titanium casting alloy

Figure 1 shows photograph of titanium casting samples. The measured sample is Ti-6Al-4V manufactured by vacuum casting method. Two samples are estimated by neutron diffraction, one is cylindrical shape and another is mechanical parts

shape. The dimension of cylindrical sample is 40 mm in diameter and 40 mm in height. The mechanical parts are 60 mm and 95 mm. After buff polishing and chemical corrosion of sample surface, the coarse crystal grains are observed on sample surface, and it could be look by the naked eye very easily.

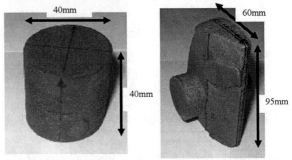

FIGURE 1. Photographs of measurement samples. The content of this sample is Ti-6Al-4V.

Figure 2 shows X-ray diffraction profile from titanium powder and from the head plan of cylindrical sample. As the crystal system of titanium is Hexagonal, it can be seen many and weak peaks in this figure.

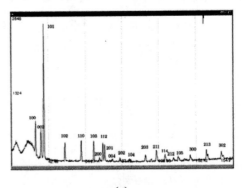

(a)

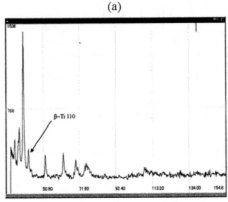

(b)

FIGURE 2. X-ray diffraction profiles from (a) Ti powder, (b) the top head plan of cylindrical sample.

Comparing these profiles, the profile from cylindrical sample changes to broad shapes and several peaks disappear in the high angle region. Furthermore, β-titanium peak appeared in side of 101 peak. As this β-titanium peak is very weak, the effect from the β-titanium phase is ignore in present stress measurement. From this profile, 101 diffraction peak has most strong intensity, and 101 diffraction peak is employed to neutron stress measurement in this study.

Condition of neutron stress measurement

Figure 3 indicates measurement positions and the sequence of neutron irradiation in titanium casting sample. The measurement was made in center position of longitudinal direction and cross section of the sample. Two measurement directions such as in figure are employed to measure the stress distribution on cross section. The measurement position is five points on each direction. The slit size is defined by the pre-measurement of the sample. If it would like to increase the analytical resolution of residual stress distributions, the small slit size is more advantage than large one.

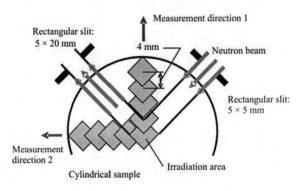

FIGURE 3. Measurement positions and the sequence of neutron irradiation in cylindrical sample.

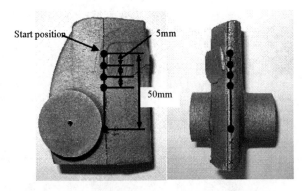

FIGURE 4. Measurement positions of mechanical parts sample. Measurement points are internal center position. (Though this figure looks like a surface, true is inside area)

However, as the diffraction intensity is very weak from titanium casting alloy in this study, the silt size was decided from the balance between beam time and

diffraction intensity. Regarding to neutron stress measurement, the measured stresses become the average stresses of irradiation cubic area, which area is commonly called "gage volume". Furthermore, about quarter of gage volume is overlapped to next area in present measurement. Figure 4 shows measurement point of mechanical parts sample. Longitudinal direction of center position is measured at intervals of 5 mm. Total measurement points is eleven points. Table 1 shows condition of neutron stress measurement. The values of Young's modulus and poisson's ratio are cited from handbook data without considering the diffraction plan dependency.

RESULTS AND DISCUSSIONS

Firstly, we have to consider a big problem of the stress free lattice spacing d_0 in here. When the stress free spacing d_0 was measured from titanium powder, the value of d_0 was about 2.276 Å. If this d_0 value is used to stress calculation in the present sample, the stress values become about 10 times larger than the tensile strength of titanium. These large stress values are not possible to accept in our common sense.

It is considered that this phenomenon causes from the difference of d_0 value between the titanium powder with the actual titanium casting manufacture. The best method to solve this problem is to make the powder from the present casting sample.

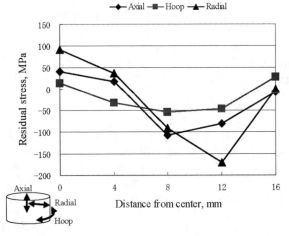

FIGURE 5. Results of stress measurement in the cylindrical sample. Radial, hoop and axial direction is also described in this figure.

However, as this procedure has to do a destructive operation, we will lose the precious titanium casting sample. Therefore in this study, it is assume that the d_0 value was defined such as the stress value of radial direction in most outside position becoming zero value. This assumption is not so deviated from true values, because residual stress of surface normal direction must be became to zero in usual. The absolute value is not so accurate, but it is possible to compare with the stress values and alterations in qualitatively tendency.

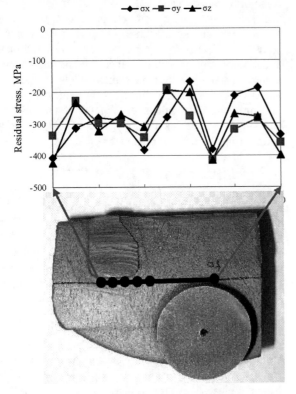

FIGURE 6. Results of stress measurement of the mechanical parts sample Three directions are also described in this figure.

Figure 5 shows the results of stress distribution in the cylindrical sample. Radial, hoop and axial direction is also described in this figure. These results show average stresses between direction 1 and direction 2 in Fig. 3.

From above mentioned assumption, the stress value of most outside position in radial direction is defined zero. Regarding to the stress alteration, the stress values in three directions is almost same tendency. The tensile residual stresses exist in the center position. After that, residual stresses change to compressive state in middle area. Finally, compressive residual stresses decrease in outer side. Comparing the each stress values, the residual stresses alteration is minimum changes in hoop stress and maximum in radial stress.

Figure 6 shows the results of stress distribution of the mechanical parts sample. Form this results, stress alterations of each direction are almost same change. All stresses are compressive residual stresses and relatively large value comparing the cylindrical sample. Unfortunately, there are some inaccurate and dispersion data in this results. These dispersions are caused from weak peak intensity by the long pass length of neutron beam in inside sample. The clearly best answer is increasing of the measurement time in one point to erase such dispersion. However, we didn't have enough beam time to try it in this case. Then, more detail and high accuracy measurement have to try in near future.

CONCLUSIONS

The large crystal grains existed in the titanium casting alloy. These large grains were distributed in whole position of cylindrical sample.

It was possible to measure the residual stress distribution from center position to near the side position in titanium casting sample. In the case of casting sample which likes a mechanical parts, residual stresses in center position is relatively large compressive residual stress in all measurement positions.

The result of three-axial stress analysis showed that the stress values in three directions were almost same tendency. Furthermore, these residual stresses changed compressive in the middle and outer part and tensile in the inner part.

REFERENCES

1. M. Nishida, T. Hanabusa, Y. Ikeuchi and N. Minakawa: *Mat-wiss u Werkstofftech* **34**, 49-55 (2003).
2. M. Nishida, M. Rifai Muslih, Y. Ikeuchi, N. Minakawa and T. Hanabusa, *Mat. Sc.i Forum* **490-491**, 239-244 (2005).
3. M. Nishida, M. Rifai, Y. Ikeuchi, N Minakawa and T. Hanabusa: *Int. J. Modern Phys.* **B 20, 25, 26 & 27**, 3668-3673 (2006).

BIOLOGY

Structural Studies of Protein-Surfactant Complexes

S. N. Chodankar, V. K. Aswal and A. G. Wagh

Solid State Physics Division, Bhabha Atomic Research Centre, Mumbai-400 085, India

Abstract. The structure of protein-surfactant complexes of two proteins bovine serum albumin (BSA) and lysozyme in presence of anionic surfactant sodium dodecyl sulfate (SDS) has been studied using small-angle neutron scattering (SANS). It is observed that these two proteins form different complex structures with the surfactant. While BSA protein undergoes unfolding on addition of surfactant, lysozyme does not show any unfolding even up to very high surfactant concentrations. The unfolding of BSA protein is caused by micelle-like aggregation of surfactant molecules in the complex. On the other hand, for lysozyme protein there is only binding of individual surfactant molecules to protein. Lysozyme in presence of higher surfactant concentrations has protein-surfactant complex structure coexisting with pure surfactant micelles.

Keywords: Small-angle neutron scattering, Protein-surfactant complex, Protein unfolding.
PACS: 61.12.Ex, 87.14.Ee, 87.15.Nn

INTRODUCTION

The function of a protein is directly dependent on its three dimensional structure and any change in its structure leads to protein denaturation causing loss in its functionality. It is known that protein denaturation can be brought about by various means and conditions, which depends upon the interplay between the different interactions among the residues responsible for its biological functionality [1]. Ionic surfactants are known to denature a protein molecule. Ionic surfactants and proteins both share the common property of charged groups and hydrophobic portions. It is believed that surfactant molecules at low concentrations undergo electrostatic binding to the protein, whereas above critical aggregation concentration hydrophobic interaction amongst surfactant molecules takes over the binding process [2-4]. At higher surfactant concentrations, the binding of surfactants to the protein macromolecules leads to their denaturation, which is considered to be the surfactant induced unfolding of the protein [5, 6]. It is the hydrophobic moieties of the surfactant which cause the protein unfolding by interacting with the nonpolar amino acids, thereby eliminating unfavorable solvent contacts.

The protein-surfactant interactions have been the subject of extensive research from the point of view of both biological and industrial applications such as in food products, cosmetics and pharmaceutical industry [7, 8]. Understanding the behavior of such protein-surfactant complexes is of vital interest and allows gaining insight into the binding mechanism between the two components and its effect on the protein structure [9, 10]. SANS is an ideal technique in elucidating the structure of such complexes. Herein, we report SANS studies performed to characterize protein-surfactant complexes formed when surfactant is added to two different protein solutions. The two proteins used are bovine serum albumin (BSA) and lysozyme which are often used as model systems in scientific studies. BSA has a molecular weight of 66.4 KDa and consists of 581 amino acids. At neutral pH, it has a negative charge and known to undergo structural changes at low and high pH values. Lysozyme on the other hand is relatively small with molecular weight of 14.6 KDa and has 129 amino acids in single polypeptide chain. It is known to have a more compact and rigid structure as compared to BSA [11].

EXPERIMENTAL METHODS

BSA and Lysozyme protein along with SDS surfactant and NaCl salt were purchased from Fluka. The samples for SANS experiments were prepared by dissolving known amount of BSA, Lysozyme, SDS and additives in D_2O. The use of D2O as solvent instead of H_2O provides better contrast in neutron experiments. The samples for BSA protein were prepared in acetate buffer solution at pH 5.0, which is close to the isoelectric pH of BSA (4.9) to minimize the interparticle interaction among BSA protein

CP989, *Neutron and X-ray Scattering in Materials Science and Biology, International Conference on Neutron and X-ray Scattering 2007*, edited by A. Ikram, A. Purwanto, Sutiarso, A. Zulfia, S. Hendrana, and Z. Nurachman
© 2008 American Institute of Physics 978-0-7354-0508-0/08/$23.00

molecules. In case of Lysozyme protein which has an isoelectric pH at 10.7, carbonate buffer was used to set the solution pH at 11.0. The ionic strength of the buffer was adjusted to 0.5M using NaCl salt. Small-angle neutron scattering experiments were performed on the SANS instrument at the Dhruva reactor, Bhabha Atomic Research Centre, Mumbai [12] and at Swiss Spallation Neutron Source, SINQ, Paul Scherrer Institut, Switzerland [13]. In particular, the samples which required low Q values to be probed were performed at the SINQ instrument. SANS data were corrected and normalized to a cross-sectional unit using standard procedures [14]. The measurements were made for 1 wt% of protein concentration and in presence of varying SDS concentrations. All the measurements were performed at fixed temperature of 30 °C.

SANS ANALYSIS

In small-angle neutron scattering experiment one measures the differential scattering cross-section per unit volume ($d\Sigma/d\Omega$) as a function of scattering vector Q. For a monodispersed macromolecular protein solution $d\Sigma/d\Omega(Q)$ can be expressed as [15]

$$\frac{d\Sigma}{d\Omega}(Q) = n(\rho_p - \rho_s)^2 V^2 P(Q)S(Q) \tag{1}$$

where n is the number density of the macromolecules, ρ_p and ρ_s, are respectively, the scattering length densities of the protein macromolecule and the solvent and V is the volume of the macromolecule. $P(Q)$ is the intra-particle structure factor and depends on the shape and size of the particle. $S(Q)$ is the inter-particle structure factor and is decided by the inter-particle distance and the particle interactions. For pure protein solution we have calculated $P(Q)$ by treating protein macromolecule as ellipsoidal. For such an ellipsoidal particle

$$P(Q) = \int_0^1 \left[\frac{3(\sin x - x\cos x)}{x^3} \right]^2 dx \tag{2}$$

$$x = Q\left[a^2\mu^2 + b^2(1-\mu^2) \right]^{1/2} \tag{3}$$

where a and b are, respectively, the semimajor and semiminor axes of the ellipsoidal protein macromolecules and μ is the cosine of the angle between the directions of a and the wave vector transfer Q. $S(Q)$ can be approximated to unity as the interparticle interactions are minimized.

The SANS data for protein-surfactant complex is analyzed considering a fractal model which considers complex consisting of micelle-like clusters randomly distributed along the unfolded polypeptide chain. The scattering intensity for such a model can be expressed

as [16].

$$I(Q) = \frac{N_1^2}{N_p N}(b_m - V_m\rho_s)^2 P(Q)S(Q) \tag{4}$$

where N_1 and N_p denote number densities of the total surfactant molecules, protein molecules in solution, respectively and N is the number of micelles on a chain. b_m is the scattering length of the surfactant molecule and V_m its volume. $P(Q)$ denotes intra-particle structure factor of the micelles. For a spherical micelle of radius R, it is given by

$$P(Q) = \left[\frac{3(\sin(QR) - QR\cos(QR))}{(QR)^3} \right]^2 \tag{5}$$

In the fractal model, the arrangement of micelle is assumed as a fractal packing of spheres. For such a case, $S(Q)$ the interparticle structure factor is written as [17].

$$S(Q) = 1 + \frac{1}{(QR)^D} \frac{D\Gamma(D-1)}{[1+(Q\xi)^{-2}]^{[(D-1)/2]}} \sin[(D-1)\tan^{-1}(Q\xi)] \tag{6}$$

where D is the fractal dimension of the micellar distribution in space and ξ the correlation length which is a measure of the extent of unfolding of the polypeptide chain.

RESULTS AND DISCUSSION

SANS data for 1 wt% BSA in presence of varying SDS concentration are shown in Figure 1. It is found that the scattering profiles has two distinct behaviors depending on the surfactant concentration range, based on which the data has been clubbed into two different sets. First set consist of SANS data (figure 1-a) for low SDS concentration (0, 0.25 and 0.5 wt%), in which case the scattering cross-section on addition of surfactant shows a similar behavior as that of pure protein solution. Whereas, SANS data at higher SDS concentration (1, 2 and 2.5 wt%) are shown in Figure 1-b, which has an entirely different feature as compared to that in figure 1-a. It is found that the protein macromolecules in pure aqueous solution have a prolate ellipsoidal shape with semimajor and semiminor axis as 70.2 and 22.2 Å, respectively. The overall scattering cross-section (Figure 1-a) increases with an increase in surfactant concentration, however the scattering profile does not seems to change much. It is believed that at low surfactant concentration binding of individual surfactant molecules due to electrostatic attraction causes increase in its volume, maintaining the ellipsoidal shape of the protein molecule. Table 1 shows changes in the dimensions of the protein on increased binding of surfactant molecules as a function of surfactant concentration. The semiminor axis remains almost same while the semimajor axes increases with increasing surfactant concentration.

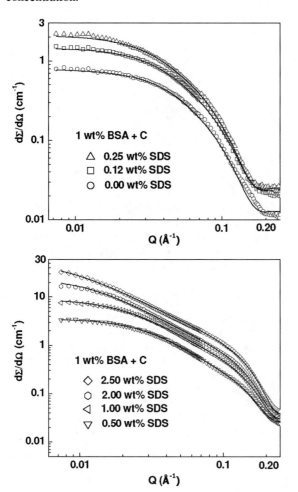

Figure 1. SANS data for 1 wt% BSA in presence varying SDS concentrations.

TABLE 1. Fitted parameters for 1 wt% BSA in presence of varying SDS concentrations.

SDS wt%	Semimajor Axis c (Å)	Semiminor Axis $a = b$ (Å)
0	22.2 ± 0.8	70.2 ± 5.1
0.25	22.2 ± 0.8	80.0 ± 6.1
0.5	23.0 ± 0.9	88.0 ± 6.4

In figure 1-b it is found that in presence of higher concentration of surfactant the scattering profile shows linearity on log-log scale in the intermediate Q range, with a Q range of linearity increasing with surfactant concentration. The linearity of the scattering profile has been fitted using a fractal structure of protein-surfactant complexes. The fractal structure is based on a necklace model considering micelle-like aggregates along an unfolded protein polypeptide chain. The slope of the linear region gives the fractal dimension, whereas the cut-offs of the linearity of the data at low and high Q value are, respectively, related to the formation of the extent of the complex and the size of the individual micelles in the complex [4]. It is observed that on increasing the SDS concentration the cut-off of linearity in the low-Q region shifts to further lower values, which is an indication of increase in the size of the complex, due to further unfolding of the protein molecule, whereas the position of high Q cut-off remains almost same. Also the slope of the linear region appears to decrease with increasing surfactant concentration, indicating decrease in the fractal dimension, suggesting further unfolding of the protein [5]. The fitted parameters for the fractal structure of the complex (Eq. 4) are presented in Table 2.

Figure 2 displays SANS data for 1 wt% Lysozyme protein in presence of varying SDS concentrations (0, 0.25, 0.5, 1, 3 and 5 wt%). It is found that lysozyme protein in pure aqueous solution is prolate ellipsoidal in shape with size dimension agreeing with that known in the literature [18]. On addition of surfactant there is increase in the scattering cross-section in the overall Q range, indicating of increase in the size of the protein macromolecule. This is due to the binding of individual surfactant molecules to the protein macromolecule as had occurred in case of BSA protein. The fitted parameters are given in table 3. On addition of surfactant it maintains its ellipsoidal shape, although it undergoes increase in its size dimension along the semimajor axis, the semiminor axis remains almost same. Interestingly, at higher surfactant concentration, the Lysozyme protein do not show the linearity features in the SANS data (indication of protein unfolding) unlike to that observed in case of BSA protein. The scattering intensity in the low Q region saturates and shows a deep in intensity on increasing the surfactant concentration further. This is believed due to the occurrence of free micelle along with the protein-surfactant complex, as binding between protein and surfactant has reached saturation stage. This is also supported by the fact that the SANS data do not fit to equation (1) or equation (4) which considers the protein solution consist of either individual macromolecules or unfolded one due to the binding of surfactant on its unfolded polypeptide chain. This clearly indicates the resistance of lysozyme to unfold in presence of surfactant as a denaturating agent.

TABLE 2. Fitted parameters for 1 wt% BSA in presence of varying SDS concentration, considering a fractal structure of micelle like clusters randomly distributed along the unfolded protein chain.

SDS wt%	Fractal Dimension D	Correlation Length ξ (Å)	Micelle Radiums R (Å)	Number of Micelles N	Aggregation Number N
1	2.15 ± 0.14	43.0 ± 2.7	18.0 ± 0.6	3	50
2	1.88 ± 0.09	87.9 ± 5.4	18.0 ± 0.6	8	43
2.5	1.71 ± 0.04	144.3 ± 7.5	18.0 ± 0.6	13	42

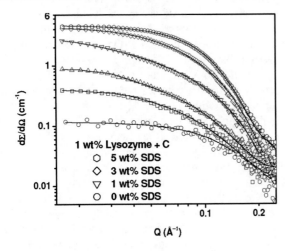

FIGURE 2. SANS data for 1 wt% Lysozyme in presence of 0, 1, 3 and 5 wt% SDS.

TABLE 3. Fitted parameters for 1 wt% lysozyme in aqueous and in presence of varying SDS concentrations

SDS wt%	Semimajor Axis c (Å)	Semiminor Axis $a = b$ (Å)
0	24.3 ± 1.3	14.0 ± 0.6
0.25	70.3 ± 6.0	14.1 ± 0.6
0.5	89.7 ± 4.2	14.1 ± 0.6
1	109.6 ± 6.0	14.1 ± 0.6

CONCLUSIONS

SANS has been used to investigate the structural evolution of the proteins BSA and lysozyme on interaction with anionic surfactant SDS. The binding of anionic surfactant to BSA protein disrupts its native structure. However there is no substantial conformational change in lysozyme protein on binding with SDS surfactant. The BSA protein on binding with SDS surfactant forms protein-surfactant complex, which exhibits a fractal structure due to formation of micelle-like aggregates along the unfolded protein chains. Increasing the SDS concentration causes increase in the size of the complex, due to further unfolding process. Though SDS binds with lysozyme protein causing increase in the size of the protein, it does not result in its unfolding. On increasing the surfactant concentration free micelles are formed along with the protein-surfactant complex. We thus show that BSA and Lysozyme proteins show entirely different behavior on binding with surfactant.

ACKNOWLEDGMENTS

We thank the Swiss Spallation Neutron Source at the Paul Scherrer Institut (PSI), Switzerland, for a beam time for performing a part of SANS work in this paper. We also thank J. Kohlbrecher for his support and useful discussion during the experiments at PSI.

REFERENCES

1. C.M. Dobson, *Nature* **426**, 884 (2003)
2. M.N. Jones, *Chem. Soc. Rev.* **21**, 127 (1992)
3. A. Stenstam, A. Khan and H. Wennerstrom, *Langmuir* **17**, 7513 (2001)
4. A.K. Morén, A. Khan, *Langmuir* **14**, 6818 (1998)
5. S.H. Chen, J. Teixeira, *Phys. Rev. Lett.* **57**, 2583 (1986)
6. S. Chodankar, V.K. Aswal, J. Kohlbrecher, R. Vavrin, A. G. Wagh. *J. Phys.: Condensed Matter.* **19**, 326102 (2007)
7. E. D. Goddard and K. P. Ananthapadmanabhan, Interactions of Surfactants with Polymers and Proteins, London, CRC Press, 1993.
8. C. Sun, J. Yang, X. Wu, X. Huang, F. Wang, S. Liu, *Biophys. J.* **88**, 3518 (2005)
9. E.L. Gelamo, R. Itri, A. Alonso, J.V. da Silva, M. Tabak, *J. Colloid Interface Sci.* **277**, 471 (2004)
10. A. Valstar, M. Almgren, W. Brown, M. Vasilescu, *Langmuir.* **16**, 922 (2000)
11. M. Vasilescu, D. Angelescu, M. Almgren, A. Valstar, *Langmuir.* **15**, 2635 (1999)
12. G.D. Wignall, F.S. Bates, *J. Appl. Cryst.* **20**, 28 (1987)
13. V. K. Aswal, P.S. Goyal, *Curr. Sci.* **79**, 947 (2000)
14. J. Kohlbrecher, W. Wagner, *J. Appl. Cryst.* **33**, 804 (2000)
15. J.B. Hayter, J. Penfold, *Colloid Polym. Sci.* **261**, 1022 (1983)
16. S.H. Chen, J. Teixeira, *Phys. Rev. Lett.* **57**, 2583 (1986)
17. J. Teixeira, *J. Appl. Cryst.* **21**, 781 (1988)
18. S. Chodankar, V.K. Aswal, *Phys. Rev.* **E72**, 041931 (2005)

X-Ray Powder Diffraction (XRD) Studies on Kenaf Dust Filled Chitosan Bio-composites

N. Muhd Julkapli and H. Md Akil

School of Materials and Mineral Resources Engineering, Engineering Campus, Universiti Sains Malaysia,
14300 Nibong Tebal, Pulau Pinang, Malaysia

Abstract. Kenaf dust filled chitosan bio-composites with various compositions of kenaf dust (i.e. 7%, 14%, 21% and 28%) were prepared using solution casting technique. The degree of relative crystallinity of the bio-composites was determined using XRD method. Two distinguishable crystalline peaks were observed in the 2θ range of 5 to 40° which indexed as 020 and 110 respectively. It was noted that the maximum peak of intensity at 020 crystalline peak increased with addition of kenaf dust as well as the second maximum peak of intensity at 110. Consequently, Fourier Transform Infrared (FTIR) analysis was done to investigate the interaction between kenaf dust and chitosan matrix. From FTIR analysis, corresponding peak of chitosan was detected at wavelength of $3233.2 cm^{-1}$ indicated that there exist intermolecular interactions between kenaf dust and chitosan matrix. These results highlighted that there are greater intermolecular forces formed in chitosan with addition of kenaf dust. Intermolecular forces were attributed to the formation of inter and intra hydrogen bonding between chitosan polymer and cellulosic kenaf dust filler.

Keywords: Chitosan, X-ray diffraction, FTIR
PACS: 61.10.Nz

INTRODUCTION

Chitin is a natural polysaccharide (N-deacetylated-2-acetamido-2-deoxy-ß-D-glucan) that exists in considerable amounts as the exoskeleton of arthropods and fungi [1]. Chemically, chitosan is composed of glucosamine and N-acetyl glucosamine linked in the ß $(1\rightarrow4)$ manner. Both chitin and chitosan, and their derivatives have many current and potential future uses because of the unusual combination of properties they possess, which include toughness, biodegradability and bioactivity; this combination makes these materials especially attractive. However, its properties, such as thermal stability, hardness and gas barrier properties are frequently not good enough to meet those wide ranges of applications. Therefore, several attempts were been made in the past in order to modify the chitosan properties in various applications. One way of modifying chitosan properties is by incorporating chitosan with rigid filler [2-3]. Recently, several attempts have also been made to incorporate natural fiber or filler into chitosan film [4]. The use of natural fiber or filler as reinforcement will ascertain that the biodegradability of chitosan will remain unaffected.

There are many natural fiber which has been reported to be suitable for usage as fiber reinforcement for instance; hemp [5], jute [6], flax [7], Kenaf and others. Most recent natural fiber which attracted considerable attention is Kenaf fiber.

Kenaf, *Hisbiscus cannabinus L*, can be processed into natural fibre with unique properties such as low density, low cost, high specific mechanical properties and biodegradability [8]. The goal of this study is to evaluate the effect of Kenaf dust as filler on the properties of chitosan as a matrix for bio-composites using XRD analysis.

EXPERIMENT AND METHOD

Chitosan powder was dissolved in 0.1 M acetic acid followed by addition of kenaf dust powder. The dispersion was then mixed using high speed homogenizer (10,000 to 29,000) rpm prior to casting and dried for 24 hours at 60°C in conventional oven.

X-ray diffractograms of powdered samples were obtained using a Bruker AXS D8 diffractometer under the following conditions: 40 kV and 40 mA with Cu $K\alpha_1$ radiation at 1.54184 Å and acceptance slot of 0.1 mm. About 20 mg of sample was spread on a sample stage, and the relative intensity was recorded in the scattering range (2θ) of 5°-40°.

Series of Fourier Transform Infra Red (FTIR) spectra were recorded with the IR spectrometer in the

CP989, Neutron and X-ray Scattering in Materials Science and Biology, International Conference on Neutron and X-ray Scattering 2007, edited by A. Ikram, A. Purwanto, Sutiarso, A. Zulfia, S. Hendrana, and Z. Nurachman

range of 4000-400 cm^{-1} at a resolution of 4 cm^{-1}. Each spectra was scanned for 4 times prior to conformation. The analyses were performed on kenaf dust, chitosan powder and composites specimens.

RESULT AND DISCUSSION

XRD Analysis

XRD provides important information typically the crystalline index and phase identification [9-10]. The summary of x-ray powder diffractogram of each composite at different percentages of kenaf dust composition is outlined in Figure 1. From the figure, there were two obvious crystalline peaks detected namely CrI$_{020}$ and CrI$_{110}$. Using the corresponding intensity value at given crystalline index, the relative crystallinity (CrI %) was determined in two ways [11]:

$$CrI_{020} = \left[\frac{I_{020} - I_{am}}{I_{020}} \times 100 \right] \quad (1)$$

$$CrI_{110} = \left[\frac{I_{110} - I_{am}}{I_{110}} \times 100 \right] \quad (2)$$

where:

I_{020} = maximum intensity below 13°
I_{am} = intensity of amorphous diffraction at 16°
I_{110} = maximum intensity at ≈ 20°

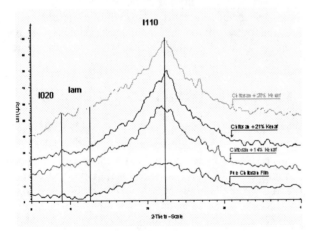

FIGURE 1. XRD analysis of; pure chitosan film, chitosan with 14 % kenaf composites, chitosan with 21 % kenaf composites and chitosan with 28 % kenaf composites

In this study, the crystalline indexes were indicated at CrI$_{110}$ and CrI$_{020}$ respectively. Following the calculation of relative crystallity using Equations 1 and 2, the crystalline indexes for all composites and pure chitosan powder is viewed in Figure 2.

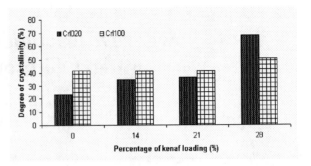

FIGURE 2. Summary of the degree of crystallinity of chitosan and its composites.

From the relative crystallinity values calculated, it can be concluded that the addition of kenaf dust increases the degree of the relative crystallinity of the chitosan film. The reason is attributed to a greater intermolecular forces exist between chitosan and cellulose (kenaf dust). As mention previously, the acetic acid was used to aid the solubility of chitosan powder. It is expected that the dissociated of carboxylic acidic ions (R-COO-), hydrions (H$^+$) and undissociated carboxylic medium (R-COOH) in acetic acid have reacted with amine groups (NH$_2$) and acetylglucosamine group (-NHCOCH$_3$) in chitosan polymer chain to form a positively charged chitosan polymer (NH^{3+}) or known as cationic chitosan polymer (Figure 3).

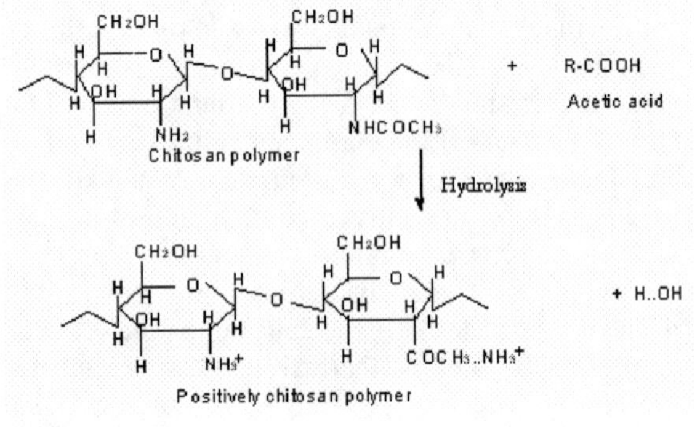

FIGURE 3. Interaction between carboxylic group in acetic acid with acetylglucosamine and glucosamine groups in chitosan polymer

Consequently, the formation of inter hydrogen bonding and intra hydrogen bonding is taken place to form cross-linked network in the composites. The proposed inter hydrogen bonding; between CH$_2$OH (functional group in cellulose polymer) and NH$_2$ (functional group in chitosan polymer) and intra hydrogen bonding; between NHCOCH$_3$ and HOCH$_2$ (functional group in chitosan polymer) are outlined in Figures 4 respectively.

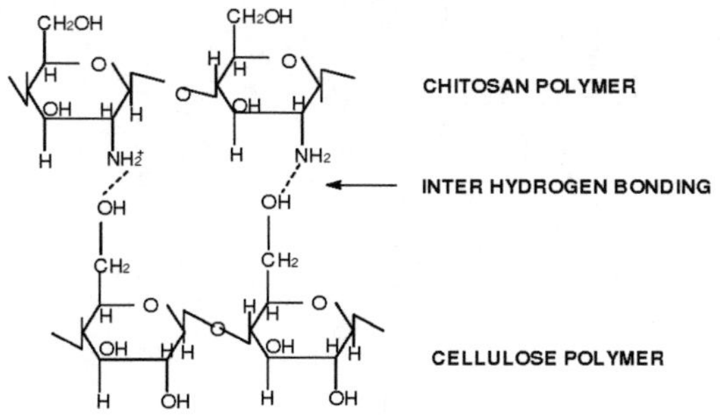

(a)

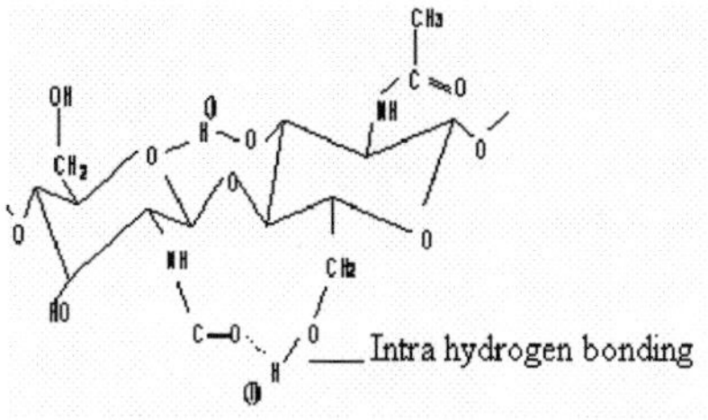

(b)

FIGURE 4. (a) Intra hydrogen bonding and (b) Inter hydrogen bonding

FTIR Analysis

FTIR analysis was used to confirm the proposed intermolecular interactions between kenaf and chitosan. Figure 5 shows the FTIR spectra of pure chitosan and chitosan with addition of 7 % to 28 % of kenaf dust. The FTIR spectra of pure chitosan film shows peak assigned to the aldeheyde group (C-CH$_3$) at 2925 to 2856 cm^{-1}, C-H bending (out of plane) at 894 cm^{-1}, the primary amine group at around 1406 to 1629 cm^{-1} and the finger print for this specimen singed at near-infrared (3233) cm^{-1} which remarks as ammonium ions (NH$^+$).

To distinguish between pure chitosan film and kenaf filled chitosan composites film spectra, chitosan with addition of 14 % kenaf dust spectra has been taken as a comparison. The FTIR spectra for chitosan mix with 14 % kenaf dust shows that, the similar peaks with pure chitosan film appears at 897 cm^{-1} singed as C-H out of plane, 1407 cm^{-1} as primary amide, 2854 to 2919 cm^{-1} remarks as two bands for –CH$_2$- groups and 3259 cm^{-1} appear as ammonium ions. Consequently, for chitosan with 14 % kenaf dust, the new peaks singed at 1549 cm^{-1} for aliphatic nitro compounds, 1151.cm^{-1} for silicones compounds and 1019 cm^{-1} corresponding to aliphatic compounds were observed.

Generally, there are less infrared bands in the 4000-1800 cm^{-1} region with many bands between 1800 and 400 cm^{-1}. The intensities of aldehyde compounds (-CH$_2$-) were also found to increase by approximately 3 % with the increasing of kenaf loading in the composites. It is expected that the hydrogen bonding was formed via the non-bonded electron pairs in aldehydes compounds.

A most distinctive difference between spectra of pure chitosan and Kenaf-filled chitosan is the intensity of reflectance for ammonium ions band, which is 96.2 % for pure chitosan and 98.3 % for 28 % kenaf filled chitosan composite. The increase in the intensity of ammonium ions is expected to offer higher chances of interaction between amine groups of chitosan with hydroxyl group (OH) of Kenaf dust.

A significant splitting of spectra was observed at wavelength between 1018 cm^{-1} to 1061 cm^{-1}, which assigned for cyclic alcohol. The percentage of reflectance between this wavelength decreases with the increasing of kenaf dust amount in composites; pure chitosan (85 %) while chitosan with 28 % kenaf dust (95.7 %).

CONCLUSIONS

From the results obtained, it can be said that the addition of kenaf dust into chitosan has significantly influenced the properties of chitosan typically the relative crystallinity level through the formation of hydrogen bonding which increases the intermolecular forces exist between chitosan and kenaf dust filler. FTIR analysis showed that there were strong interactions occurred between kenaf dust and chitosan in the composite.

ACKNOWLEDGEMENT

The authors wish to thank CIDB, USM, MARDI and AMREC for their assistance that has resulted in this article.

REFERENCES

1. R. Marguerite, *Prog. Polym. Sci.* **31**, 603-632 (2006)
2. P.C. Srivinasa, M.N. Ramesh, K.R. Kumar, R.N. Tharanathan, *Carbohydrate Polym.* **53**, 431-438 (2003)
3. Robert Y.M. Huang, R. Pal and G.Y. Moon, *J. Membrane Sci.* **160**,17-30 (1999)
4. Takashi N.; Koichi H.; Masaru K.; Katsuhiko N. and Hiroshi I.; *Composites Sci. Techn.* **63**, 1281-1286 (2003)
5. Aziz S. H. and Ansell M. P., *Composites Sci. Tech.* **64**, 1231-1238 (2004)
6. Rouison D.; Sain M., and Couturier M.,*Composites Sci. Tech.* **64** , 629-644 (2004)
7. Charles L; Webber III and Venita K. Bledsoe, *ASHS Press*, 348-357 (2002)
8. Liu C. and Bai R., *J. Membrane Sci.* **267**, 68-77 (2005)

9. Prashanth K.V.H; Lakshman K.; Sharmala T.R.; and Tharanathan R.N., *Intl. Biodeterior and Biodegradation* **56**, 115- 120 (2005)

10. Klung P.H., and Alexander E.L., *A Wiley Interscience Publication*, USA (1954)

11. Zhang Y; Xue C.; Xue Y.; Gao R., and Zhang X., *Carbohydrate Res.* **340**, 1914-1917 (2005)

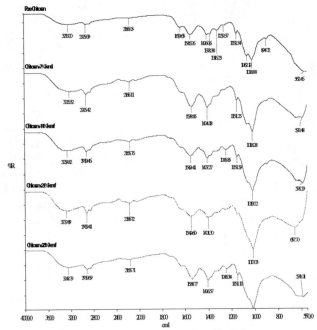

Reflectance (%) correspond to Kenaf content

(a)

Composition of Kenaf dust (%)	Reflectance (%)		
	A (aldehyde group)	B (primary amine)	C (ammonium ions)
0	95.5	92.2	96.2
7	96.2	95.2	97.0
14	97.2	96.3	98.3
21	98.2	97.7	98.7
28	98.3	97.5	93.0

(b)

FIGURE 5. (a) FTIR spectra of pure chitosan, chitosan with 7 % of Kenaf dust, chitosan with 14 % of Kenaf, chitosan with 21 % of Kenaf and chitosan with 28 % of Kenaf, (b) Corresponding functional groups as identified by FTIR.

CERAMICS

Ferroelectric Properties of $Ba_2Bi_4Ti_5O_{18}$ Doped with Pb^{2+}, Al^{3+}, Ga^{3+}, In^{3+}, Ta^{5+} Aurivillius Phases

A. Rosyidah[1], D. Onggo[1], Khairurrijal[2] and Ismunandar[1]

[1]*Inorganic & Physical Chemistry Research Division, Institut Teknologi Bandung, Indonesia*
[2]*Physics of Electronic Materials Research Division, Institut Teknologi Bandung, Indonesia*

Abstract. In recent years, bismuth layer structured ferroelectrics (BLSFs) have been given much attention because some materials, such as $Ba_2Bi_4Ti_5O_{18}$, are excellent candidate materials for nonvolatile ferroelectric random access memory (FRAM) applications. BLSFs are also better candidates because of their higher Curie points. Recently, we have carried out computer simulation in atomic scale in order to predict the energies associated with the accommodation of aliovalent and isovalent dopants (Pb^{2+}, Al^{3+}, Ga^{3+}, In^{3+}, Ta^{5+}) in the Aurivillius structure of $Ba_2Bi_4Ti_5O_{18}$. In this work, the predicted stable phases were synthesized using solid state reactions and their products then were characterized using powder X-ray diffraction method. The cell parameters were determined using Rietveld refinement in orthorhombic system with space group of $B2cb$. The cell parameters for $Ba_2Bi_4Ti_5O_{18}$ doped with Pb^{2+}, Al^{3+}, Ga^{3+}, In^{3+}, Ta^{5+} were $a = 5.5006(6)$ $b = 5.4990(5)$ $c = 50.5440(7)$ Å; $a = 5.5012(4)$ $b = 5.4986(8)$ $c = 50.5449(7)$ Å; $a = 5.5006(3)$ $b = 5.4999(3)$ $c = 50.5437(9)$ Å; $a = 5.5007(4)$ $b = 5.4989(7)$ $c = 50.5446(6)$ Å; and $a = 5.5000(5)$ $b = 5.4995(8)$ $c = 50.5436(6)$ Å. Results from the ferroelectric properties measurement for $Ba_2Bi_4Ti_5O_{18}$ doped with Pb^{2+}, Al^{3+}, Ga^{3+}, In^{3+}, Ta^{5+} were $P_r = 16.7$ $\mu C/cm^2$, $E_c = 35.1$ kV/cm; $P_r = 15.9$ $\mu C/cm^2$, $E_c = 33.8$ kV/cm; $P_r = 15.6$ $\mu C/cm^2$, $E_c = 34.2$ kV/cm; $P_r = 15.3$ $\mu C/cm^2$, $E_c = 34.0$ kV/cm; $P_r = 16.9$ $\mu C/cm^2$, $E_c = 35.6$ kV/cm.

Keywords: Aurivillius phase, Rietveld refinement, ferroelectric properties, $Ba_2Bi_4Ti_5O_{18}$
PACS: 77.80.-e, 61.10.Nz

INTRODUCTION

Bismuth layer-structured ferroelectrics (BLSFs) is the common acronym for the Aurivillius phase materials that are ferroelectric. The Aurivillius family[1], with general formula of (Bi_2O_2) $(A_{m-1}B_mO_{3m+1})$ can be described as the combination of regular stacking between the $(Bi_2O_2)^{2+}$ slabs and perovskite-like $(A_{m-1}B_mO_{3m+3})^{2-}$ blocks. The integer, m, describes the number of sheets of corner-sharing BO_6 octahedra forming the ABO_3-type perovskite blocks. The 12 coordinate perovskite-like A-site could be a mono-, di- or trivalent element (or combination) with large cation such as Na^+, K^+, Ca^{2+}, Sr^{2+}, Ba^{2+}, Pb^{2+}, Bi^{3+} or Ln^{3+} and the 6-coordinate perovskite-like B-site could be filled by smaller cations such as Fe^{3+}, Cr^{3+}, Ti^{4+}, Nb^{5+} or W^{6+}. Whereas the perovskite blocks offer large possibilities in terms of compositional flexibility due to numerous possible combinations of A and B cations, the cation sites in the $(Bi_2O_2)^{2+}$ layers are almost exclusively occupied by Bi^{3+}. BLFs ceramics are characterized by high Curie points, low dielectric constants, low dielectric losses, low aging, high dielectric breakdown strengths, strong anisotropic electromechanical coupling factors and low temperature coefficients of resonant frequency[2,3].

$Ba_2Bi_4Ti_5O_{18}$ is one such BLSFs material. The crystal structure of $Ba_2Bi_4Ti_5O_{18}$ is composed of a pseudo-perovskite $(Ba_2Bi_2Ti_5O_{16})^{2-}$ block interleaved with $(Bi_2O_2)^{2+}$ layers along the **c**-axis. $Ba_2Bi_4Ti_5O_{18}$ with an odd number of BO_6 octahedra is expected to exhibit spontaneous polarization along the **c**-axis. The $Ba_2Bi_4Ti_5O_{18}$ polycrystalline ceramic has been reported[4] to have a dielectric permittivity of 850 at 325 °C (the Curie temperature) and 360 at room temperature; and the electrical properties in the single crystal also has been reported[5].

One interesting feature of the Aurivillius phases resides in the compositional flexibility of the perovskite blocks which allows incorporating various cations. It is thus possible to modify the ferroelectric properties[6,7] according to the chemical composition. Although this phenomenon was observed since many years, its structural origin not yet clearly elucidated. Dopants are added into a wide variety of Aurivillius in order to modify their properties. The goal in some cases is to create or enhance desirable properties, while in other this is to eliminate or reduce undesirable

CP989, Neutron and X-ray Scattering in Materials Science and Biology, International Conference on Neutron and X-ray Scattering 2007, edited by A. Ikram, A. Purwanto, Sutiarso, A. Zulfia, S. Hendrana, and Z. Nurachman
© 2008 American Institute of Physics 978-0-7354-0508-0/08/$23.00

effects. Subbarao[7] and Newnham[8] showed that it to be a ferroelectric with the highest known Curie point in the bismuth layer-structured ferroelectrics family at that time of 940 °C. However, the information about the effect of doping in this material is still limited.

This paper will present results of systematics doping in $Ba_2Bi_4Ti_5O_{18}$ with the aim that a better understanding of doping effect will be gained. The discussion will be restricted to the substitutional isovalent and aliovalent cation in $Ba_2Bi_4Ti_5O_{18}$.

EXPERIMENTAL METHOD

The polycrystalline sample of $Ba_2Bi_4Ti_5O_{18}$ with aliovalent and isovalent dopants (Pb^{2+}, Al^{3+}, Ga^{3+}, In^{3+}, Ta^{5+}) were prepared by the standard solid-state reaction method. Stoichiometric quantitaties of Bi_2O_3, TiO_2, $BaCO_3$, PbO, Al_2O_3, Ga_2O_3, In_2O_3 and Ta_2O_5 (Aldrich Chem. Co.), all with a purity of 99.99%, were thoroughly mixed and ground, and heated in an alumina crucibles at elevated temperature until phase purity was establised. Typical reaction conditions were heated for 24 h at 700 °C, 24 h at 850 °C, 24 h at 1000 °C, and a further 24 h at 1100 °C, with intermediate re-grindings between each stage. The sample was slowly cooled to room temperature in air.

The purity of the product was monitored by powder X-ray diffraction using monochromatized Cu $K\alpha_1$ radiation λ = 0.1541 nm. Unit cell parameters were least squares refined by the RIETICA program[9]. The ferroelectric properties of $Ba_2Bi_4Ti_5O_{18}$ with aliovalent and isovalent dopants were evaluated from the P-E hysteresis curves, using a high-voltage test system (Model RT-66A, Radiant Technologies, Albuquerque, NM)

RESULTS AND DISCUSSION

The X-ray diffraction pattern of the as-prepared powders showed that structures of $Ba_2Bi_{3.95}A_{0.05}Ti_5O_{18}$ (A = Pb, Al, Ga, In), $Ba_2Bi_4Ti_{4.95}Ta_{0.05}O_{18}$ are orthorombic with space group $B2cb$. Preliminary examination of the raw X-ray powder diffraction data for the doped composition suggested that these powders are single phase compounds and also can be indexed with the orthorombic symmetry. The Rietveld refinement with orthorombic and $B2cb$ space group were then carried out and proceed without incident. Typical Rietveld plot are shown in Figure 1.

The powder X-ray diffraction patterns of all composition $Ba_2Bi_{3.95}A_{0.05}Ti_5O_{18}$ (A = Pb, Al, Ga, In) and $Ba_2Bi_4Ti_{4.95}Ta_{0.05}O_{18}$ suggest that all doped samples are isostructure with the parent $Ba_2Bi_4Ti_5O_{18}$[10,11]. The cell parameters of the

different as prepared materials are listed in Table 1. Refining the site mixing between the Bi^{3+} an A-sites proved to be impossible without introducing a refinement constraint. Because it was necessary to allow for atoms to fractionally occupy one site, a constraint was set up to ensure that the total site occupancy added to unity. In order to accomplish the atoms mixing, a new fictive atom was created by splitting one atom into two atoms. Oxygen anion parameters were not refined. From the development of the cell parameters, several distinct effects can be deduced depending on the fraction of iso- and alio-valent substitutions. For all groups of materials, the cell volume remains increase because of the doping.

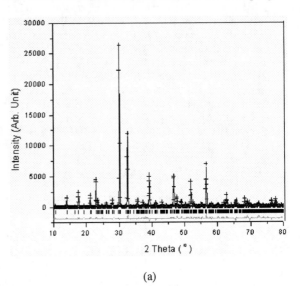

(a)

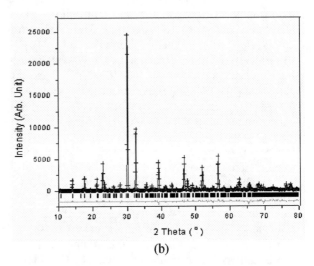

(b)

FIGURE 1. Rietveld refinement plot showing the observed (+), calculated (solid line) and difference for $Ba_2Bi_{3.95}Pb_{0.05}Ti_5O_{18}$ (a) and $Ba_2Bi_{3.95}Al_{0.05}Ti_5O_{18}$ (b). The tick marks show the positions of the allowed Bragg reflections in space group $B2cb$.

TABLE 1. Cell parameters for $Ba_2Bi_{3.95}A_{0.05}Ti_5O_{18}$ (A = Pb, Al, Ga, In) and for $Ba_2Bi_4Ti_{4.95}Ta_{0.05}O_{18}$ were determined using Rietveld refinement applying orthorhombic system, in space group $B2cb$.

Parameters	$Ba_2Bi_4Ti_5O_{18}$*)	Pb^{2+}	Al^{3+}	Dopant Ga^{3+}	In^{3+}	Ta^{5+}
a (Å)	5.4985(3)	5.5006(6)	5.5012(4)	5.5005(3)	5.5007(4)	5.5000(5)
b (Å)	5.4980(4)	5.4990(5)	5.4986(8)	5.4999(3)	5.4989(7)	5.4995(8)
c (Å)	50.3524(8)	50.5440(7)	50.5449(7)	50.5437(9)	50.5446(6)	50.5436(6)
V (Å^3)	1522.3(1)	1528.88(1)	1528.94(5)	1528.86(4)	1528.89(8)	1528.83(6)
$\frac{2(a-b)}{(a+b)}$	1.82×10^{-4}	2.91×10^{-4}	4.73×10^{-4}	1.09×10^{-4}	3.27×10^{-4}	0.91×10^{-4}
$t=\frac{r(A)+r(O)}{\sqrt{2}[r(B)+r(O)]}$	1.006	1.002	0.991	0.992	0.995	1.006
R_p	4.37	7.94	9.73	6.46	7.28	8.25
R_{wp}	5.51	3.85	4.66	3.05	3.23	5.76
R_{exp}	1.22	1.25	1.23	1.26	1.30	1.21
R_{Bragg}	2.94	8.62	10.80	8.98	8.38	7.54
GOF	1.97	2.36	4.72	3.91	4.67	4.85

$R_p = 100\sum|y_{obs}-y_{cal}|/|y_{obs}|$; $R_{wp} = 100\{\sum w_i(y_{oi}-y_{ci})^2/\sum w_i(y_{oi})^2\}^{1/2}$; $R_{Bragg} = 100\sum|I_o-I_c|/|I_o|$; GOF = R_{wp}/R_{exp}
*) Ismunandar *et al.* (2004)

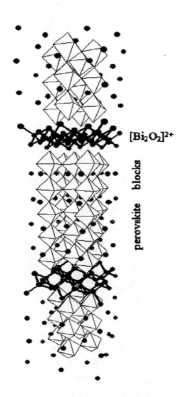

FIGURE 2. Structure of $Ba_2Bi_4Ti_5O_{18}$ consist of $[Bi_2O_2]^{2+}$ layer interleaved with perovskite-like $[Ba_2Bi_2Ti_5O_{16}]^{2-}$ blocks.

Atomistic simulation techniques have been employed to investigate the Aurivillius oxides phases: Bi_3TiNbO_9, $Bi_4Ti_3O_{12}$, $BaBi_4Ti_4O_{15}$ and $Ba_2Bi_4Ti_5O_{18}$[12,13]. The simulation suggested that M^{3+} dopants (Al, Ga, In) and Pb^{2+} are favorable in

Bi(2) sites in $(Bi_2O_2)^{2+}$ block and unlikely to occur at any of the perovskite sites (Bi sites and Ti sites). This seems in contrary to the expected, since in term of coordination preference and ionic size these three cations are suitable to substitute Ti^{4+}. However, this reconfirms that vacancy creation in perovskite block needs very high energy. On the other hand, the calculation on Ta^{5+} dopant shows that substitution is likely to occur at any of the perovskite sites on energetic grounds. The formation single phase compounds and the similarity of the diffraction patterns of $Ba_2Bi_{3.95}A_{0.05}Ti_5O_{18}$ (A = Pb, Al, Ga, In) and $Ba_2Bi_4Ti_{4.95}Ta_{0.05}O_{18}$ gave further evidence of these site preferences.

The observed cell parameters are larger than $Ba_2Bi_4Ti_5O_{18}$ in agreement with the larger ionic radii of Ba^{2+} and Bi^{3+}, as could be seen on the VIII coordinated Bi^{3+} and Ba^{2+} (XII coordinated Bi^{3+} ionic radii is unavailable). Compared the degree of the orthorombic splitting with $Ba_2Bi_4Ti_5O_{18}$, given by $\frac{2(a-b)}{(a+b)}$ obtained for $Ba_2Bi_{3.95}Ga_{0.05}Ti_5O_{18}$ and $Ba_2Bi_4Ti_{4.95}Ta_{0.05}O_{18}$ compounds are smaller than those obtained $Ba_2Bi_{3.95}A_{0.05}Ti_5O_{18}$ (A = Pb, Al, In). These can be explained using tolerance factor, t, defined as $t=\frac{r(A)+r(O)}{\sqrt{2}[r(B)+r(O)]}$ where r_A is the radius of the 12-coordinate A^{2+} cation, r_O, the radius of the 4 coordinate oxygen anion and r_B the radius of the six-coordinate Ti^{4+} cation. As the size of the A-type cation increases through the series Al (0.53) < Ga (0.61) < In (0.76) < Ta (0.78) < Pb (1.49) then t increases, from less then unity in $Ba_2Bi_{3.95}Al_{0.05}Ti_5O_{18}$ to greater than 1 in $Ba_2Bi_{3.95}Pb_{0.05}Ti_5O_{18}$ and $Ba_2Bi_4Ti_{4.95}Ta_{0.05}O_{18}$.

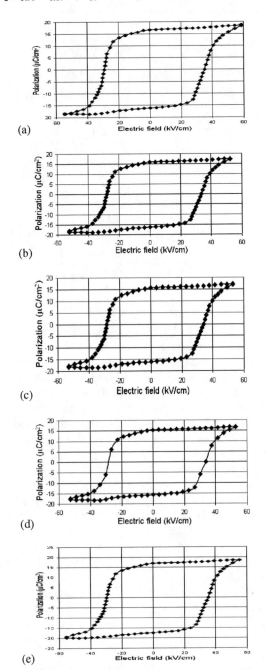

(a)

(b)

(c)

(d)

(e)

FIGURE 3. The P-E hysteresis loops of the $Ba_2Bi_{3.95}A_{0.05}Ti_5O_{18}$ (A = Pb, Al, Ga, In) a-d and e for $Ba_2Bi_4Ti_{4.95}Ta_{0.05}O_{18}$ ceramics.

The bismuth barium-titanat, $Ba_2Bi_4Ti_5O_{18}$ have the orthorhombic structure as shown in Figure 2. The structure of $Ba_2Bi_4Ti_5O_{18}$ is thus built up of $[Bi_2O_2]^{2+}$ layer betwen which $[Ba_2Bi_2Ti_5O_{16}]^{2-}$ layers are inserted. In the $Ba_2Bi_2Ti_5O_{16}$ units, Ba/Ti ions are enclosed by oxygen octahedra which are linked through corners

forming O-Ba/Ti-O octahedra. Thus $Ba_2Bi_2Ti_5O_{16}$ units pose a remarkable similarity to the perovskite-type structure. The height of the perovskite-type layer sandwiched between Bi_2O_2 layers in $Ba_2Bi_4Ti_5O_{18}$ is equal to 10 for O-Ba/Ti-O distances or approximately to m = 5 ABO_3 perovskites.

Aurivillius ceramics are interesting ferroelectrics. The layered structure makes this kind of ferroelectrics have good fatigue endurance[13]. We would like to investigate, what are the effects of doping on the ferroelectric properties. Figure 3 shows the P-E hysteresis loops of the $Ba_2Bi_{3.95}A_{0.05}Ti_5O_{18}$ (A = Pb, Al, Ga, In) a-d and e for $Ba_2Bi_4Ti_{4.95}Ta_{0.05}O_{18}$ ceramics, which indicates the ferroelectricity in this compound.

The remanent polarization (P_r) of the different as prepared materials are listed in Table 2. The disadvantage of the layer-structure perovskite materials for high-temperature piezoelectric applications is their relatively high ferroelectricity. This ferroelectricity is electronic-type and therefore, can be suppressed by doping[15-16].

The contribution of each constituent ion to the total spontaneous ferroelectric Polarization is calculated

$$P_s = \sum_i \frac{(m_i \times \Delta x_i \times Q_i e)}{V} \qquad (3)$$

where m_i is the site multiplicity, Δx_i is the atomic displacement along the a-axis from the corresponding position in the tetragonal structure, $Q_i e$ is the ionic charge of the ith constitute ion, and V is the volume of the unit cell. Figure 4 compares the Polarization calculation results for $Ba_2Bi_{3.95}Al_{0.05}Ti_5$ and $Ba_2Bi_4Ti_5O_{18}$. This clearly shows that the substitution of Al for Bi in the perovskite enhances the Polarization. This is in contrast with the result of similar substitution in two layer Aurivillius, where the substitution results in reducing the total Polarizations[17]. Although it should be noted that considering the estimated standart deviation and the contribution of other ions, the total Polarizations could be reduced. However, the precise oxygen atoms positions would be needed, i.e. neutron powder diffraction experiment is need.

$Ba_2Bi_{3.95}A_{0.05}Ti_5O_{18}$ (A = Pb, Al, Ga, In) and $Ba_2Bi_4Ti_{4.95}Ta_{0.05}O_{18}$ samples for examination of the ferroelectric properties show the saturated hysteresis curves at room temperature along the **ab**-axis. The ferroelectric properties based on the saturated hysteresis curves, in which the remanent polarization (P_r) and the coercive field (E_c) are in the saturated states. Generally, ferroelectrics depend on Curie temperature have a atomic displacement, which leads to a spontaneous polarization and a switching field[18,19]. In addition, the increase in the strain energy of the BO_6 octahedra from the bismuth layer also is assumed to be increased in $Ba_2Bi_{3.95}A_{0.05}Ti_5O_{18}$ (A = Pb, Al, Ga, In) and $Ba_2Bi_4Ti_{4.95}Ta_{0.05}O_{18}$, which allows easy movement of the octahedral cations in the electric-field direction. These findings are consistent with those expected for BLSF materials with odd- and evennumbered m, respectively. Polarization reversal occurs by the spontaneous polarization vector "rocking" up to ~10° [5,20]. Therefore, the P_r and E_c values along the c-axis are smaller than those along the ab-axis. This theory can be applied to $Ba_2Bi_{3.95}A_{0.05}Ti_5O_{18}$ (A = Pb, Al, Ga, In) and $Ba_2Bi_4Ti_{4.95}Ta_{0.05}O_{18}$. Therefore, in $Ba_2Bi_{3.95}A_{0.05}Ti_5O_{18}$ (A = Pb, Al, Ga, In) and $Ba_2Bi_4Ti_{4.95}Ta_{0.05}O_{18}$, the P_r and E_c values along the c-axis were smaller than those along the ab-axis.

TABLE 2. The remanent polarization (P_r) of the $Ba_2Bi_{3.95}A_{0.05}Ti_5O_{18}$ (A = Pb, Al, Ga, In) and $Ba_2Bi_4Ti_{4.95}Ta_{0.05}O_{18}$.

Parameters	$Ba_2Bi_4Ti_5O_{18}$[5]	Dopant				
		Pb^{2+}	Al^{3+}	Ga^{3+}	In^{3+}	Ta^{5+}
P_r ($\mu C/cm^2$)	12.0	16.7	15.9	15.6	15.3	16.9
E_c (kV/cm)	30.0	35.1	33.8	34.2	34.0	35.6

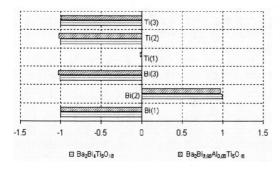

FIGURE 4. The Polarization contribution in $Ba_2Bi_{3.95}Al_{0.05}Ti_5O_{18}$ compared with those in $Ba_2Bi_4Ti_5O_{18}$.

In summary this experiment has shown that structure of new five layers Aurivillius compound $Ba_2Bi_4Ti_5O_{18}$ has determined. The non Bi cations which exclusively occupy the inner of perovskite layers result in enhancement of ferroelectricity.

CONCLUSIONS

The combined X-ray diffraction and Rietveld refinement confirm that series of $Ba_2Bi_{3.95}A_{0.05}Ti_5O_{18}$ (A = Pb, Al, Ga, In) and $Ba_2Bi_4Ti_{4.95}Ta_{0.05}O_{18}$ adopt orthorhombic system, in space group $B2cb$. The the P-E hysteresis loops of the ceramics indicates the ferroelectricity in this compound.

ACKNOWLEDGMENT

AR thank for financial support from Direktorat Jenderal Pendidikan Tinggi (Dikti), as well as Chemistry Department, Institut Teknologi Bandung.

REFERENCES

1. B. Aurivillius, *Arkiv foer Kemi* **1**, 463 (1949)
2. T. Takenaka, K. Sakata, *J. Appl. Phys.* **55**, 1092 (1984)
3. B. Frit, J.P. Mercurio, *J. Alloys Compd.* **188**, 27 (1992)
4. B. Aurivillius, P.H. Fang, *Phys. Rev.* **126**, 893 (1962)
5. H. Irie, M. Miyayama, T. Kodo, *J. Am. Ceram. Soc.* **83**, 2699 (2000)
6. Y. Wu, M.J. Forbess, S. Seraji, S.J Limmer, T.P. Chou, C. Nguyen, G. Cao, *J. App. Phys.* **90**, 5296 (2001)
7. E.C. Subbarao, *J. Am. Ceram. Soc.* **45**, 564 (1962)
8. R.W. Wolfe, R.E. Newnham, D.K. Jr. Smith, M.I. Kay, *Ferroelectrics* **3**, 1 (1971)
9. C.J. Howard, B.A. Hunter, Lucas Heights Research Laboratories, NSW, Australia, 1998, pp 1-27.
10. Ismunandar, T. Kamiyama, A. Hoshikawa, Q. Zhou and B.J. Kennedy, *J. Neutron Res.* **13**, 183 (2005)
11. Ismunandar, T. Kamiyama, A. Hoshikawa, Q. Zhou, B.J. Kennedy, Y. Kubota and K. Kato, *J. Solid State Chem.* **177**, 4188 (2004)
12. A. Rosyidah, D. Onggo, Khairurrijal, Ismunandar, Prosiding Seminar Nasional Kimia dan Kongres Nasional Himpunan Kimia Indonesia, (2006), 321-328.
13. A. Rosyidah, D. Onggo, Khairurrijal, Ismunandar, *J. Chin. Chem. Soc.* (2007), Accepted.
14. J.G. Thompson, A.D. Rae, R.L. Withers, D.C. Craig, *Acta Cryst.* **B 47**, 174 (1991)
15. H.S. Shulman, M. Testorf, D. Damjanovic, N.J. Setter, *J. Am. Ceram. Soc.* **79**, 3124 (1996)
16. A. Voisard, D. Damjanovic, N. Setter, *J. Eur. Ceram. Soc.* **19**, 1251 (1999)
17. Y. Shimakawa, Y. Kubo, Y. Nakagawa, S. Goto, T. Kamiyama, H. Asano, *Phys. Rev.* **B. 61**, 6559 (2000)
18. S.K. Kim, M. Miyayama, H. Yanagida, *J. Ceram. Soc. Jpn.* **102**, 722 (1994)
19. H. Irie, M. Miyayama, T. Kudo, *Jpn. J. Appl. Phys.* **38**, 5958 (1999)
20. S.E. Cummins, L.E. Cross, *J. Appl. Phys.* **39**, 2268 (1968)

Electrical Characteristics of NTC Thermistor Ceramics Made of Mechanically Activated Fe$_2$O$_3$ Powder Derived from Yarosite

D. Gustaman Syarif[1] and A. Ramelan[2]

[1]Nuclear Technology Center for Materials and Radiometry-BATAN,
Jl. Tamansari 71, Bandung, 40132, Indonesia
[2]Material Study Program, ITB, Jl. Ganesha 10, Bandung, Indonesia

Abstract. Electrical characteristics of ceramics for NTC thermistor made of mechanically activated Fe$_2$O$_3$ powder which was derived from yarosite mineral has been studied. The powder of Fe$_2$O$_3$ was derived from yarosite mineral by precipitation and calcination. The powder was mechanically activated by blending using an electric blending machine. The ceramics were produced by pressing the calcined and activated powders to produce pellets and sintering the pellets at 1100°C and 1200°C for 1 hour in air. Electrical characterization was done by measuring electrical resistivity of the ceramics at various temperatures (25°C-100°C). Microstructure and structural analyses were carried out by using a Scanning Electron Microscope (SEM) and X-ray diffraction (XRD), respectively. The XRD analyses showed that the sintered ceramics had crystal structure of hexagonal (hematite). The presence of second phase could not be identified from the XRD analyses. From SEM data, it was known that the ceramics from activated powder had larger grains due to small size of the activated powder. According to the electrical data, it was known that the ceramics made of the activated powder had lower thermistor constant (B) and room temperature electrical resistivity (ρ_{RT}). The value of B and ρ_{RT} of the produced ceramics fitted market requirement.

Keywords: Thermistor, NTC, Fe$_2$O$_3$, yarosite, powder activation.
PACS: 82.45.Xy

INTRODUCTION

NTC thermistors are widely used in the world due to their potential use for many applications such as temperature measurement, circuit compensation, suppression of in rush-current, flow rate sensor and pressure sensor in many sectors (http://www.betatherm.com). It is well known that most NTC thermistors are produced from spinel ceramics based on transition metal oxides with general formula of AB$_2$O$_4$ where A is metal ion in tetrahedral position and B is metal ions in octahedral position[1-9]. Many studies have been done to improve the characteristics of the spinel based-NTC thermistors[5, 6, 10]. The study on the production of Fe$_2$O$_3$ based-NTC ceramics from yarosite mineral is seldom[11]. Especially, the study on the effect of mechanically powder activation on characteristics of Fe$_2$O$_3$ based-NTC ceramics from yarosite mineral has not been reported yet.

In order to get capability in production of NTC thermistor from abundant material in Indonesia, we have studied a fabrication of NTC ceramics from yarosite mineral[11]. In that work, it was known that the NTC ceramics could be well produced from the yarosite mineral. However, the room temperature resistivity was still high; making the application of the ceramics is rather limited. Many ways can be considered to decrease the room temperature resistivity and one of them is powder activation especially using a mechanical method namely blending one. The size of powder particle can be change to be smaller using the blending method. The small size powder has large surface area, so it is easy to sinter. The sintering powder having large surface area may produce large grains which are considered decreasing the room temperature resistivity since the ceramics with large grains have small number of scattering center for charge carrier. This work studied the difference in electrical characteristic between thermistor ceramics made from mechanically activated Fe$_2$O$_3$ powder derived from yarosite mineral and those made from the non-activated one (initial powder).

CP989, *Neutron and X-ray Scattering in Materials Science and Biology, International Conference on Neutron and X-ray Scattering 2007*, edited by A. Ikram, A. Purwanto, Sutiarso, A. Zulfia, S. Hendrana, and Z. Nurachman

METHODOLOGY

Powder of yarosite mineral (From P.D. Karat Pertambangan) was dissolved in HCl acid and precipitated using NH$_4$OH solution. The precipitation was washed and dried at 80°C for 24 hours. The dried powder was calcined at 700°C for 2 hours in air to get Fe$_2$O$_3$ powder which then called as initial powder. An amount of calcined powder was mechanically activated by blending for 30 minutes using an electrical blending machine from Karl Korb. This powder was called as activated powder. The initial and activated powders were then pressed with pressure of 4 Ton/cm^2 into green pellets. The green pellets were sintered at 1100°C and 1200°C for 1 hour in air.

The crystal structure of the sintered pellets was analyzed with X-ray diffraction (XRD) using K_α radiation at 40KV and 25mA. The microstructure of the pellets was investigated by using SEM. For this examination, some pellets were fractured for the samples (Fractography). The opposite-side surfaces of the sintered pellets were coated with Ag paste. After the paste was dried at room temperature, the Ag coated-pellets were heated at 750°C for 10 minutes. The resistivity was measured at various temperatures from 25 to 100°C in steps of 5°C. Thermistor constant (B) was derived from curves of ln resistivity vs. $1/T$ where T is temperature in Kelvin.

RESULTS

Figure 1 shows an appearance of a typical ceramic. Ceramics made from the initial powder were generally visually good. However, those made from the activated powder were containing some protrudes on the surface as shown in Figure 1. The protrudes on the surface is caused by a pressure from volatile material as a reaction product between impurities contained in the ceramic and Fe$_2$O$_3$ during sintering. The chemical composition of the powder is shown in Table 1. Figure 2 and Figure 3 show the XRD profiles of pellet ceramics from the initial and activated powders sintered at 1100°C. Meanwhile, Figure 4 and Figure 5 show the XRD profiles of pellet ceramics from the initial and activated powders sintered at 1200°C. As shown in the Figure 2-5, the profiles are similar. There is no difference in crystal structure can be observed due to different sintering temperatures and mechanical powder treatment. The XRD profiles show that the structure of the ceramics is the same namely hexagonal after being compared to the XRD standard profile of hematite from JCPDS No. 33-0664. No peaks from second phases were observed. It may be due to the small concentration of impurities which smaller than the precision limit of the X-ray diffractometer used.

FIGURE 1. Visual appearance of a ceramic of Fe$_2$O$_3$ sintered at 1200°C from activated powder.

TABLE 1. Chemical composition of Fe$_2$O$_3$ powder derived from Yarosite.

No.	Oxide	Weight %
1.	Fe$_2$O$_3$	93.86
2.	SiO$_2$	2.26
3.	Al$_2$O$_3$	2.63
4.	TiO$_2$	1.04
5.	K$_2$O	0.036
6.	Na$_2$O	0.083
7.	MnO	0.091

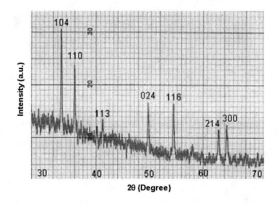

FIGURE 2. XRD profile of Fe$_2$O$_3$ ceramic sintered at 1100°C from initial powder.

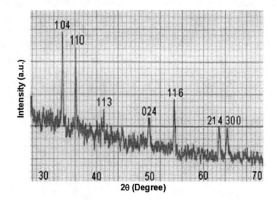

FIGURE 3. XRD profile of Fe$_2$O$_3$ ceramic sintered at 1100°C from activated powder.

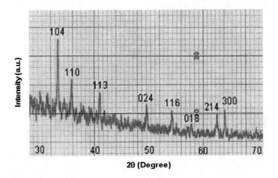

FIGURE 4. XRD profile of Fe_2O_3 ceramic sintered at 1200°C from initial powder.

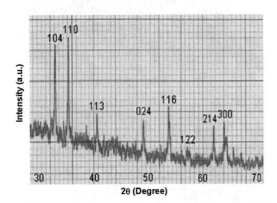

FIGURE 5. XRD profile of Fe_2O_3 ceramic sintered at 1200°C.

Figure 6 shows the microstructures of the ceramics sintered at 1100°C and 1200°C from the initial and activated powders. As shown in Figure 6, a higher temperature sintering produces pellet having larger grains. This is a consequence of high atomic mobility at a high temperature[12]. It can be seen from the Figure 6 also, that the pellet made of activated powder has larger grains than that made of initial powder. This means that the activated powder is easier to sinter because it has small particle size and large surface area. The powder with large surface area is easy to sinter. The large surface area will promote sintering resulting in large grains and small number of grain boundaries.

$$\rho = \rho_0 \exp(B/T) \tag{1}$$

where, ρ = electrical resistivity, ρ_o = electrical resistivity at infinite temperature, B is the thermistor constant and T is the temperature in Kelvin.

$$\alpha = B/T^2 \tag{2}$$

where, α = sensitivity. In this case, T is 300 K (Room temperature).

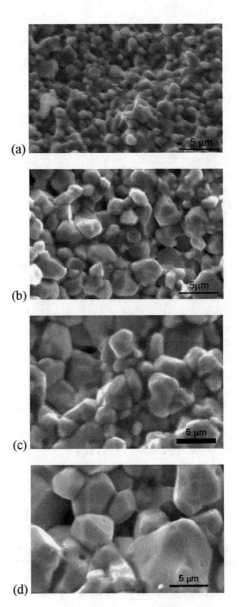

(a)

(b)

(c)

(d)

FIGURE 6. SEM micrographs of sintered pellets a) Sintered at 1100°C from initial powder, b) Sintered at 1100°C from activated powder, c) Sintered at 1200°C from initial powder, d) Sintered at 1200°C from activated powder.

The electrical data of Figure 7 shows that the electrical characteristics of the ceramics follow the NTC tendency expressed by Eq. 1. As shown in Table 1, the powder treatment (activation) decreases the room temperature resistivity (ρ_{RT}) and thermistor constant (B). The room temperature resistivity (ρ_{RT}) is the resistivity at $T = 300K$, the B constant is the gradient of curves of Figure 7 and α is the sensitivity which is calculated using equation (2). Compared to the (B) value for market requirement where $B \geq 2000K$, the value of B for our ceramics is larger and it means better.

TABLE 2. Electrical characteristics of NTC Fe_2O_3 based-ceramics.

No.	Sintering Temp. (°C)	Powder Treatment	B (K)	α (%/K)	ρ (Ohm.cm)
1	1100°C	No	4944	5.49	895139
2	1100°C	30 minutes Blending	3523	3.92	13211
3	1200°C	No	3924	4.36	44279
4	1200°C	30 minutes Blending	3149	3.50	7618

As can be seen from Table 2, the decrease of the room temperature resistivity due to the powder activation by 30 minutes blending is larger than that due to the increase of the sintering temperature from 1100°C to 1200°C. The electrical resistivity of the pellet from activated (blended) powder sintered at 1100°C is smaller that that from the initial powder sintered at 1200°C.

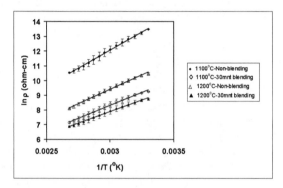

FIGURE 7. *Ln* resistivity (ρ) vs. $1/T$ of Fe_2O_3 ceramics.

DISCUSSION

The increase of the grain size is caused by small size of the powder that makes the powder more reactive. The reactive powder is easy to sinter, producing large grained-ceramics. The change of the electrical characteristics is controlled by the change of the microstructure. The ceramics having large grains have less scattering centers for charge carrier, so therefore they have high conductivity or low resistivity as shown by the results in this work.

In decreasing the room temperature resistivity, the powder activation by 30 minutes blending is more effective than the increase of sintering temperature

from 1100°C to 1200°C. The powder activation causes a more intimate contact among grains, though the increase of sintering temperature of 100°C (from 1100°C to 1200°C) yields relatively larger grains. Besides that, there is a possibility that micro cracks are present in the pellet made from the initial powder sintered at 1200°C resulting in a relatively higher resistivity.

CONCLUSIONS

The grain size of the Fe_2O_3 based-NTC ceramics made from yarosite increases after powder activation (blending), causing a decrease in room temperature resistivity and thermistor constant. The decrease of the room temperature resistivity makes the ceramics is more applicable. The value of (ρ_{RT}) and (B) of the ceramics made in this work fits the market requirement. The powder treatment is very simple, so it can be easily adopted in a production route of NTC thermistor.

ACKNOWLEDMENT

The authors express their thank to Mr. M. Yamin for his help in pellet fabrication dan Mr. Yudi Setiadi for his help in providing XRD data.

REFERENCES

1. E. S. Na, U. G. Paik, S. C. Choi, *J. Ceram. Proces. Res.* **2,** 31 (2001).
2. Y. Matsuo, T. Hata, T. Kuroda, US Patent No.4, 324, 702 (1982).
3. H. J. Jung, S. O. Yoon, K. Y. Hong, J. K. Lee, US Patent No.5, 246,628 (1993).
4. K. Hamada, H. Oda, US Patent No. 6,270,693 (2001).
5. K. Park, *Mater. Sci. Eng.*, **B104,** 9 (2003).
6. K. Park, D. Y. Bang, *J. Mater. Sci. Mater. Electronics.* **14,** 81 (2003).
7. S. G. Fritsch, J. Salmi, J. Sarrias, A. Rousset, S. Schuurman, A. Lannoo, *Mater. Res. Bull.*, **39,** 1957 (2004).
8. R. Schmidt, A. Basu, A.W. Brinkman, *J. Eur. Ceram. Soc.*, **24,** 1233 (2004).
9. K. Park, I.H. Han, *Mater. Sci. Eng.* **B119,** 55 (2005).
10. Windartun, D. Gustaman, The Proceeding of the International Conference on Mathematics and Natural Sciences (ICMNS), ITB, Bandung, (2006) pp. 809-813.
11. D. Gustaman, Guntur D.S., M. Yamin, The Proceeding of the National Seminar of Science and Techniques, Bandung, BATAN, (2005) pp. 344-352.

The Effect of SiO$_2$ Addition on the Characteristics of CuFe$_2$O$_4$ Ceramics for NTC Thermistor

Wiendartun[1] and D. Gustaman Syarif[2]

[1]*Department of Physics, Education University of Indonesia, UPI*
Jl. Setiabudi, Bandung, Indonesia
[2]*Nuclear Technology Center for Materials and Radiometry-BATAN,*
Jl. Tamansari 71, Bandung, 40132, Indonesia

Abstract. The effect of SiO$_2$ addition on the characteristics of CuFe$_2$O$_4$ ceramics for NTC thermistors has been studied. The ceramics were produced by pressing a homogeneous mixture of CuO, Fe$_3$O$_4$ and SiO$_2$ (0-0.75 w/o) powders in appropriate proportions to produce CuFe$_2$O$_4$ based ceramics and sintering the pressed powder at 1100°C for 2 hours in air. Electrical characterization was done by measuring electrical resistivity of the ceramics at various temperatures (25°C-100°C). Microstructure and structural analyses were also carried out by using optical microscopy and X-ray diffraction (XRD), respectively. The XRD analyses showed that the CuFe$_2$O$_4$ and SiO$_2$ added-CuFe$_2$O$_4$ ceramics have crystal structure of cubic spinel. The presence of second phase could not be identified from the XRD analyses. According to the electrical data, it was known that the SiO$_2$ addition increased the thermistor constant (B) and the room temperature electrical resistivity (ρ_{RT}). The value of B and ρ_{RT} of the produced CuFe$_2$O$_4$ ceramics namely B = 2548-3308°K and ρ_{RT} = 291-9400 ohm's, fitted market requirement.

Keywords: Thermistor, NTC, CuFe$_2$O$_4$, SiO$_2$.
PACS: 82.45.Xy

INTRODUCTION

NTC thermistors are widely used in the world due to their potential use for many applications such as temperature measurement, circuit compensation, suppression of in rush-current, flow rate sensor and pressure sensor in many sectors [1]. It is well known that most NTC thermistors are produced from spinel ceramics based on transition metal oxides with general formula of AB$_2$O$_4$ where A is metal ion in tetrahedral position and B is metal ions in octahedral position [2-10]. Many studies have been done to improve the characteristics of the spinel based-NTC thermistors[6, 7,11]. However, the study on the effect of SiO$_2$ addition on the characteristics of CuFe$_2$O$_4$ spinel ceramics for NTC thermistor has not been reported yet.

Generally, the CuFe$_2$O$_4$ ceramics used as soft magnet [12-15] as well as catalyst [16-18], however, potentially, the CuFe$_2$O$_4$ ceramics have capability of being NTC thermistors due to its semi conductive property. According to CuO-Fe$_2$O$_3$ phase diagram [19], there is an area where the ceramic composing of CuO and Fe$_2$O$_3$ heated at 1100°C will have a microstructure containing liquid phase. In room temperature, the liquid phase may be a boundary material. This boundary material theoretically will influence the characteristics of the ceramics, especially the electrical characteristics. Since an additive such as SiO$_2$ is added, the characteristics of the CuFe$_2$O$_4$ may change because two conditions may happen. The conditions are, the first, the SiO$_2$ dissolves in the CuFe$_2$O$_4$ by substituting Cu ions or Fe ions, the second, the SiO$_2$ does not dissolve but segregated at grain boundaries and it may react with the liquid phase that originally exists.

Since the first condition happens, the CuFe$_2$O$_4$ ceramics may have a lower electrical resistivity when the substitution of Fe^{3+} and /or Cu^{2+} creating free electron in the conduction band. Meanwhile, since the second one happens, the electrical resistivity may be higher because the segregated SiO$_2$ may change the microstructure. In our previous study [12], it was known that TiO$_2$ tend to increase the room temperature resistivity and thermistor constant. This work is to know the effect of SiO$_2$ addition on the characteristics of the CuFe$_2$O$_4$ ceramics for NTC thermistors, especially the electrical characteristic based on the above mentioned hypothesis. The results were compared to our previous study.

CP989, Neutron and X-ray Scattering in Materials Science and Biology, International Conference on Neutron and X-ray Scattering 2007, edited by A. Ikram, A. Purwanto, Sutiarso, A. Zulfia, S. Hendrana, and Z. Nurachman

METHODOLOGY

Powders of CuO, Fe_3O_4 and SiO_2 were weighed in appropriate proportions to fabricate SiO_2 added-$CuFe_2O_4$ ceramics where the SiO_2 were 0, 0.25, 0.50 and 0.75 weight %. The mixture of powders was calcined at 800°C for 2 hours. After calcination, the powder was crushed and sieved with a siever of < 38 μm. The sieved powder was then pressed with pressure of 4 ton/cm² into green pellets. The green pellets were sintered at 1100°C for 2 hours in air.

The crystal structure of the sintered pellets was analyzed with X-ray diffraction (XRD) using K_α radiation at 40KV and 25mA. After grinding, polishing, etching the pellets, the microstructure of the pellets was investigated by an optical microscope. The opposite-side surfaces of the sintered pellets were coated with Ag paste. After the paste was dried at room temperature, the Ag coated-pellets were heated at 750°C for 10 minutes. The resistivity was measured at various temperatures from 25 to 100°C in steps of 5°C.

RESULTS

Figure 1 shows appearance of typical SiO_2 Added-CuF_2O_4 ceramics. The ceramics are visually good. Figures 2, 3, and 4 show the XRD profiles of $CuFe_2O_4$ ceramic added with 0, 0.25 and 0.75 weight % SiO_2, respectively. As shown in the figure 2 - 4, the profiles are similar. The XRD profiles show that the structure of the ceramics is cubic spinel after being compared to the XRD standard profile of $CuFe_2O_4$ from JCPDS No. 22-1012. No peaks from second phases are observed. It may be due to the small concentration of SiO_2 added which smaller than the precision limit of the X-ray diffractometer used. The added SiO_2 may be dissolved or not. It cannot be concluded from the XRD profiles in this works. The microstructure and electrical data may be used to evaluate whether the SiO_2 added was dissolved or not.

FIGURE 1. Visual appearance of typical SiO_2 Added-CuF_2O_4 ceramic

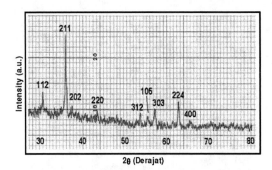

FIGURE 2. XRD profile of $CuFe_2O_4$ ceramic

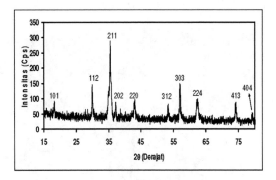

FIGURE 3. XRD profile of 0.25 w/o SiO_2 added-CuF_2O_4 ceramic

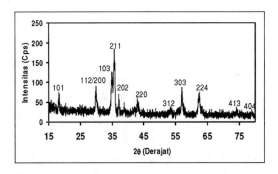

FIGURE 4. XRD profile of 0.75 w/o SiO_2 added-CuF_2O_4 ceramic

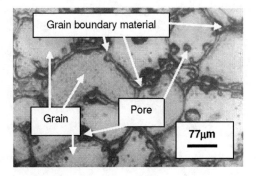

FIGURE 5. Microstructure of the $CuFe_2O_4$ ceramic

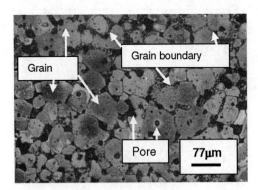

FIGURE 6. Microstructure of the 0.25 w% SiO_2 added-$CuFe_2O_4$ ceramic

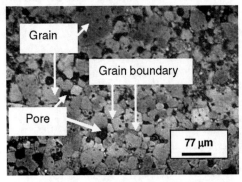

FIGURE 7. Microstructure of the 0.75 w% SiO_2 added-$CuFe_2O_4$ ceramic

Figure 5, Figure 6 and Figure 7 are microstructures of the $CuFe_2O_4$ ceramics added with 0, 0.25, and 0.75 weight % SiO_2, respectively. From Figures 5-7, it is clearly seen that the addition of SiO_2 decreases the size of grains. The decrease of the grain size is due to the segregation of the SiO_2 at grain boundaries. Since the SiO_2 dissolved in $CuFe_2O_4$ the grains should become larger when Si^{4+} creates iron vacancy. Since the accommodation of Si^{4+} creates electron as the compensation, it will be no effect on the grain growth.

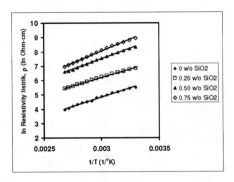

FIGURE 8. *Ln* resistivity (ρ) vs. $1/T$ of SiO_2 added-$CuFe_2O_4$ ceramics

The electrical data of Figure 8 shows that the electrical characteristics of the ceramics follow the NTC tendency expressed by Eq. 1. As shown in Table 1, the addition of SiO_2 increases the room temperature resistivity (ρ_{RT}) and thermistor constant (B). The room temperature resistivity is the resistivity at $T = 300°K$, the B constant is the gradient of curves of Figure 8 and α is calculated using equation (6). Compared to the (B) value for market requirement where $B \geq 2000°K$, the value of B for our ceramics is larger and it means better.

$$\rho = \rho_o \exp(B/T) \qquad (1)$$

where, ρ = Electrical resistivity, ρ_o = Electrical resistivity at infinite temperature, B is the thermistor constant and T is the temperature in Kelvin.

$$\alpha = B/T^2 \qquad (2)$$

where, α = Sensitivity, B = Thermistor constant and T = Temperature in Kelvin degree. In this case, T is $300°K$ (Room temperature).

TABLE 1. Electrical characteristics of the SiO_2 added-$CuFe_2O_4$ ceramics.

No.	Additive SiO_2 (w/o)	B (°K)	α (%/°K)	ρ_{RT} (ohm-cm)
1.	0	2548	2,83	291
2.	0,25	2358	2,62	1079
3.	0,50	2884	3,20	4788
4.	0,75	3308	3,68	9400

DISCUSSION

The decrease of the grain size is caused by the segregation of the added SiO_2. The segregated SiO_2 inhibits the grain growth during sintering. This data confirms that the added-SiO_2 is not dissolved in the $CuFe_2O_4$ ceramics which could not be concluded from the XRD data.

The change of the electrical characteristics is controlled by the change of the microstructure due to the SiO_2 addition. Since the SiO_2 dissolved in $CuFe_2O_4$ by substituting Cu and/or Fe which results in electron as the compensation, the room temperature resistivity (ρ_{RT}) and thermistor constant (B) should decrease. The increase (ρ_{RT}) and (B) as shown by our data indicates that the added SiO_2 is not dissolved and tends to segregate at the grain boundaries.

The ionic radius difference between Si^{4+} (54 pm[20]) and Cu^{2+} (87 pm[20]) is too large, so the possibility of Si to substitute Cu is small. The ionic radius difference between Si^{4+} and Fe^{3+} (69 pm[20]) is smaller than that between Si^{4+} and Cu^{2+}, however, this difference is still large. So the possibility of Si to substitute Fe is also small. The microstructure data showed that the SiO_2 addition decreased the grain size. So, therefore it can be concluded that the added SiO_2 is segregated at the grain boundaries and the electrical characteristic is controlled by the microstructure.

Compared to the previous study about the effect of TiO_2 on electrical characteristic of $CuFe_2O_4$[11], there is a significant difference in electrical resistivity. The room temperature resistivity of the SiO_2 added-$CuFe_2O_4$ ceramics is larger than that of the TiO_2 added-$CuFe_2O_4$ ones[11]. It may be due to the difference effect of the different additive. There is a high possibility that a part of added TiO_2 was dissolved in the $CuFe_2O_4$ ceramics, so that the resistivity of the TiO_2 added-$CuFe_2O_4$ lower than that of the SiO_2-added $CuFe_2O_4$ ceramics in the current work. Considering the ionic radius of the Ti^{4+} ion i.e. 74.5 pm [20] which is close to that of Fe^{3+} ion (69 pm[20]), it is possible that TiO_2 dissolved in $CuFe_2O_4$ where the Ti^{4+} substitutes the Fe^{3+} decreasing the resistivity of the ceramics.

CONCLUSION

The grain size of the $CuFe_2O_4$ ceramics decreases by addition of SiO_2 because the added SiO_2 segregated at grain boundaries and inhibited grain growth during sintering. The addition of SiO_2 increased the room temperature resistivity (ρ_{RT}) and the thermistor constant (B) of the $CuFe_2O_4$ ceramics through changing the microstructure. The value of (ρ_{RT}) and (B) of the $CuFe_2O_4$ ceramics made in this work fits the market requirement.

ACKNOWLEGMENT

The authors wish to acknowledge their deep gratitude to DIKTI, Department of National Education of Indonesian Government for financial support under hibah PEKERTI program with contract No. 014/SPP/PP/DP2M/II/2006.

REFERENCES

1. BetaTHERM Sensors [on line]. Available: http://www.betatherm.com.
2. E.S. Na, U.G Paik, S.C. Choi, *J. Ceram. Process. Res.* **2**, 31-34 (2001)
3. Y. Matsuo, T. Hata, T. Kuroda, "Oxide thermistor composition", US Patent 4,324,702, April 13, 1982
4. H.J. Jung, S. O. Yoon, K. Y. Hong, J.K. Lee, "Metal oxide group thermistor material", US Patent 5,246,628, September 21, 1993.
5. K. Hamada, H. Oda, "Thermistor composition", US Patent 6,270,693, August 7, 2001.
6. K. Park, *Mater. Sci. Eng.* **B104**, 9-14 (2003)
7. K. Park, D.Y. Bang, *J. Mater. Sci. Mater. Electronics* **14**, 81-87 (2003)
8. S.G. Fritsch, J. Salmi, J. Sarrias, A. Rousset, S. Schuurman, A. Lannoo, *Mater. Res. Bull.* **39**, 1957-1965 (2004)
9. R. Schmidt, A. Basu, A. W. Brinkman, *J. Eur. Ceram. Soc.* **24**, 1233-1236 (2004).
10. K. Park, I. H. Han, *Mater. Sci. Eng.* **B119**, 55-60 (2005)
11. The Effect of TiO_2 Addition on the Characteristics of $CuFe_2O_4$ Ceramics for NTC Thermistors, International Conference on Mathematics and Natural Sciences (ICMNS) 2006, ITB, Bandung, October 2006.
12. J.Z. Jiang, G.F. Goya, H.R. Rechenberg, *J. Phys. Condens. Mater* **11**, 4063 (1999).
13. G.F. Goya, H.R. Rechenberg, J.Z. Jiang, *J. Magn. Magn. Mater.* **218** 221 (2000).
14. G.F. Goya, H.R. Rechenberg, *J. Appl. Phys.* **84** (2), 1101 (1998)
15. C.R. Alves, R. Aquino, M.H. Sousa, H.R. Rechenberg, G.F. Goya, F.A. Tourinho, J. Depeyrot, *J. Metastable Nanocryst. Mater.* **20-21**, 694 (2004).
16. K. Satoshi, T. Toyokazu, T. An, *Catal. Lett.* **100**, 89-93 (2005)
17. W.F. Shangguan, Y. Ternaoka, S. Kagawa, *Appl. Catal.* **B16**, 149-154 (1998)
18. R.C. Wu, H.H. Qu, H. He, Y.B. Yu, *Appl. Catal.* **B48** (1), 49 (2004)
19. Anonymous, Phase diagram for Ceramics, ASTM.
20. M. Barsoum, Fundamental of Ceramics, McGraw-Hill, 1997.

Fractal Studies on Titanium-Silica Aerogels using SMARTer

E. Giri Rachman Putra[1], A. Ikram[1], Bharoto[1], E. Santoso[1], T. Chiar Fang[2],
N. Ibrahim[2], A. Aziz Mohamed[3]

[1] Neutron Scattering Laboratory, BATAN, Kawasan Puspiptek Serpong, Tangerang 15314, Indonesia
[2] Department of Physics, Faculty of Science Universiti Teknologi Malaysia (UTM), 81310 Skudai, Johor, Malaysia
[3] Materials Technology Group, Industrial Technology Division Agensi Nuklear Malaysia, 43000 Kajang, Malaysia

Abstract. Power-law scattering approximation has been employed to reveal the fractal structures of solid-state titanium-silica aerogel samples. All small-angle neutron scattering (SANS) measurements were performed using 36 meters SANS BATAN spectrometer (SMARTer) at the neutron scattering laboratory (NSL) in Serpong, Indonesia. The mass fractal dimension of titanium-silica aerogels at low scattering vector q range increases from -1.4 to -1.92 with the decrease of acid concentrations during sol-gel process. These results are attributed to the titanium-silica aerogels that are growing to more polymeric and branched structures. At high scattering vector q range the Porod slope of -3.9 significantly down to -2.24 as the roughness of particle surfaces becomes higher. The cross over between these two regimes decreases from 0.4 to 0.16 nm^{-1} with the increase of acid concentrations indicating also that the titanium-silica aerogels are growing.

Keywords: SANS, aerogels, fractals, power-law scattering
PACS: 25.40. Dn, 82.70. Gg, 83.85. Hf

INTRODUCTION

Aerogel is one of the most interesting solid materials that has very wide-range applications, recently. Because of its low density and large specific surface area, aerogels have been applied for absorber of dust[1], a host matrix for long life nuclear waste[2] or a carrier for active compound or pharmaceuticals in drug-delivery system[3]. A spesific application of aerogles based on their optical properties is for Cherenkov radiation detectors[4] and optical sensors. However, the most recent application of aerogles is for catalysts[5].

Unlike conventional foam, aerogel is a special porous materials with extreme porosity in a nano-to-micrometer scale. It is composed of individual particles with a few nanometer in size which are linked in three dimensional structure. Aerogels are typically prepared by synthesizing metal-alkali precursors in an excess solvent by sol-gel process compared to traditional glass melting or ceramics powder methods. The solvent is removed from the pores under supercritical conditions thereby the large capillarity force and shrinkage that always occurs when liquid is removed from small pores will be avoided.

Porous materials which behave as disordered solid materials has to be characterized to obtain its structure information such as pore dimension, surface structures, etc. Fractal concept has been found[6] to be important in many structure studies of disordered materials since many processes as well as the structures can be described and understood in terms of non-integral dimensions.

Small-angle scatterings of neutron (SANS) and X-ray (SAXS) have proved to be well suited and direct techniques to investigate the microsructure of fractals and other disordered systems on nanometer scale, such as polymers[7,8] biological molecules, aggregates[7], mineral rocks[9] and gels[7,10]. The theoretical and experimental small-angle scattering techniques have been developed and well established[11]. A characteristic feature of the small-angle scattering from fractal scatterers and many other disorder solids is often obeys a "power-law" scattering in the magnitude of the scattering vector q given as

$$I(q) \propto q^{-D} \qquad (1)$$

where D related to the structure of the scatterer is positive constant. Eq. (1) applies only in the regime $\xi \gg q^{-1} \gg a$, where a is a typical chemical or bond

CP989, Neutron and X-ray Scattering in Materials Science and Biology, International Conference on Neutron
and X-ray Scattering 2007, edited by A. Ikram, A. Purwanto, Sutiarso, A. Zulfia, S. Hendrana, and Z. Nurachman

distance related to local structure and ξ is the correlation length or average diameter of a scatter.

In discussion of small-angle scattering from fractals, there are two types of fractal systems, i.e. mass fractals and surface fractals. Mass fractals are aggregates of primary particles and the distribution of mass $M(r)$ within a sphere of radius r is given by

$$M(r) \propto r^{D_M} \qquad (2)$$

where D_M is the mass fractal dimension and has a non-integer value for random or disorder object, $1 \le D_M \le 3$. However, for rods, disks, and spheres, D_M respectively is 1, 2, and 3. In contrast to mass fractals, for surface fractals the object is uniform, so that no scattering occurs from the bulk. In this case the magnitude of scattering vector is given by

$$I(q) \propto q^{D_s - 6} \qquad (3)$$

where D_s is the three-dimensional surface fractal dimension which relates surface area, A, to length. If the surface is within a sphere of radius r, then

$$S(r) \propto r^{2 - D_s} \qquad (4)$$

The meaningful values of D_s lie in the interval $2 \le D_s \le 3$. When a surface is smooth rather than fractal or non-fractal, $D_s = 2$ and scattering intensity $I(q)$ decays as q^{-4} in Eq. (3) as the surface becomes independent of r. This behaviour is commonly called Porod's law. For fractal surface, Porod slopes between -3 and -4 are expected. At large scattering vector q, the intensity $I(q)$ for scattering from a thin disk has slope of -2 and -1 from a thin rod, Eq. (1).

In this paper, we report the fractal structure change of titanium-silica aerogels prepared by sol-gel process method with various acid concentrations and then analyzed the small-angle neutron scattering data as a direct technique to determine the fractal dimension.

EXPERIMENTAL METHOD

The titanium-silica aerogels with various acid concentrations from 1 N to 1.75 N were synthesized with sol-gel process method at Universiti Teknologi Malaysia, Malaysia[12]. The acid concentration of sulfuric acid was preferred in producing aerogels materials effectively. The sample then was poured into 1 mm thick quartz cell and then covered with other 1 mm thick quartz cell as a sandwich with sample thickness of 1 mm.

The SANS experiments were performed using the 36 meter SANS-BATAN spectrometer (SMARTer) at the neutron scattering laboratory (NSL) – BATAN in Serpong, Indonesia. The rotational speed of the mechanical velocity selector was adjusted to 5000 rpm

with tilting angle of 0° to generate neutron wavelength of 0.39 nm with the spread of 0.135. Three different sample-to-detector distances of 1.5, 4 and 13 m have been set up covering the scattering vector q-range of $0.03 < q \ (\mathrm{nm}^{-1}) < 3$. The length of collimation or sample-to-source distance was set for 4 and 8 m. The exposure time for each aerogel sample and each sample-to-detector distance was 2 hours. The scattering data were corrected for empty cell as a background, electronic background as a noise, sample transmission and detector efficiency using the SMARTer data reduction software.

For data correction, empty cell, electronic background, and incoherent scatterings from water were collected for 10, 12 and 10 hours, respectively. During the experiment, the temperature was maintained at room temperature.

RESULTS AND DISCUSSION

Figure 1 shows a typical power-law scattering profiles from silica aerogles which have two power-law regimes[13]. At high q range or short-length, scattering from smooth particles results in a slope of -4, while at intermediate q range the scattering slope is -1.93, representing a thin disk shape, describing the particle aggregation in a polymeric structure by a mass-fractal morphology close to $D \sim 2$. It is consistent with Eqs. (1) and (3), with $D = 2$ and $D_s = 2$, respectively. However, at low q-range the curve does not show the independent scattering on scattering vector q since the dimension of particle is larger than ξ and it is in uniform structure[13].

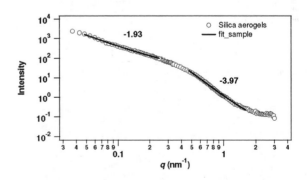

FIGURE 1. SANS profiles from pure silica aerogels were obtained in the wide scattering vector q-range.

Figure 2 shows the power-law scattering profiles from silica aerogels containing titanium as a function of acid concentrations. A similar scattering profile with silica aerogles is shown from titanium-silica aerogel sample at the lowest acid concentration of 1 N, Figure 2a. The change of slopes at high and intermediate q range is related to changing of particle structures as the effect of the acid concentration. The particles become rougher on the surface as the slope decreases from -3.97 to -3.93, while the aggregate has more elongated shape as the magnitude of scattering intensity $I(q) = q^{-1.5}$ compare to that of the silica aerogels, $I(q) = q^{-1.93}$.

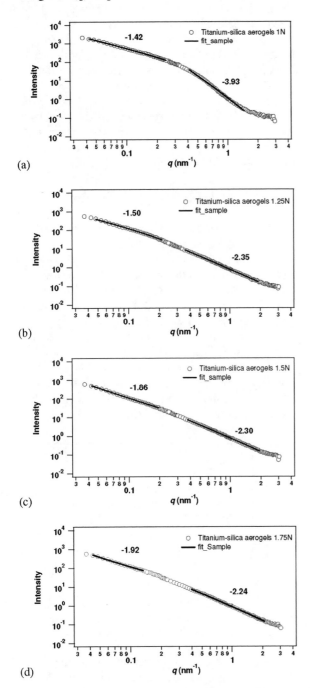

(a)

(b)

(c)

(d)

FIGURE 2. SANS profiles from titanium-silica aerogels with various acid concentrations during sol-gel process (a) 1 N; (b) 1.25 N; (c) 1.5 N and (d) 1.75 N. The slopes were determined by least square fitting

Figure 2 also shows the slopes of titanium-silica aerogles with acid concentrations of 1 N (Figure 2a) to 1.25 N (Figure 2b) at high q-range that changed dramatically from -3.9 to -2.35. This indicates that increasing pH of 0.25 N from 1 N has altered the surface structure from relatively smooth particles to highly rough particles or relatively dense polymeric structure[13] and the surface fractals is vanished. Meanwhile, by further increasing the acid concentrations, the slope does not change as much as the first one. It is also pointed out that the slopes at intermediate q-range show another fashion where they decrease from -1.42 to -1.92 gradually when the acid concentration increases from 1 N to 1.75 N, Figure 2a-d. These changes also indicate that the titanium-silica aerogels are growing to more polymeric and branched structures with the magnitute scattering intensity $I(q) = q^{-2}$ by decreasing the acid concentration. All data measurements on aerogles samples were given in Table 1.

TABLE 1. Data measurements of aerogel samples

Aerogels materials	Slope (high q)	Slope (low q)	Cross over (nm^{-1})	Size (nm)
Pure silica	-3.97	-1.93	0.40	17.7
Titanium-silica 1.0N	-3.93	-1.42	0.25	25.1
Titanium-silica 1.25	-2.35	-1.50	0.20	31.4
Titanium-silica 1.5	-2.30	-1.86	0.19	33.1
Titanium-silica 1.75	-2.24	-1.92	0.16	39.3

The cross over between two regimes for the titanium-silica aerogels at various concentrations can be related to the size of an agglomeration of particles in the aerogels with the order of $2\pi/q_o$ as in the Bragg relation. As the pH increases from 1 N to 1.75 N, a breaking point which roughly marks the change of slope in the curve decreases from 0.4 nm^{-1} to 0.16 nm^{-1} which corresponds to particle size of 17.7 nm to 39.3 nm. These results also indicate that the titanium-silica aerogels are growing with the increase of pH. Similar to the fractal structures, the mass fractal structure was indentified in the tintanium-silica aerogels and changed by increasing the pH.

CONCLUSION

An excellent performance in revealing the fractal structures of titanium-silica aerogels has been accomplished by SANS BATAN spectrometer (SMARTer). Power-law scattering on titanium-silica aerogels at high and intermediate q-range changes as a function of acid concentrations. The acid concentration affected the changing of fractal structure of aerogels more rather than the impurities in titanium. Mass fractal structures have been found with the dimension D_M between 1.42 and 1.92.

REFERENCES

1. K. L. Yeung, N. Yao, S. Cao, US Patent 11/288, 259, 29 November 2005; http://eetd.lbl.gov/ECS/aerogels/satoc.htm

2. T. Woignier, J. Reynes, J. Phalippou and J.L. Dussossoy, *J. Sol-Gel Sci. Techn.* **19**, 833-837 (2000).

3. I. Smirnova, Synthesis of silica aerogles and their application as a drug delivery system, PhD thesis, Technishen Universitat Berlin, 2002; I. Smirnova, S. Suttiruengwong, W. Arlt, *J. Non-Cryst. Solids* **350**, 54-60 (2004)

4. E. Nappi, *Nucl. Phys. B - Proceedings Supplements,* 61, Supplement 2, 270-276 (1998)

5. B. C. Dunn, et al. *Appl. Catal.* **A 278**, 233-238 (2005)

6. B. B. Mandelbrot, Fractals, Form and Chance, Freeman, San Fransisco, 1977; D. Avnir, D. Farin, P. Pfeifer, Nature 308, 1984, 5956; P. Pfeifer, D. Avnir, *J. Chem. Phys.* **79,** 3558 (1983)

7. J. E. Martin, A. J. Hurd, J. Appl. Cryst. 20, 61-78 (1987)

8. F. Horkay, P. J. Basser, A. M. Hecht, E. Geissler, *Polymer* **46**, 4242-4247 (2005)

9. D. Sen, S. Mazumder, S. Tarafdar, *J. Mater. Sci.* **37**, 941-947 (2002)

10. I. M. Miranda, F. M. A Margaca, J. Teixeira, *J. Mol. Struc.* **383**, 271-276 (1996)

11. P. G. Hall, R. T. Williams, *J. Colloid Interface Sci.* **104**, 151-174 (1985); P. W. Schmidt, *J. Appl. Cryst.* **24**, 414-435 (1991)

12. Tan Chiar Fang, Master Thesis, Department of Physics, Universiti Teknologi Malaysia, Malaysia, 2005.

13. D. W. Schaefer, K. D. Keefer, *Phys. Rev. Lett.* **56**, 2199-2202 (1986); C. I. Merzbacher, J. G. Barker, K. E. Swider, D. R. Rolison, *Adv. Colloid Interface Sci.* **76**, 57-69 (1998)

Microstructure and Optical Properties of $Al_xGa_{1-x}N$/GaN Heterostructure Thin Films Grown on Si(111) Substrate by Plasma Assisted Metalorganic Chemical Vapor Deposition Method

H. Sutanto[1,2], A. Subagio[1,2], E. Supriyanto[1,3], P. Arifin[1],
M. Budiman[1], Sukirno[1], M. Barmawi[1]

[1]*Laboratory of Electronic Material Physics, Bandung Institute of Technology, Bandung*
Jl. Ganesha No. 10 Bandung 40132, Indonesia
[2]*Department of Physics, Diponegoro University, Semarang*
Kampus MIPA-UNDIP, Jl. Prof. Soedharto, SH, Tembalang-Semarang 50275, Indonesia
[3]*Department of Physics, Jember University, Jember*
Jl. Kalimantan III/25Kampus Tegalboto-Jember, Indonesia

Abstract. Microstructure and optical properties of $Al_xGa_{1-x}N$/GaN heterostructures thin films grown on Si(111) substrate by Plasma Assisted-Metalorganic Chemical Vapor Deposition (PA-MOCVD) were investigated. The surface morphology and crystal orientation of the films were determined by scanning electron microscope (SEM) and X-ray diffractometer (XRD), respectively. The content of Al in $Al_xGa_{1-x}N$ films (x) was determined by means of NIR-UV visible optical reflectance spectroscope. The surface morphology of films depends significantly on the content of Al. Films with higher value of x showed the smaller grain size and the smoother surface. Films with x = 0.29 and x = 0.36 showed crystal orientation of (10$\underline{1}$0) plane, while films with x = 0.12 have two crystal orientation of (10$\underline{1}$0) and (10$\underline{1}$1) planes. The optical reflectance spectra showed that the ordered of oscillation depend on the smoothness of the film surface, while the number of oscillation related to the thickness of films. The calculated band gap was 3.34 eV for GaN and in the range of 3.34 to 6.20 eV for $Al_xGa_{1-x}N$ depending on the x values.

Keywords: $Al_xGa_{1-x}N$/GaN, MOCVD, crystal structure, optical reflectance.
PACS: 78.55.Cr, 78.66.Fd

INTRODUCTION

In recent year, GaN and related III-V nitrides have attracted substantial scientific and industrial interest mainly due to its excellent properties. Specially $Al_xGa_{1-x}N$/GaN heterostructure field-effect transistors (HFETs) are being rapidly developed for high-power, high-temperature operation at microwave frequencies[1–3]. The growth of high-quality thin films GaN and their alloys thin films on sapphire is usually difficult due to the large lattice mismatch and the large difference in thermal expansion coefficient between GaN and the sapphire substrate. GaN thin films also have been grown on various other substrates, such as Si, GaAs, 6H-SiC, etc[4-6]. Silicon is an attractive substrate because of its high crystal quality, large area size, low cost and the potential application in integrated circuits. Therefore, GaN and related

materials and structures grown on Si substrate are promising for developing new generation of devices by combination of Si and III-N based materials and technologies in the 21st century. However, due to the even larger difference in lattice constant (~17%) and thermal expansion coefficient (~56%) between GaN and the silicon substrate compared to that between GaN and sapphire, it is even more difficult to grow high-quality GaN films and structures on Si substrates than on sapphire.

Buffer layers between GaN and silicon substrates have been grown to overcome the problem of lattice mismatch and to improve the quality of the GaN thin film. These include AlN[7], carbonized silicon[8], nitridized GaAs[9], oxidized AlAs[10], and so on. Recently, we have successfully made efforts on the growth and characterization of GaN films on silicon

CP989, Neutron and X-ray Scattering in Materials Science and Biology, International Conference on Neutron and X-ray Scattering 2007, edited by A. Ikram, A. Purwanto, Sutiarso, A. Zulfia, S. Hendrana, and Z. Nurachman
© 2008 American Institute of Physics 978-0-7354-0508-0/08/$23.00

substrates with low temperature GaN buffer layer by plasma assisted MOCVD method[11-12].

Optical reflectance spectroscopy is one of the experimental methods which enable the measurement of thickness layer and band gap energy of each layers material[13]. The advantages of this technique were that this method was contactless and non destructive character.

In this paper we report the preliminary study of microstructure and optical properties of $Al_xGa_{1-x}N$/ GaN heterostructure thin films grown on Si(111) substrates by PA-MOCVD method

EXPERIMENTAL METHOD

$Al_xGa_{1-x}N$/GaN films were grown on Si(111) substrates by Plasma Assisted-Metal Organic Chemical Vapor Deposition (PA-MOCVD). Trimethylgallium (TMGa), trimethylaluminum (TMAl) and N_2 gas plasma were used as precursors of Ga, Al and N, respectively. The purified H_2 by passing it through a heated palladium cell was used as the carrier gas. A low power downstream plasma cavity supplied the reactive N plasma from N_2 gas and reactive H plasma from H_2 gas. The plasma was generated by 2.45 GHz microwave plasma source of 200 watt.

The silicon (Si) substrates were prepared by cleaving a 2 inch into several 1.0 x 1.5 cm^2 pieces with a diamond scribe. The substrate were sequentially cleaned by acetone and methanol for 10 minutes each, followed by washing in de-ionized water (DI-water) and etched in solution of H_2O_2 : H_2SO_4 : DI-water = 1 : 3 : 1 at 70°C for 5 minutes. Then they were dipped into 2% HF solution for 5 minutes to remove native SiO_2 on the Si substrate. Finally, the substrate were put into running DI-water and then blown with dry nitrogen. The substrate was immediately introduced into the growth reactor and heated up to 650°C for thermal cleaning in the H_2 ambient to remove native oxide on the surface of Si. *In situ* hydrogen plasma cleaning on substrates was carried out for 10 minutes with the H_2 gas flow rate of 50 sccm. GaN buffer layer was grown with TMGa of 0.19 sccm and N_2 flow of 90 sccm at 500°C, which produces about 25 nm thick of layer. After the deposition of buffer layer, the substrate temperature was raised to the growth temperature of 680°C for the growth of GaN films with same parameter as buffer layer and temperature of 700°C for the growth of $Al_xGa_{1-x}N$ films with the flow rate of TMGa and N_2 gas were 0.19 sccm and 90 sccm, respectively. The flow rates of TMAl were varied between 0.05 and 0.13 sccm.

The surface morphology and cross-section images of $Al_xGa_{1-x}N$/GaN films were observed by scanning electron microscope (SEM, JEOL JSM-6360 LA). The structural properties of the $Al_xGa_{1-x}N$/GaN films were determined by X-Ray Diffractometer (Cu K_α (λ = 1.54056 Å), PAN Analytical). The optical properties of the films were measured by means of optical reflectance spectroscopy in the range of 200 – 2400 nm (NIR-UV Visible Spectroscope, PC101).

RESULTS AND DISCUSSION

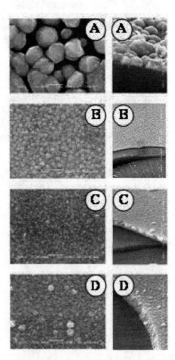

FIGURE 1. Surface morphology (left) and cross-sectional SEM images (right) of $Al_xGa_{1-x}N$/GaN thin-films. (A) without buffer layer; with buffer layer and various TMAl flow rate of (B) 0.05 sccm; (C) 0.07 sccm and (D) 0.13 sccm.

Fig. 1 shows the surface morphology and cross-sectional SEM images of $Al_xGa_{1-x}N$/GaN thin-films. Films grown without buffer layer (Fig. 1. A) are dominated by hexagonal islands with rough surface morphology, large grains size, inhomogeneous and separated films nucleation. This is caused by high lattice mismatch and high interface energy between GaN and the substrate. In addition, the large grain size of films indicates that the film nucleation is low. In contrast, films grown with GaN buffer layer (sample B; C and D) show smoother surface morphology and smaller grain size as the Al content increased. The ionic radius of element is responsible to the increase of surface quality since the ionic radius is a dominant factor on the growth of film grain. The size of ionic radius of Al is 0.5 Å, which is smaller than that of Ga (0.62Å).

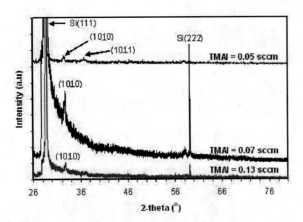

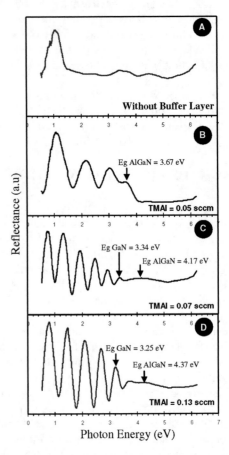

FIGURE 2. The XRD patterns of $Al_xGa_{1-x}N$/GaN thin-films; (A) without buffer layer; with buffer layer and various TMAl flow rate of (B) 0.05; (C) 0.07 and (D) 0.13 sccm.

Fig. 2 shows the XRD patterns of $Al_xGa_{1-x}N$/GaN thin-films. In conjunction with $Al_xGa_{1-x}N$ crystallographic plane, the X-ray diffraction peaks of sample (B) are observed at 2-theta of $32.46°$ and $37.14°$ which correspond to (10$\underline{1}$0), and (10$\underline{1}$1) planes, respectively. While, the X-ray diffraction peaks of sample (C) and sample (D) show one peak each, at 2-theta of $32.66°$ and $32.72°$ which correspond to (10$\underline{1}$0) plane. For (10$\underline{1}$0) plane, the peak is shifted to the higher diffraction angle as the Al content is increased. The crystallinity of films can be identified from the FWHM (*Full Width at Half Maximum*). The better film crystallinity is associated to the smaller FWHM value. The FWHM values of (10$\underline{1}$0) peak were $0.373°$; $0.285°$ and $0.375°$ for samples (B), (C) and (D), respectively.

Fig. 3 shows the room temperature reflectance spectra of $Al_xGa_{1-x}N$/GaN thin-films. The film with rough surface morphology (Fig 3.A) shows disorder oscillation patterns. The rough surface morphology of the films results in peculiar scattering of incident beam to any directions which lead to un-perfect excitonic and band-to-band absorption of the films. Optical reflectance spectra of the films with smoother surface (Fig. 3. B, C and D) are in good order. Oscillation pattern of the optical reflectance spectra is related to the quality of the films. The thickness (d) of the film can be calculated from the interference equation below:[14]

$$d = \frac{N}{2}\left[\frac{\lambda_1\lambda_2}{n_1\lambda_2 - n_2\lambda_1}\right] \qquad (1)$$

where (N+1) is the number of maxima from λ_1 to λ_2, n_i is the refractive index at the maximum λ_i. The thickness of $Al_xGa_{1-x}N$/GaN films calculated using this method are 261 nm, 511 nm, and 600 nm, for sample B, C and D, respectively. These results are in good agreement with the thicknesses calculated from cross-section SEM analysis (Fig. 1).

FIGURE 3. The optical reflectance spectra of $Al_xGa_{1-x}N$/GaN thin-films (A) without buffer layer and with buffer layer and flow rate of (B) 0.05 sccm, (C) 0.07 sccm, and (D) 0.13 sccm.

The last oscillation peak in the optical reflectance spectra (shown in dashed line at Fig. 3) related to the band-to-band transition which corresponds to the band-gap energy (E_g). The similar results are shown by other researches[15-16].

The content of Al in $Al_xGa_{1-x}N$ (x) can be obtained by combining the E_g calculated from optical reflectance and Vegard's law:[17]

$$E_g (Al_xGa_{1-x}N) = x.E_g(AlN) + (1-x).E_g(GaN) - b.x.(1-x) \qquad (2)$$

where x is Al molar fraction of $Al_xGa_{1-x}N$ thin-film; E_g(GaN) = 3.34 eV; E_g(AlN) = 6.20 eV and b = bowing parameter.

Using the bowing parameter b = -0.012[11], the calculated Al content (x) are 0.12; 0.29 and 0.36 for sample B, C and D, respectively. Even though the band gap of GaN is difficult to be determined from optical reflectance spectra of samples A and B, the estimated E_g (GaN) of sample C is 3.34 eV, which is in good agreement with that obtained by other workers[17]. However, estimated E_g (GaN) of sample D is smaller than that of sample C. This result shows that GaN in sample C is in relaxation form, while GaN in sample D is in the strained form, due to the lattice mismatch between GaN and $Al_xGa_{1-x}N$.

CONCLUSIONS

The surface morphology of $Al_xGa_{1-x}N$/GaN heterostructure thin-films grown on Si(111) by PA-MOCVD method depends on Al content (x). Films with higher x exhibit the smaller grain size and the smoother surface morphology. Films grown without buffer layer have rough surface morphology compared to that grown with buffer layer. $Al_xGa_{1-x}N$ films with x = 0.29 and x = 0.36 have single crystal orientation of $(10\underline{1}0)$ plane, while film with x = 0.12 has two crystal orientations of $(10\underline{1}0)$ and $(10\underline{1}1)$ planes. The optical reflectance spectra show that the order of oscillation depends on the smoothness of film surface, while the number of oscillation related to the thickness of films. The band gap energy of $Al_xGa_{1-x}N$ can be deduced from the last oscillation peak in the optical reflectance spectra.

ACKNOWLEDGMENTS

This work was supported by The Ministry of Research and Technology, Republic of Indonesia, through the project of "Riset Unggulan Terpadu" (RUT XII), contract No. ITB-INDONESIA 01/Perj/Dep.III/RUT/PPKI/II/2005.

REFERENCES

1. W. Knap, C. Skierbiszewski, K. Dybko, J. Lusakowski, M. Siekacz, I. Grzegory, S. Porowski, *J. Cryst. Growth* **281**, 194-201 (2005)
2. W.S. Tan, H.L. Cai, X.S. Wu, S.S. Jiang, W.L. Zheng and Q.J. Jia, *J. Alloys and Compounds* **397**, 231-235 (2005)
3. J.R. Juang, D.R. Hang, M.G. Lin, T.-Y. Huang, G.H. Kim, C.T. Liang, Y.F. Chen, W.K. Hung, W.H. Seo, Y. Lee and J.H. Lee, *Chin. J. Phys.* **42**, 629-636 (2004)
4. P. Kung, A. Saxler, X. Zhang, D. Walker, T.C. Wang, I. Ferguson and M. Razeghi, *Appl. Phys. Lett.* **66**, 2958-2960 (1995)
5. A. Trampert, O. Brandt, H. Yang and K.H. Ploog, *Appl. Phys. Lett.* **70**, 583-585 (1997)
6. B.N. Sverdlov, G.A. Martin, H. Morkoc and D.J. Smith, *Appl. Phys. Lett.* **67**, 2063-2065 (1995)
7. S. Guha and N. Bojarczuk, *Appl. Phys. Lett.* **72**, 415-417 (1998)
8. A.J. Steckl, J. Devrajan, C. Tran and R.A. Stall, *Appl. Phys. Lett.* **69**, 2264-2266 (1996)
9. J.W. Yang, C.J. Sun, Q. Chen, M.Z. Anwar, M.A. Khan, S.A. Nikishin, G.A. Seryogin, A.V. Qsinsky, L. Chernyak, H. Temkin, C. Hu and S. Mahajan, *Appl. Phys. Lett.* **69**, 3566-3568 (1996)
10. N.P. Kobayashi, J.T. Kobayashi, P.D. Dapkus, W.J. Choi, A.E. Bond, X. Zhang and D.H. Rich, *Appl. Phys. Lett.* **71**, 3569-3571 (1997)
11. H. Sutanto, A. Subagio, E. Supriyanto, B. Mulyanti, P. Arifin, M. Budiman, Sukirno and M. Barmawi, Proceedings of 2005 Asian Physics Symposium, ISBN No: 979-98010-2-8, Dept. of Physics, Bandung Institute of Technology, Indonesia, 2005, pp.52-55.
12. H. Sutanto, A. Subagio, E. Supriyanto, B. Mulyanti, P. Arifin, M. Budiman, Sukirno and M. Barmawi, Proceedings of 3rd Kentingan Physics Forum, ISBN N0: 979-97651-1-0, Dept. of Physics, Sebelas Maret University Surakarta, Indonesia, 2005, pp.133-136.
13. J. Misiewicz, P. Sitarek, G. Sek and R. Kudrawiec, *Mater. Sci.* **21**, 263-318 (2003).
14. R. Palomino-Merino, A. Conde-Gallardo, M. Garcia-Rocha, I. Hernandez-Calderon, V. Castano and R. Rodriquez, *Thin Solid Films* **401**, 118-123 (2001)
15. R. Kudrawiec, M. Syperek, J. Misiewicz, R. Paskiewicz, B. Paskiewicz and M. Tlaczala, *Superlattices and Microstructures* **36**, 633-641 (2004)
16. A.T. Winzer, R. Goldhahn, G. Gobsch, A. Dadgar, H. Witte, A. Krtschil and A. Krost, *Superlattices. and Microstructures* **36**, 693-700 (2004)
17. I. Vurgaftman and J.R. Meyer, *J. Appl. Phys.* **89**, 5815-5875 (2001)

Dislocations in P-MBE Grown ZnO Layers Characterized by HRXRD and TEM

A. Setiawan[1], I. Hamidah[1], S. Maryanto[2], S. Aisyah[3], and T. Yao[4]

[1]Department of Mechanical Engineering and Graduate School of Science Education
Indonesia University of Education, Setiabudhi 229 Bandung 40154, Indonesia
[2]Geological Research and Development Centre, Diponegoro 57 Bandung 40122, Indonesia
[3]Department of Chemistry Education Indonesia University of Education
Setiabudi 229 Bandung 40154, Indonesia
[4]Center for Interdisciplinary Research, Tohoku University Aramaki, Aoba-ku, Sendai, 980-8578, Japan

Abstract. We have characterized dislocations in ZnO layers grown on c-sapphire (α-Al$_2$O$_3$) by plasma-assisted molecular-beam epitaxy (P-MBE) with and without MgO buffer layer. ZnO without MgO buffer was grown three-dimensionally (3D), while ZnO with MgO buffer was grown two-dimensionally (2D). Mosaic spread (tilt and twist angles), types and density of dislocations in the layers were studied by both high-resolution X-ray diffraction (HRXRD) and transmission electron microscopy (TEM). HRXRD experiments reveal that screw dislocation densities in the ZnO layer are 8.1×10^8 cm^{-2} and 6.1×10^5 cm^{-2}, for ZnO with and without MgO buffer, respectively, while edge dislocation densities are 1.1×10^{10} cm^{-2} and 1.3×10^5 cm^{-2}, for ZnO with and without MgO buffer, respectively. HRXRD and TEM data showed the same result that the major dislocations in the ZnO layers are edge type dislocations running along c-axis. Therefore, HRXRD technique can be applied to characterize dislocations in ZnO layers.

Keywords: ZnO, dislocations, HRXRD, TEM
PACS: 81.05.Dz, 61.72.Lk, 61.72.Dd, 68.37.Lp

INTRODUCTION

ZnO is a direct band energy gap semiconductor (E_g=3.37 eV at RT) with a wurtzite structure. The most outstanding feature of ZnO is its large exciton binding energy, 60 meV, which is about three times larger than that of ZnSe and of GaN. Recent reports on the lasing mechanisms of ZnO have shown that ZnO is a promising photonic material for exciton devices in the wavelength ranging from blue to ultraviolet[1-3].

Because of its low cost, large size, and high quality, c-sapphire has been extensively used as substrate for ZnO epitaxy. The greatest disadvantages of ZnO epitaxy on c-sapphire are the large lattice misfits (18%) and thermal mismatch (13%) between ZnO and c-sapphire and the formation of 30°-rotated domains[4]. Consequently, ZnO layers grown on c-sapphire showed a rough surface morphology and poor crystalline quality [4].

In order to overcome the problems caused by the large mismatch between ZnO and c-sapphire substrate, insertion of a buffer layer material which can reduce lattice misfit between ZnO and c-sapphire seems to be a key to obtain high quality ZnO. The growth of double buffer layers consisting of low temperature (LT)-MgO buffer and LT-ZnO buffer followed by a high temperature (HT) annealing has been utilized with success by P-MBE[5]. Here, the highly mismatched heterointerface of ZnO/α-Al$_2$O$_3$ (18%) is broken up into two much more slightly mismatched interfaces of ZnO/MgO (9%) and MgO/α-Al$_2$O$_3$ (8%) by inserting a MgO layer, which eventually leads to surface adhesion and lateral epitaxial growth. In the present paper, we investigate detailed structural quality of ZnO layers grown on c-sapphire with and without MgO buffer layer.

TEM and HRXRD are greatly powerful tools that are utilized in this study. Here, HRXRD has been used to assess the crystal quality of heteroepitaxial layers including tilt and twist angles, and dislocation densities, while TEM has been used to determine types and density of dislocations. Tilt and twist angle, and hence dislocation densities, can be evaluated by a series of HRXRD measurements[6-8].

CP989, Neutron and X-ray Scattering in Materials Science and Biology, International Conference on Neutron and X-ray Scattering 2007, edited by A. Ikram, A. Purwanto, Sutiarso, A. Zulfia, S. Hendrana, and Z. Nurachman

EXPERIMENT

ZnO layers were grown on c-sapphire by P-MBE either with or without MgO buffer. The substrates were degreased in acetone and methanol in ultrasonic cleaner and then chemically etched in a H_2SO_4 (96%): H_3PO_4 (85%)= 3:1 solutions for 15 minutes at 160 °C. Prior to growth, the substrates were thermally cleaned at 750 °C in the preparation chamber for 1 hour. The substrates were then treated in oxygen plasma at 650 °C for 30 minutes in the growth chamber to produce an oxygen terminated c-sapphire surface. Oxygen flow rate and plasma power were set to 2.5 sccm and 300 W, respectively. The sample structure was as follows. Firstly, a LT-ZnO buffer was grown at 490 °C on thermally cleaned c-sapphire followed by annealing at 750 °C for 3 min. Then, HT-ZnO was grown at 700 °C. In the case of ZnO with MgO buffer, LT-MgO buffer was grown at 490 °C followed by LT-ZnO at 490 °C. The thickness of ZnO layers was 450 nm. The whole growth process was monitored by in-situ RHEED. Structural characterizations were carried out by HRXRD and cross-sectional TEM. HRXRD experiments were carried out with a Phillips X'Pert MRD diffractometer.

RESULTS AND DISCUSSION

Surface morphology

Figure 1 shows RHEED patterns and the corresponding top view AFM images of ZnO layers (a) without and (b) with MgO buffer. ZnO without MgO buffer shows spotty RHEED patterns, indicating 3D growth mode, while ZnO with MgO buffer shows streaky and specular spot RHEED patterns, indicating 2D growth mode.

The rms values of surface roughness at 1 μm^2 scan area are 13 nm and less than 1 nm for ZnO layer grown without and with MgO buffer, respectively.

Structural quality addressed by HRXRD

In order to address the defect structures of wurtzite ZnO by HRXRD, 0002 Ω- and 10-11 Φ- and Ω- rocking curves measurements were performed. Note that the broadening of 0002 Ω- and 10-11 Φ- rocking curves represents lattice disordering along the growth direction (out of plane) and in-plane disordering, respectively.

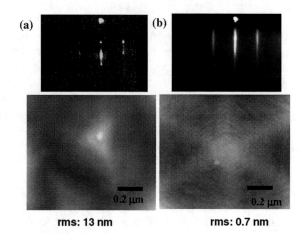

FIGURE 1. RHEED-patterns and corresponding AFM images of ZnO layers grown on c-sapphire (a) without and (b) with MgO buffer.

Figure 2 provides a comparison of (a) 0002 Ω and (b) 10-11 Ω rocking curves of ZnO layers grown with and without MgO buffer. FWHM values of 0002 Ω scans are 565 arcsec and 18 arcsec, for ZnO grown without and with MgO buffer, respectively. FWHM values of 10-11Ω scan are 1346 arcsec and 1076 arcsec, for ZnO films grown without and with MgO buffer, respectively.

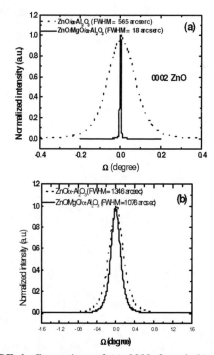

FIGURE 2. Comparison of (a) 0002 Ω and (b) 10-11 Ω scans of a ZnO layer grown without (*dotted curve*) and with (*solid curve*) MgO buffer.

It should be noted here that FWHM of the 0002 Ω scan was greatly reduced by employing an MgO buffer layer. This directly indicates a small tilt in the c-plane because of the extreme ordering along the growth direction of ZnO(0001) as a consequence of well-controlled layer-by-layer epitaxial growth. Significant broadening of the 10-11 reflection as compared to the 0002 is an indicative of the presence of high edge dislocation density. Note that all type of dislocations (edge, screw, and mixed) broaden the 10-11 reflection, whereas the 0002 reflection is only sensitive to screw and mixed type of dislocations. Furthermore, the FWHM value of the 10-11 Ω scan of the ZnO grown with MgO buffer is smaller than that without MgO, indicating a much lower edge dislocation density.

Mosaic spreads (tilt and twist angles)

Mosaic crystals can be characterized by means of tilt and twist angles and average size of the mosaic blocks. The tilt describes the out-of-plane rotation of the blocks and the twist describes the in-plane rotation. Figure 3 shows the typical Williamson-Hall (W-H) plots for ZnO samples with and without MgO buffer. W-H plot with a linier fit is performed using full width at half maximum (FWHM) of (0002), (0004), (0006) Ω scans, where "FWHM x $\sin(\theta)/\lambda$" is plotted against "$\sin(\theta)/\lambda$" in order to determine a tilt angle[8-10]. Here θ is a diffraction angle and λ is a wavelength of CuK_α (0.154056 nm). In the W-H plots, a tilt angle is the slope of the fitted line[8]. From Fig. 3, the tilt angles are determined to be $0.1541°$ and $0.0056°$ for the ZnO samples without and with MgO buffer, respectively.

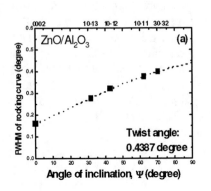
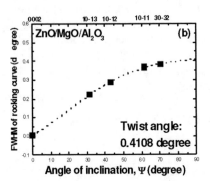

FIGURE 3. Williamson-Hall plots for a ZnO layer on c-sapphire (a) with and (b) without MgO buffer

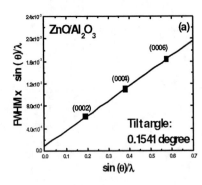
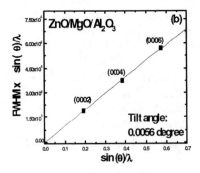

FIGURE 4. FWHMs of *(hkl)* Ω rocking curve for reflections (marked at upper X-axis) as a function of inclination angle Ψ of the reflecting lattice planes with respect to the sample surface. (a) ZnO/α-Al₂O₃ and (b) ZnO/MgO/Al₂O₃.

In order to determine the twist angle, FWHMs of (0002), (10-13), (10-12), (10-11), (30-32) Ω scans are plotted as a function of the inclination angles. Figures 4(a) and 4(b) show the plots for ZnO layers without and with MgO buffer, respectively. A twist angle of the samples is determined by estimating the FWHM at an inclination angle of $90°$ by extrapolation[7]. From Fig.4, the twist angles are determined to be $0.4387°$ and $0.4108°$, for ZnO without and with MgO buffer, respectively.

Types and densities of dislocations accessed by HRXRD

From the determined tilt and twist angles, <0001> screw and 1/3<11-20> edge dislocations densities are evaluated. Dislocation densities are determined using formalism by Ayers[11], which has been successfully used to determine the dislocation densities in GaN epilayers on sapphire[8,9]. In this formalism, Gaussian shape rocking curve and Gaussian distribution of the orientations of the mosaics are assumed. Dislocation density d is determined using equation of $d = \alpha^2/(4.35b^2)$, where α is the mosaic angle and **b** is the Burgers vector. For the screw dislocation density, the tilt angle and the Burgers vector of <0001> are used, while the twist angle and the Burgers vector of 1/3<11-20> are used for the edge dislocation density. From the determined tilt angle, screw dislocation densities are determined to be 6.1×10^8 cm^{-2} and 8.1×10^5 cm^{-2}, for ZnO without and with MgO buffer, respectively. We note that the screw dislocation density of ZnO is greatly reduced, about three-order of magnitude, by employing a MgO buffer. From the determined twist angles, edge dislocations densities are determined to be 1.3×10^{10} cm^{-2} and 1.1×10^{10} cm^{-2}, for ZnO without and with MgO buffer, respectively. The edge dislocation density is slightly reduced by employing a MgO buffer. Low screw dislocation density in ZnO layer with MgO buffer implies well ordering in the growth direction while the high edge dislocation density means high disordering in the c-plane. The high edge dislocation density is caused by high in plane lattice misfit between ZnO layer and MgO buffer.

Types and densities of dislocations accessed by TEM

Types of dislocation in the ZnO samples were characterized by cross-sectional TEM under two-beam condition, as shown in Figure 5. The samples were observed near the [2-1-10] zone axis with diffraction vectors **g** = [0006] (Fig. 5(a) and Fig. 5(c)) and **g** = [03-30] (Fig 5(b) and Fig. 5(d)). By invisibility criterion, screw-type dislocations should be visible under **g** = 0006 and invisible under **g** = 03-30. In the contrast, edge-type-dislocations should be invisible under **g** = 0006 and visible under **g** = 03-30. Mixed-type dislocations should be visible under both of the **g** vectors. By averaging several pictures, for ZnO layer with MgO buffer, threading dislocations were roughly distributed as 31% of screw-type (Burgers vectors **b** = [0001]), 61% of edge-type (Burgers vectors **b** = 1/3<11-20>), and 8% of mixed-type (Burgers vectors **b** = 1/3<11-23>) dislocations. For ZnO with MgO buffer, threading dislocations were distributed as 98% of edge-type and 2% of screw-type and mixed type dislocations. Here, major threading dislocations running along c-axis are edge-type dislocation with Burgers vector of 1/3<11-20> in both the samples. Comparison of Fig. 5(a) and Fig. 5(c) reveals the screw dislocations are greatly reduced by introducing the MgO buffer. From plane-view TEM images(data are not shown here), total dislocation densities were was determined to be 6×10^{10} cm^{-2} and 2×10^{10} cm^{-2}, for ZnO layers without and with MgO buffer, respectively.

ZnO/Al$_2$O$_3$

ZnO/MgO/Al$_2$O$_3$

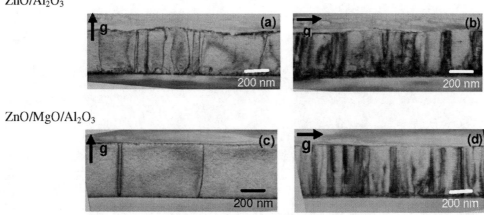

FIGURE 5. Two beam bright-field cross-sectional electron micrographs of the ZnO/α-Al$_2$O$_3$ and ZnO/MgO/α-Al$_2$O$_3$ near the [2-1-10] zone axis with g = 0006 ((a) and (c)) and g = 03-30 ((b) and (d)).

Figure 5 shows another important fact. At the interface region, a high density of interfacial threading dislocations was observed. Surprisingly, the dislocation density rapidly decreases beyond 20 nm. Furthermore, by increasing layer thickness, the density of threading dislocation in ZnO with MgO buffer decreases faster than that of without MgO buffer. It can only be understood if these threading dislocations are not along the c-axis so that they strongly interact with each other and annihilated quickly[12]. Lattice misfit between ZnO and MgO (9%) is smaller than between ZnO and c-sapphire (18%). Therefore, the density of interfacial defect in ZnO with MgO buffer is smaller that that of without MgO buffer.

CONCLUSIONS

We have studied dislocations in ZnO layers grown on α-Al_2O_3 by P-MBE with and without MgO buffer layer. Mosaic spreads (tilt and twist angles), type and density of dislocations in layers were characterized by HRXRD and the results were compared with TEM data. Screw dislocation densities in ZnO layer determined by HRXRD are 8.1×10^8 cm^{-2} and 6.1×10^5 cm^{-2}, for ZnO with and without MgO buffer, respectively, while edge dislocation densities in ZnO layer are 1.1×10^{10} cm^{-2} and 1.3×10^{10} cm^{-2}, for ZnO with and without MgO buffer, respectively. HRXRD and TEM data showed the same results that the major dislocations in the ZnO layers are edge type dislocations running along c-axis. Those values of dislocation densities are agree with TEM data. Therefore, HRXRD technique can be applied to characterize dislocations in of ZnO layers.

ACKNOWLEDGMENT

The author would like to thank Prof. S.K. Hong, Dr. H. J. Ko, Dr. Yefan Chen and Prof. M. W. Cho for their useful discussions on P-MBE growth and characterization. The author also would like to thank Directorate general of Higher Education of Indonesia for its financial support under Fundamental Research Project (contract number: 014/DP2M/II/2006).

REFERENCES

1. D.M. Bagnall, Y.F Chen, Z. Zhu, T. Yao, S. Koyama, M. Y. Shen, T. Goto, *Appl. Phys. Lett.* **70**, 2230-223 (1997)
2. D.M. Bagnall, Y.F Chen, Z. Zhu, T. Yao, M.Y. Shen, T. Goto, *Appl. Phys. Lett.* **73**, 1038-1040 (1998)
3. H. J. Ko, Y. F Chen, K. Miyajima, A. Yamamoto, T. Goto, *Appl. Phys. Lett.* **77**, .537-539 (2000)
4. P. Fons, K. Iwata, A. Yamada, K. Matsubara, S. Niki, K. Nakahara, T. Tanabe, H. Takasu, *Appl. Phys. Lett.* **77**, 1801-1803 (2001)
5. Y.F. Chen, H.J. Ko, S.K. Hong, T. Yao, *Appl. Phys. Lett.* **76**, 559-561 (2000)
6. V. Srikant, J. S. Speck, D. R. Clarke, *J. Appl. Phys.* **82**, 4286-4295 (1997)
7. H. Heinke, V. Kirchner, S. Einfeldt, D. Hommel, *Appl. Phys. Lett.* **77**, 2145-2147 (2000)
8. T. Metzger, R. Hopler, E. Born, O. Ambacher, M. Stutzmann, R. Stommer, M. Stultzmann, R. Stommer, M. Schuster, H. Gobel, S. Cristiansen, M. Albrecht, H.P. Strunk, *Phil. Mag. A* **77**, 1013-1025 (1998)
9. G. K. Williamson, W.H. Hall, *Acta Metall.* **1**, 22 -31 (1953)
10. S. K. Hong (Private communication)
11. J. E. Ayers, *J. Cryst. Growth* **135**, 71-77 (1994)
13. Y. F. Chen, S. K. Hong, H. J. Ko,V. Kirshner H. Wenis, T. Yao, K. Inaba, Y. Segawa, *Appl. Phys. Lett.* **78**, 3352-3354 (2001)

Fabrication of TiO$_2$ Thick Film for Photocatalyst from Commercial TiO$_2$ Powder

S. Fuji Asteti[1] and D. Gustaman Syarif[2]

[1]Department of Chemistry, UNJANI, Cimahi, Bandung, Indonesia
[2]Nuclear Technology Center for Materials and Radiometry-BATAN,
Jl. Tamansari 71, Bandung, 40132, Indonesia

Abstract. Photocatalytic activity of TiO$_2$ thick film ceramics made of commercial TiO$_2$ powder has been studied. The TiO$_2$ powder was nano sized one that was derived from dried TiO$_2$ suspension. The TiO$_2$ suspension was made by pouring some blended commercial TiO$_2$ powder into some amount of water. The paste of TiO$_2$ was made by mixing the nano sized TiO$_2$ powder with organic vehicle and glass frit. The paste was spread on a glass substrate. The paste was dried at 100°C and heated at different temperatures (400°C and 500°C) for 60 minutes to produce thick film ceramics. The photocatalytic activity of these films was evaluated by measuring the concentration of a solution of methylene blue where the thick films were inside after being illuminated by UV light at various periods of times. The initial concentration of the methylene blue solution was 5 ppm. Structural analyses were carried out by X-ray diffraction (XRD). The XRD analyses showed that the produced thick film ceramic had mainly crystal structure of anatase. According to the photocatalytical data, it was known that the produced thick film ceramics were photocatalyst which were capable of decomposing an organic compound such as the methylene blue.

Keywords: Photocatalyst, TiO$_2$, nano sized powder, thick film.
PACS: 82.45.Jn

INTRODUCTION

Environment pollution has become world problem especially water pollution that has attracted much attention recently. Organic material (VOM: Volatile organic materials) [1-7], metal ions [5] and micro organism like bacteria [8] are some of water pollutants. Some efforts can be done to solve the water pollution problem. One effort is the application TiO$_2$ photocatalyst for water purification and disinfection [1-8]. The TiO$_2$ photocatalyst receives a great attention because chemically stable [9] and harmless to human [10].

The TiO$_2$ photocatalyst is divided into two kinds i.e. mobile [1-6,8,11,12] and immobile [7-10,13-16]. The mobile photocatalyst is generally in the form of powder. The powder photocatalyst is very effective in destroying trace organic material as a pollutant such as dyes (methylene blue etc.) due to its large surface contact area. However, recollecting the powder is a big problem. It is more serious when the powder is applied to drinking water purification. So, the immobilization of the photocatalyst is necessary to eliminate this problem. The immobilization of the TiO$_2$ can be made using various methods. One of them is by converting the TiO$_2$ photocatalyst powder into a thick film. The advantage of the thick film technology is its ease and simplicity. The thick film can be applied not only for water purification but also for air purification.

The thick film was generally produced by applying screen printing method using a paste that made of a mixture of ceramic powder and organic vehicle. Substrate is required for the thick film such as glass, alumina or metal. The kind of substrate depends on the application of the thick film. For TiO$_2$ thick film photocatalyst the nano sized powder is the important part. The nano sized powder can be produced by various methods such as sol gel and coprecipitation from some kinds of precursors. The nano sized powder can also be produced by mechanical treatment such as blending or crushing. A production of nanosized TiO$_2$ powder using a combination of mechanical and colloidal solution treatment has not been reported yet. This work is a report of a route to produce a thick film TiO$_2$ photocatalyst from nano sized powder using commercial TiO$_2$ powder as the initial powder and of photocatalytic activity of the thick film.

CP989, Neutron and X-ray Scattering in Materials Science and Biology, International Conference on Neutron and X-ray Scattering 2007, edited by A. Ikram, A. Purwanto, Sutiarso, A. Zulfia, S. Hendrana, and Z. Nurachman
© 2008 American Institute of Physics 978-0-7354-0508-0/08/$23.00

METHODOLOGY

An amount of commercial TiO$_2$ powder (E. Merck) was blended in an electrical blending machine using water as a medium for about 45 minutes. The blended powder and some water were poured into a beaker glass. A suspension was formed. The upper part of the solution containing colloidal powder was taken and poured into another beaker glass. This suspension was dried by heating it at 95°C for several hours. After drying, the powder was taken out. An X-ray diffraction analyses was carried out to evaluate the crystal structure and the particle size of the powder. A TiO$_2$ paste was made from the powder by mixing it with glass frit and organic vehicle composed of 90 % alpha terpineol and 10% ethyl cellulose. The ceramic powder part of the paste was composed of 95 % TiO$_2$ powder and 5 % glass frit. The paste was spread on a glass slide as a substrate. The film was dried at 100°C for about 2 hours and heated at 400 and 500°C for one hour to produce good adhered TiO$_2$ thick film ceramics. The thick film was subjected to the XRD analyses to know the crystal structure of the film. The photocatalytic activity of the thick film was evaluated by measuring the concentration a methylene blue solution where the thick film was inside after being illuminated by a UV light at different period of times.

RESULTS

From the calculation using Scherrer method[17], it was confirmed that the particle size of the initial TiO$_2$ powder was 83 nm. Figure 1 shows a visual appearance of a typical TiO$_2$ thick film ceramic. The ceramics were generally visually good and were relatively well adhered on the glass substrates and strong enough when immersed in a solution of methylene blue for photocatalytic activity testing.

Figure 2 shows the XRD profile of the TiO$_2$ initial powder and Figure3 shows the XRD profiles of the thick film ceramics heated at different temperatures. As shown in the Figure 2 and 3, the profile of the initial powder and those of thick films fired at 400°C and 500°C are similar. Majority of the peaks of the XRD profiles fit those of the XRD standard profile of TiO$_2$ anatase from JCPDS No. 21-1272. However, as can be seen in the profile of the initial powder, some peaks from brookite TiO$_2$ (compared to XRD standard profile for brookite from JCPDS No. 29-1360) are observed. The same feature is also observed in the profile of the thick film fired at 400°C. This indicates that until 400°C the TiO$_2$ thick film was composed of anatase as the main part and brookite as the minor part. At 500°C, the brookite transforms to anatase. Then,

anatase is then the only one structure of the TiO$_2$ fired at 500°C.

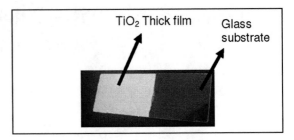

FIGURE 1. Visual appearance of typical TiO$_2$ thick film ceramic.

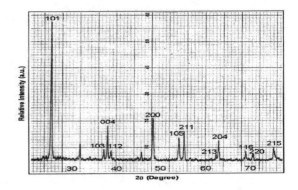

FIGURE 2. XRD profile of TiO$_2$ powder.

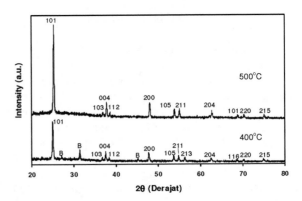

FIGURE 3. XRD profile of TiO$_2$ thick film ceramics heated at 400°C and 500°C.

Figure 4 shows the microstructures of the thick film ceramics fired at 400°C and 500°C. As shown in Figure 4, the microstructure is nearly the same and difficult to differentiate. The grains size of both thick films is nearly the same i.e. about 150 nm.

The photocatalytic activity data of Figure 4 shows that the concentration of methylene blue decreases after irradiated by UV. The decrease of the concentration is larger for the methylene blue solution with the TiO$_2$ thick film inside. Here, the role of the photocatalyst as the organic compound decomposer is seen. The photocatalyst capability of degrading organic compound is identified by measuring the degradation degree after one hour UV irradiation which is generally about 60-70%. In this work the degradation degree of the films fired at 400 and 500°C are 60% and 73% respectively (Table 1). These are good enough. The degradation degree here is defined as the initial concentration of methylene blue minus the concentration of the methylene blue after one hour UV irradiation divided by the initial concentration of the methylene blue in %.

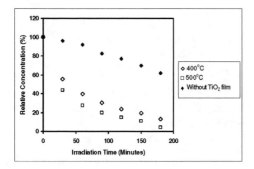

FIGURE 4. Concentration of the methylene blue as a function of irradiation time.

The degradation degree of the film fired at 500 °C is higher than that of the film fired at 400 °C. Small grains are theoretically sensitive to interact with the methylene blue due to large surface contact area. However, the grain size of both films fired at 400 °C and 500 °C is nearly the same and difficult to differentiate. So, the different capability of degrading methylene blue of the films fired at 400 °C and 500 °C may be caused by the different Ti oxide composition of the films. The microstructures of the thin film heated at 400 °C and 500 °C are shown in Figure 5. The film fired at 500°C is free from brookite and composed of anatase only. The film composed of anatase only has a better photocatalytic activity. Degradation mechanism of methylene blue is as follow.

When a TiO_2 thick film is irradiated by UV light, electron transfer from the valence band to the conduction band will happen and in the valence band holes are formed. The reaction can be written as [17]:

$$TiO_2 + h\upsilon \ (h\upsilon > E_g) \rightarrow TiO_2 \ (e^- + h^+) \qquad (1)$$

where, $h\upsilon$ = photon energy of UV, E_g = band gap energy, e^- = electron and h^+ = hole. For anatase, the E_g is 3.0 eV. Figure 6 shows the scheme of the TiO_2 anatase band gap.

Electron and hole will react with water and oxygen forming radicals through reactions below[17]:

$$h^+ + H_2O \rightarrow OH + H^+ \qquad (2)$$
$$e^- + H^+ + O_2 \rightarrow HO_2^{\bullet} \qquad (3)$$
$$HO_2^{\bullet} + e^- + H^+ \rightarrow H_2O_2 \qquad (4)$$
$$H_2O_2 + e^- \rightarrow OH^- + {}^{\bullet}OH \qquad (5)$$

The radicals will be strong oxydant which is capable of decomposing organic compound such as methylene blue. The general reaction can be written as:

$$Radicals + MB \rightarrow mineralized\ products \qquad (6)$$

where, MB is methylene blue. A visual appearance of a typical methylene blue solution after irradiated with UV light for 5 hours is depicted in Figure 7.

The photocatalitical capability of the TiO_2 thick film produced here is the consequence of the small (nano size) grains of the film which allow the excited electrons in conduction band to interact with the methylene blue solution easily and here the electron - hole recombination chance is relatively small. Theoretically the photocatalytical activity of the films can be enhanced by doping as schematically shown in Figure 6. Enhancing photocatalytical activity of the TiO_2 films by doping is one of our future works on photocatalyst.

(a)

(b)

FIGURE 5. The microstructures of the thick film ceramics heated at (a) 400 °C and (b) 500 °C.

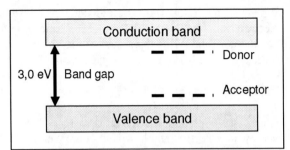

FIGURE 6. The TiO$_2$ anatase bandgap.

TABLE 1. Degradation degree of TiO$_2$ thick films.

No.	Firing Temperature (°C)	Degradation degree (%)
1	400	60
2	500	73
3	Without TiO$_2$	8

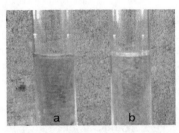

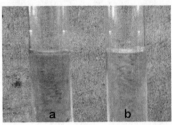

FIGURE 7. Typical solution of methylene blue after irradiated with UV for 5 hours. a) Without TiO$_2$ thick film, b) With TiO$_2$ thick film.

CONCLUSIONS

TiO$_2$ thick film photocatalyst can be well produced from a commercial TiO$_2$ powder using a method as a combination of mechanical and solution treatment.

The thick film fired at 400°C has TiO$_2$ anatase crystal structure containing brookite as a minor part while that fired at 500°C has only TiO$_2$ anatase crystal structure (free from brookite).

The produced TiO$_2$ thick films have a good photocatalytic capability for decomposing methylene blue.

ACKNOWLEDGMENTS

The authors thanks to Mr. M. Yamin from Materials Physics Group PTNBR-BATAN for his help during experiment.

REFERENCES

1. J.A. Byrne, B.R. Eggins, W. Byers, Norman M.D. Brown, *Appl. Catal. B: Environ.* **20,** L85 (1999).
2. H. Lachheb, E. Puzenat, A. Houas, M. Ksibi, E. Elaloui, C. Guillard, J.M. Herrmann, *Appl. Catal. B: Environ.* **39,** 75-90 (2002).
3. Y. Ohko, I. Ando, C. Niwa, T. Tatsuma, T. Yamamura, T. Nakashima, Y. Kubota, A. Fujishima, *Environ. Sci. Techn.* **35,** 2355 (2001).
4. H. Yamashita, K. Maekawa, H. Nakao, M. Anpo, *Appl. Surf. Sci.* **237,** 393-397 (2004).
5. M.A. Barakat, Y.T. Chen, C. P. Huang, *Appl. Catal. B: Environ.* **53,** 13 (2004).
6. C.H. Ao, S.C. Lee, J.Z. Yu, J.H. Xu, *Appl. Catal. B: Environ.* **54,** 41 (2004).
7. Y. Zu, L. Zhang, W. Yao, L. Cao, *Appl. Surface Sci.* **158,** 32 (2000).
8. A.G. Rincon, C. Pulgarin, *Appl. Catal. B: Environ.* **49,** 99 (2004).
9. P. Chrysicopoulou, D. Davazoglou, Chr. Tripalis, G. Kordas, *Thin Solid Films,* **323,** 188 (1998).
10. M. Terashima, N. Inoue, *Appl. Surface Sci.* **169-170,** 535 (2001).
11. X. Zhang, M. Zhou, *Appl. .A:Gener.* **282,** 285 (2006).
12. M. Mrowiz, W. Baicerski, A.J. Colussi, M.R. Hoffmann, *J. Phys. Chem. B,* **108,** 17269 (2004).
13. N. Negishi. K. Takeuchi, T. Ibusuki, *Appl. Surface Sci.* **121/122,** 417 (1997).
12. D.G. Syarif, A. Miyashita, T. Yamaki, T. Sumita, Y.S. Choi, Hisayoshi Itoh, *Appl. Surface Sci.* **193,** 287-292 (2001).
13. Rodrigues, M. Gomez, S. E. Lindquist, C. G. Granqvist, *Thin Solid Films* **360,** 250 (2000).
14. Y. Cheng, D. Dinonysious, *Appl. Catal.* **62,** 255 (2005).
15. Y.Paz, A.Heller, *J. Mater. Res.* **12,** 2759 (1997) .
16. H. Lin, H. Kozuka, T. Yoko, *Thin Solid Films* **315,** 111 (1998).
17. Q. Wu, D. Li, Y. Hou, L. Wu, X. Fu, X. Wang, *Mater. Chem. Phys.***102,** 53 (2007).

Structure and Morphology of Neodymium-doped Cerium Oxide Solid Solution Prepared by a Combined Simple Polymer Heating and D.C.-Magnetron Sputtering Method

I. Nurhasanah[1], M. Abdullah[2], and Khairurrijal[2]

[1]*Department of Physics, Faculty of Mathematics and Natural Sciences,*
Diponegoro University, Semarang, 50275, Indonesia
[2]*Physics of Electronics Materials Research Division, Faculty of Mathematics and Natural Sciences,*
Institut Teknologi Bandung, Bandung 40132, Indonesia

Abstract. Neodymium-doped Cerium Oxide (NDC) solid solution is attractive alternative material to replace yttria-stabillized zirconia (YSZ) used as an electrolyte for solid-oxide fuel cells (SOFCs). In this study Nd-CeO$_2$ nanoparticles with Nd of 3, 6 and 9 at./at.-% were synthesized by simple polymer heating. The NDC thin films were deposited on silicon substrates by using target made from the nanoparticles. Deposition process was carried out by D.C.-magnetron sputtering at temperature as low as 375°C. XRD pattern was used to confirm solid solubility and structural properties of the films. The results indicated that all samples are single phase solid solution with cubic fluorite structure. Their lattice parameters increase with increasing Nd content. It was also found that the mean grain size decrease with increasing Nd content. SEM analysis showed that NDC thin films have dense and uniform thickness. These results revealed that the nanoparticles and thin films of NDC solid solution are successfully prepared by a combined simple polymer heating and D.C.-Magnetron Sputtering method at low temperature.

Keywords: Cerium oxide, D.C. magnetron sputtering, nanoparticles, polymer heating.
PACS: 68.55.-a

INTRODUCTION

Solid oxide fuel cell (SOFC) has been studied extensively as the future generation of electrical energy device because it is advantages such as high energy conversion and little air pollutant. The main component of the SOFC is the electrolyte which determines the cell operating temperature. The established solid oxide fuel cell (SOFC) currently employs an yttria-stabilized zirconia (YSZ) as an electrolyte. This SOFC is operated at about 1000°C to reach the required level of ionic conductivity. The relatively high temperature would decrease the efficiency and stability of the cell. Thus the use of YSZ in the SOFC is not sufficient at reduced operating temperature. As a consequence of these limitations, the combination of two approaches has been adopted to decrease the cell operating temperature, i.e. reducing the electrolyte thickness and searching the materials with higher ionic conductivity at lower temperature [1-4].

Rare earth-doped CeO$_2$ (RDC) has attracted attention due to its higher ionic conductivity than YSZ

even though at low temperature. RDC powders are typically produced by solid-state reaction starting from individual component oxides, which requires repeated mechanical mixing and extensive heat treatment at high temperature (~ 1300°C for 10 h). The densified powders usually have large grain size, low surface area and poor chemical homogeneity. The large grain shows highly resistive grain boundaries for ionic conduction and blocks the ionic current in the electrolyte. Suzuki et al. reported that ionic conductivity of GDC nanocrystalline increased as the grain size decreased (≤ 100 nm). They explained that phenomenon by reduction of the activation energy due to the segregation of impurities in the grain boundary volume [5]. This may imply that typical grain size of thin film electrolytes to be used in low-temperature SOFC is to be below 100 nm. In addition to achieve high ionic conductivity, the film must have dense and uniform microstructure.

Several kinds of wet-chemical methods have been developed and successfully used for low temperature synthesis of RDC and produced powders with fine grain/particle size. In this paper, Neodymium-doped

CP989, Neutron and X-ray Scattering in Materials Science and Biology, International Conference on Neutron
and X-ray Scattering 2007, edited by A. Ikram, A. Purwanto, Sutiarso, A. Zulfia, S. Hendrana, and Z. Nurachman
© 2008 American Institute of Physics 978-0-7354-0508-0/08/$23.00

Cerium Oxide (NDC) thin films were deposited on silicon substrate by employing D.C.-magnetron sputtering method. The sputtered targets were made from nanoparticles synthesized by simple polymer heating method. The combination of these two methods is very promising to obtain thin film electrolyte for low temperature SOFC (LTSOFC) due to simplicity, fast, and readily scaled up for industrial applications. We studied the deposited thin films by X-ray diffraction and scanning electron microscopy to identify the phase purity, crystal structure, lattice parameter and sample morphology. The deposited thin films show cubic fluorite structure as well as dense and uniform morphology which indicate that the thin films are applicable as solid electrolyte for then LTSOFC.

EXPERIMENTAL METHOD

The NDC nanoparticles with neodymium (Nd) contents of 3, 6 and 9 at./at.-% were synthesized by heating a mixed solution of Ce-nitrate and Nd-nitrate at 600° for 30 hours. They were pressed to form disks with diameters of 25 mm and thicknesses of 5 mm. The disks were sintered at 1000°C for 4 hours in air atmosphere and then used as sputtering targets. The NDC thin films with various Nd contents were deposited by using DC-magnetron sputtering in argon gas at a temperature of 375 °C for 90 min. The total pressure of the chamber is 40 mTorr. The DC power was kept at a constant value of about 30 W on the target. The structure and morphology of the NDC nanoparticles, targets and thin films were investigated by using the Phillips PW 1710 X-ray diffractometer (XRD) and JEOL JSM-6360LA scanning electron microscope (SEM).

RESULTS AND DISCUSION

The samples were analyzed using x-ray diffraction to identify solubility Nd in CeO_2 and to confirm CeO_2 phase. The XRD pattern of NDC nanoparticles and targets with different dopant contents are shown in Figure 1. This figure shows that all NDC samples have several diffraction peaks. These diffraction peaks correspond to (111), (200), (220), (311) and (222) of pure CeO_2. In contrast to the NDC nanoparticles and targets, the NDC thin films show different XRD patterns as depicted in Figure 2. Peaks of (111) (200), and (311) are observed in the thin film containing 3 at./at.-% of Nd while the thin film with 6 at./at.-% of Nd has (200) and (311) diffraction peaks and for the film with 9 at./at.-% of Nd the diffraction peak is only at (200).

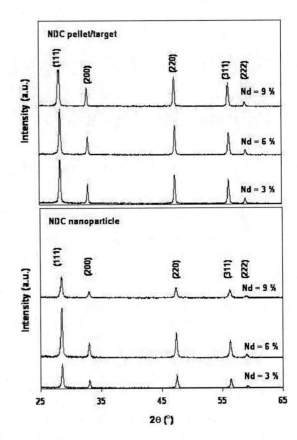

FIGURE 1. XRD Pattern of NDC nanoparticles and sputtering target

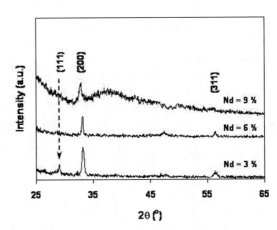

FIGURE 2. XRD pattern of NDC thin films

The lattice constants of the NDC nanoparticles and targets were calculated from the (111) peak, while the (200) peak was used to calculate lattice constant of the NDC thin films. As shown in Table 1, the calculated lattice constants confirm that all samples have cubic fluorite structure. These are consistent with the result reported previously that the NDC has the CeO_2 fluorite structure in the limiting value of Nd up to 50 at./at.-% [6,7]. It also reveals that the NDC structure is not appreciably altered and the Nd dopant is incorporated (substituted) into the CeO_2 lattice. It seems that the solubility of Nd_2O_3 in CeO_2 is high. Noting that conditions that are favorable for the formation of solid solutions are when the sizes of ions differ by less than about 15% [8], these results are acceptable when considering the atomic radii of Nd^{+3} (0.11 nm) and Ce^{+4} (0.097 nm) which differ around 12%.

TABLE 1. Lattice Constant and Crystallite Size

Nd (%)	NDC Nanoparticles		NDC Targets		NDC Thin Films	
	a (nm)	D (nm)	a (nm)	D (nm)	a (nm)	D (nm)
3	0.5407	29.40	0.5421	30.21	0.5399	29.44
6	0.5409	26.39	x0.5439	26.69	x0.5405	19.27
9	0.5414	23.49	0.5445	19.94	0.5447	17.68

We also find that the lattice constant increases slightly when the concentration of Nd dopant increases as shown in Table 1 can be explained as follows. The presence of Nd atom increases the effective radius of lattice basis, which results in a slight increase in the distance between atoms (lattice constant). The increase in the lattice constant implies the dissolution of Nd_2O_3 in CeO_2 has completed at relatively low temperature (< 800°C).

The crystallite size is found to decrease as the Nd fraction increases. The mechanism behind this result might be a slight prevention of crystallite growth due to the presence of Nd ions in the sample/precursor due to the difference in valence of two ions [8].

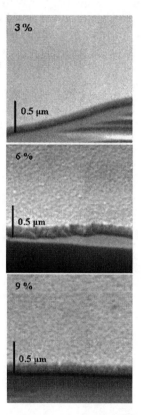

FIGURE 4. SEM image of NDC thin films

(a)   (b)

FIGURE 3. SEM images of NDC (a) nanoparticles and (b) targets

Using the Scherrer equation, the crystallite size of NDC samples is estimated and tabulated in Table 1.

The morphologies of NDC nanoparticles, targets and thin films were investigated by using SEM. Figure 3 shows the morphologies of NDC nanoparticles and targets NDC. The microstructures among NDC nanoparticles could not be distingusihed, but the nanoparticles are mainly composed of nearly spherical shape with uniform size (Figure 3. (a)). As shown in Figure 3.(b), the 3 at./at.-% NDC target is more porous with pores are continuous and open. But, in the 6 at./at.-% and 9 at./at.-% NDC targets there are closed, isolated pores and the grain growth less pronounced and the targets are relatively dense.

The micrographs of NDC thin films are shown in Figure 4. All the thin films are composed of circular grains distributed homogeneously. From the cross-sectional view, it is seems that the films have good adhesion with the substrate and dense without pores or voids among grains. It can be seen that the films have very uniform thickness. The thicknesses of the NDC thin films are 91 nm, 136 nm and 182 nm for 3, 6 and 9 at./at.-% Nd contents, respectively.

CONCLUSIONS

The structure and morphology of NDC thin films prepared by a combined simple polymer heating and D.C.-Magnetron Sputtering methods has been investigated. The dissolution of Nd_2O_3 in CeO_2 of samples including nanoparticle, target and thin film is examined based on diffraction peaks and lattice parameter. It is found that the dissolution of Nd_2O_3 in CeO_2 is completed at low temperature and the samples have fluorite structure. The Nd doping prevents the grain growth and expands lattice constant. The NDC thin films are show uniform and dense microstructure.

ACKNOWLEGMENTS

Research Grants from Hibah Pasca Sarjana IV from Directorate of Higher Education, Ministry of National Education Republic of Indonesia are gratefully acknowledged.

REFERENCES

1. H. Inaba, and H. Tagawa, *Solid State Ionics* **83**, 1(1996)
2. S. C. Singhal, *MRS Bulletin*, March, 16-21 (2000)
3. B.C.H. Steele and A. Heinzel, *Nature* **414**, 345-352 (2001)
4. B. Zhu, X.T. Yang, J. Xu, Z.G. Zhu, S.J. Ji, M.T. Sun, and J.C. Sun, *J. Power Sources* **118**, 47-53 (2003)
5. T. Suzuki, I. Kosacki, and H. U. Anderson, *Solid State Ionics* **151**, 111-121 (2002)
6. H. Nitani, T. Nakagawa, M. Yamanouchi, T. Osuki, M. Yuya, and T. A. Yamamoto, *Mater. Lett.* **58**, 2076 (2004)
7. S.V. Chavan, M.D. Mathews, and A.K. Tyagi, *Mater. Res. Bull.* **40**, 1558 (2005)
8. W.D. Kingery, H.K. Bowen, and D.R. Uhlmann, *Intoduction to Ceramics,* 2nd Ed., Wiley, 1976.

The Influence of a Cu-doped on the Structure of La$_{0.5}$Ca$_{0.5}$Mn$_{1-x}$Cu$_x$O$_3$ (0 < x < 0.2)

Y. E. Gunanto[1], B. Kurniawan[1], A. Purwanto[2], W. A. Ardhi[2]

[1]*Material science study program, Department of Physics, University of Indonesia, Depok, 16424, Indonesia*
[2]*Centre of Technology for Nuclear Industrial Material, National Nuclear Energy Agency (BATAN),
Serpong, Tangerang, 15341, Indonesia*

Abstract. We investigated a Cu–doped on a Mn-site at a compound of La$_{0.5}$Ca$_{0.5}$MnO$_3$ for an understanding of the influence on a structure and temperature transition of a magnetic. Characterization of the structure La$_{0.5}$Ca$_{0.5}$MnO$_3$ compound has been performed. The result of the X-ray diffraction XRD measurement shows that material at room temperature is a single phase with the orthorhombic structure and a space group *Pnma* with lattice parameters a = 5.41792 Å, b = 7.63906 Å, c = 5.42265 Å and $\alpha = \beta = \gamma = 90°$. From the energy dispersive X-ray spectroscopy EDAX, we get the composition is La$_{0.493}$Ca$_{0.528}$MnO$_{2.57}$. The compound of a La$_{0.5}$Ca$_{0.5}$MnO$_3$ with a Cu-doped (La$_{0.5}$Ca$_{0.5}$Mn$_{1-x}$Cu$_x$O3, with 0≤x≤0.2) is in characterization process. From the previous experiment, a Cu was doped on a Mn-site at the La$_{0.1}$Ca$_{0.9}$MnO$_3$ compound (La$_{0.1}$Ca$_{0.9}$Mn$_{0.9}$Cu$_{0.1}$O$_3$), we obtained that it also crystalilizes in the orthorhombic *Pnma*, but the lattice parameters is decreasing if it is compared without a Cu doping. This might be due to a Jahn-Teller distortion, which change the average of an (Mn,Cu) – O bond length and an (Mn,Cu) – O – (Mn,Cu) bond angle.

Keyword: XRD, Cu-doped, La$_{0.5}$Ca$_{0.5}$MnO$_3$, EDAX, La$_{0.493}$Ca$_{0.528}$MnO$_{2.57}$.
PACS: 61.12.Ld, 61.10.Nz

INTRODUCTION

Research on AMnO$_3$ (A=La, Ca) system has been done by a number of people and there is a wide research interest in La$_{1-x}$A$_x$MnO$_3$ (A = alkaline-earth metal) manganites with a distorted perovskite structure. This is due to the fact that their properties are attractive. The properties of interest are the magnetic and transport properties such as: CMR, Metal – Insulator Properties (MIT) and the other [1-8]. La$_{0.5}$Ca$_{0.5}$MnO$_3$ represents the boundary between competing ferromagnetic (FM) and charge-ordered antiferromagnetic (CO-AFM) ground state, a favorable scenario for a phase separation phenomena [9]. The compound of La$_{0.5}$Ca$_{0.5}$MnO$_3$ being a paramagnetic at room temperature, changes on cooling to a mainly FM metallic phase at $T_C \approx 220$ K, and subsequently to charge-ordered antiferromagnetic (CO-AFM) phase at $T_{CO} \approx 150$ K (180 K upon warming) [10]. Recently, a substitution of Mn with another atom has also been done which dramatically gives some effect on the scale of magnetic and transport of Manganese perovskites [11-24].

In this measurement, we report that the change of lattice parameters can be caused by a Cu-doped at Mn position. Although the change of lattice parameters occurs, but the structure is the same for all compound.

EXPERIMENTAL METHOD

Samples La$_{0.5}$Ca$_{0.5}$Mn$_{1-x}$Cu$_x$O$_3$ ($0 \leq x \leq 0.2$) were prepared by a solid-state reaction method. A stoichiometric mixture of La$_2$O$_3$, CaCO$_3$, MnO$_2$ and CuO with purity more than 99%. That compound were mixed thoroughly in a ball mill 10 hours, then further heated at 1350°C for 6 hour. After that the sample is processing again with the ball mill 5 hour and heated at 1100°C for 24 hour to obtain the single-phase compound. An X-ray diffraction of the powders was carried out at a room temperature (RT) using Philips PW 1710 ABD 3520, and refining data with the Rietan program.

RESULTS AND DISCUSSION

Figure 1 shows an X-ray diffraction of La$_{0.5}$Ca$_{0.5}$Mn$_{1-x}$Cu$_x$O$_3$ ($0 \leq x \leq 0.2$), all samples are single phase with the space group *Pnma* orthorhombic and the lattice parameters $b \approx \sqrt{2}\,a \approx \sqrt{2}\,c$.

CP989, Neutron and X-ray Scattering in Materials Science and Biology, International Conference on Neutron and X-ray Scattering 2007, edited by A. Ikram, A. Purwanto, Sutiarso, A. Zulfia, S. Hendrana, and Z. Nurachman
© 2008 American Institute of Physics 978-0-7354-0508-0/08/$23.00

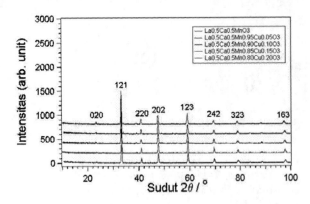

FIGURE 1. XRD patterns of $La_{0.5}Ca_{0.5}Mn_{1-x}Cu_xO_3$ ($0 \le x \le 0.2$)

The Cu-doped increases (x = 0.05 – 0.15), lattice volume decreases, except for x = 0.20 the volume lattice increase. This result is the same with the compound $La_{0.1}Ca_{0.9}Mn_{1-x}Cu_xO_3$ (x = 0.1), where the volume of lattice is smaller than x = 0. The Cu-doped could be increasing the transition temperature of magnetic from 12 K (x = 0) to 80 K (x = 0.1)[25].

Figure 2 shows a neutron scattering pattern for a compound of $La_{0.1}Ca_{0.9}Mn_{1-x}Cu_xO_3$ (x = 0 and 0.1) with several temperature. If the temperature decreases, the intensity increases. The new peak arise at a temperature of T = 80 K with the Bragg angle θ = 12° (2θ = 24°) and more obvious at T = 12 K. The magnetic characteristic rise at T = 80 K.

TABLE 1. Lattice parameters for the compound of $La_{0.5}Ca_{0.5}Mn_{1-x}Cu_xO_3$ ($0 \le x \le 0.20$)

Lattice parameter	x				
	0	**0.05**	**0.10**	**0.15**	**0.20**
a(Å)	5.4179	5.4237	5.4235	5.4192	5.4390
b(Å)	7.6391	7.6878	7.6455	7.6644	7.6664
c(Å)	5.4227	5.4106	5.4386	5.4214	5.4513
$\alpha = \beta = \gamma$	90°	90°	90°	90°	90°
Vol	224.43	225.60	225.51	225.17	227.30

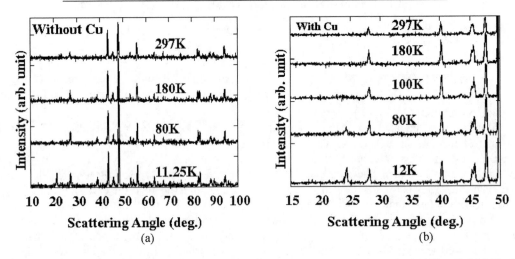

FIGURE 2. Neutron scattering pattern of (a) $La_{0.1}Ca_{0.9}MnO_3$ at temperatures 11.25, 80, 180, and 298 K (b) La0.1 $Ca_{0.9}Mn_{0.9}Cu_{0.1}O_3$ at temperatures T = 12, 80, 100, 180, and 297 K.

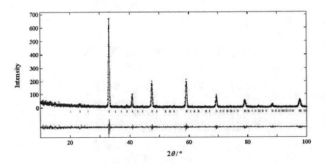

FIGURE 3. XRD scattering pattern of $La_{0.5}Ca_{0.5}MnO_3$

From EDAX analysis at room temperature, the polycrystal was to have a form of $La_{0.493}Ca_{0.528}MnO_{2.57}$ not $La_{0.5}Ca_{0.5}MnO_3$, this may be caused by the furnace temperature so high that more oxygen are burning. Fig.3 shows the pattern of XRD $La_{0.5}Ca_{0.5}MnO_3$ is a single phase with a refinement of a Rietan program.

The result of a pattern XRD after refining with Rietan program for a compound of $La_{0.5}Ca_{0.5}Mn_{1-x}Cu_xO_3$ ($0.05 \le x \le 0.2$) is shown at figures 4 - 7. All samples are single phase too.

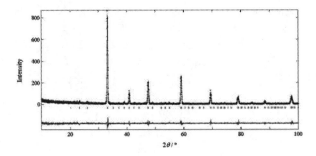

FIGURE 4. XRD scattering pattern of La$_{0.5}$Ca$_{0.5}$Mn$_{0.95}$Cu$_{0.05}$O$_3$

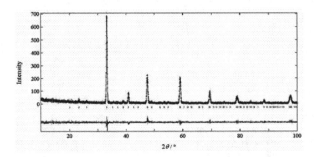

FIGURE 5. XRD scattering pattern of La$_{0.5}$Ca$_{0.5}$Mn$_{0.90}$Cu$_{0.10}$O$_3$

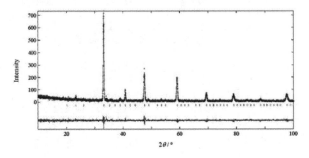

FIGURE 6. XRD scattering pattern of La$_{0.5}$Ca$_{0.5}$Mn$_{0.85}$Cu$_{0.15}$O$_3$

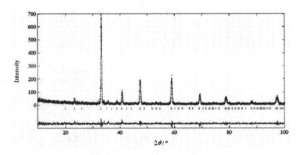

FIGURE 7. XRD scattering pattern of La$_{0.5}$Ca$_{0.5}$Mn$_{0.80}$Cu$_{0.20}$O$_3$

Although all samples have the structure of orthorombic, the space group of *Pnma*, but the bond length and bond angle is differs (Fig.8. (a) – (e)). This is caused by Jahn-Teller distortion.

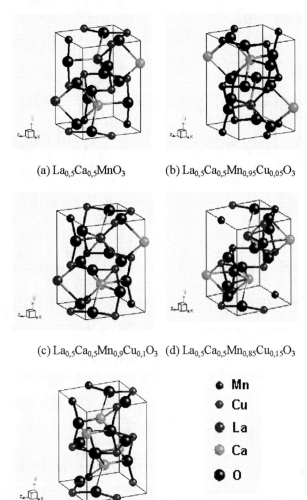

(a) La$_{0.5}$Ca$_{0.5}$MnO$_3$ (b) La$_{0.5}$Ca$_{0.5}$Mn$_{0.95}$Cu$_{0.05}$O$_3$

(c) La$_{0.5}$Ca$_{0.5}$Mn$_{0.9}$Cu$_{0.1}$O$_3$ (d) La$_{0.5}$Ca$_{0.5}$Mn$_{0.85}$Cu$_{0.15}$O$_3$

• Mn
• Cu
• La
• Ca
• O

(e) La$_{0.5}$Ca$_{0.5}$Mn$_{0.80}$Cu$_{0.20}$O$_3$

FIGURE 8. Crystal system of La$_{0.5}$Ca$_{0.5}$Mn$_{1-x}$Cu$_x$O$_3$ ($0 \leq x \leq 0.20$)

The effect of a Cu-doped for x = 0.05 and 0.20 makes distortion larger than the undoped Cu (x =0), but for x = 0.10 smaller than the undoped Cu. M. Pissas *et. al.* found for compound La$_{1-x}$Ca$_x$MnO$_3$ (x = 0.80, 0.85), the distortion parameter are 1.61 x 10^{-5} and $<d>$ = 1.917 for x = 0.80 and the distortion parameter 2.6 x 10^{-5} and $<d>$ = 1.913 for x = 0.85 [26].

TABLE 2. Selected bond length (Å) and bond angles (deg) for $La_{0.5}Ca_{0.5}Mn_{1-x}Cu_xO_3$ ($0 \leq x \leq 0.20$) samples at T = 300 K. The average bond length $<d> = (1/6)\Sigma_{i=1,6}d_i$, the distortion parameter of the MnO_6 octahedra $S_d^2 = (1/6)\Sigma_{i=1,6}[(d_i - <d>)/<d>]^2$.

	x				
	0	**0.05**	**0.10**	**0.15**	**0.20**
Bond (A) Mn – O(2)	1.8781	1.7657	1.9149	1.8848	1.8357
Bond (A) Mn – O(1)	1.9778	1.9733	1.9552	1.9649	1.9869
Bond (A) Mn – O(2)	1.9945	2.0909	1.9617	2.0086	2.0600
Angle (°) O(2)-Mn-O(1)	89.3	94.0	94.2	89.1	91.8
Angle (°) O(2)-Mn-O(2)	89.3	89.1	90.7	90.4	91.3
<d>	1.9501	1.9433	1.9439	1.9528	1.9609
S_d^2	6.9442×10^{-4}	4.7865×10^{-3}	1.1340×10^{-4}	6.8917×10^{-4}	2.2689×10^{-3}

CONCLUSIONS

The present work based on analysis of an X-ray diffraction data shows that all compound $La_{0.5}Ca_{0.5}Mn_{1-x}Cu_xO_3$ ($0 \leq x \leq 0.20$) samples have an orthorhombic structure with the space group *Pnma*. The effect of a Cu-doped makes the volume lattice larger than the undoped Cu. The Cu-doped increases (x = 0.05 – 0.15), lattice volume decreases, except for x = 0.20, the volume lattice increases. The distortion parameters are larger (x = 0.05, 0.20), but smaller (x = 0.10) than the undoped Cu.

REFERENCES

1. M.S. Kim, J.B. Yang, Q. Cai, X. D. Zhou, W. J. James, W.B. Yelon, P.E. Parris, D. Buddhikot, S.K. Malik. 2005. arXiv:cond-mat/0502225.
2. L. Sudheendra and C.N.R. Rao. *J. Phys. Condens. Matter* **15**, 3029-3040 (2003)
3. M. Roy, J.F. Mitchell, P. Schiffer. 2000. arXiv:cond-mat/0001064.
4. L. Sudheendra, A.R. Raju and C.N.R. Rao. 2002. arXiv:cond-mat/0205370.
5. J. J. Hamilton, E. L. Keatley, H. L. Ju, A. K. Raychaudhuri, V. N. Smolyaninova, and R. L. Greene, *Phys. Rev.* **B 54**, 14926 (1996)
6. A. Ignotov, S. Khalid, R. Sujoy and N. Ali., *J. Synchrotron Rad.* **8**, 898-900 (2001)
7. Von Helmolt et al., *J. Appl. Phys.* **76**, 6925-6928 (1994).
8. Jin.S.S et al, *Science* **264**, 413-415 (1994)
9. A. Moreo, S. Yunoki, and E. Dagotto, *Science* **283**, 2034 (1999)
10. P. Schiffer, et al, *Phys. Rev. Lett*. **75**, 3336 (1995)
11. C. Martin, A. Maignan, and B. Raveau, *J. Mater. Chem.* **6**, 1245 (1996)
12. J. Blasco, J. Garcia, J.M. De Teresa, M.R. Ibarra, J. Perez, P.A. Algarabel, and C. Marquina, *Phys. Rev.* **B 55**, 8905 (1997)
13. K.H. Ahn, X.W. Wu, K. Liu, and C.L. Chien, *J. Appl. Phys.* **81**, 5505 (1997)
14. S.L. Yuan, Y. Jiang, G. Li, J.Q. Li, Y.P. Yang, MX.Y. Zhang, P. Tang, Z. Huang, *Phys. Rev.* **B 61**, 3211 (2004)
15. J. Yang, W.H. Song, Y.Q. Ma, R.L.Z hang, B.C. Zhao, Z.G. Sheng, G.H. Zheng, J.M. Dai, and Y.P. Sun. 2004. arXiv:cond-mat/0408303.
16. C. Zhang, J.S. Kim, B.H. Kim, and Y. W. Park, *Phys. Rev.* **B 70**, 024505 (2004).
17. E. Granado, Q. Huang, J.W. Lynn, J. Gopalakrishnan, and K. Ramesha. 2004. arXiv:cond-mat/0412621.
18. M. Pissas and G. Kallias, *Phys. Rev.* **B 68**, 134414 (2003)
19. X.G. Li, R.K. Zheng, G. Li, H.D. Zhou, R. X. Huang, J.Q. Xie, and Z.D. Wang. 2002. arXiv:cond-mat/0212012.
20. R.K. Zheng, R. X. Huang, A.N. Tang, G. Li, X.G. Li, J.N. Wei, J.P. Shui, and Z. Yao. 2002 arXiv:cond-mat/02012013.
21. K.Y. Wang et al., *Phys. Stat. Sol.* **184**, 515-522 (2001)
22. S.L. Yuan et al., *J. Phys. Condens. Matter* **12** (2000)
23. F. Rivadulla, M.A. Lopez-Quintela, L.E. Hueso, P. Sande, J. Rivas, and R.D.S anchez. *Phys. Rev.* **B 62**, (2000)
24. J. Sacanell, F.P arisi, P. Levy, and L. Ghivelder. 2004. arXiv:cond-mat/0412499.
25. Y.E. Gunanto, A. Purwanto, B. Kurniawan, A. Fajar, and H. Mugirahardjo, 2006, Proceedings of ICMNS 2006, 1075 – 1077.
26. M. Pissas, G. Kallias, M. Hofmann, and D.M. Tobbens. *Phys. Rev.* **B**, 064413 (2002)

The Synthesis and Characterization of Titania Nanotubes Formed at Various Anodisation Time

S. Sreekantan, L. M. Hung, Z. Lockman, Z. A. Ahmad and A. F. Mohd Noor

School of Materials & Mineral Resource Engineering, Universiti Sains Malaysia
14300, Nibong Tebal, Seberang Perai Selatan, Malaysia

Abstract. One-dimensional (1D) nanostructured titania such as nanowires, nanorod and nanotubes have attracted considerable attention recently due to their unique physical properties and their potential application in water photoelectrolysis, photocatalysis, gas sensing, and photovoltaic. In this work, a simple anodisation method has been developed to fabricate titania nanotubes in 1M Na_2SO_4 containing various amount of NH_4F. The dimension of the titania nanotube produced depend on the electrochemical process parameter: composition of the electrolyte, pH of the electrolyte and time of anodisation. As for this paper, the effect of fluoride content and anodisation time on the formation of titania nanotube was discussed in detail. The nanotubes formed were analyzed by field emission scanning electron microscope (FESEM) and X-Ray diffraction (XRD). The minimum fluoride content that is required to form nice well ordered nanotube for sample anodized for 30 minutes is 0.3g whereas for 120 minutes is 0.1g.

Keywords: titania nanotube, fluoride, anodization
PACS: 61.46.Fg

INTRODUCTION

Titania is gaining considerable interest due to unique and excellent properties due in optics, electronics, photochemistry and biology [1]. Recently, studies have indicated that for these applications, well-arrayed titania nanotubes [2-6] are of interest because nanotubes have high surface-to-volume ratios which increases surface area for reaction to produce higher efficiency devices. Furthermore, nanotubes contain free spaces in their interior that can be filled with active materials enabling them to be engineered to produce advanced materials of multi functionality. Anodization is a simple cost effective method which is commonly used to form nanotubular structure. The morphologies and the dimension of the materials produced via anodization are associated to the electrochemical process parameter. Therefore, this paper investigates the formation of titania nanotubes as a function of fluoride content and the anodisation time.

EXPERIMENTAL METHOD

High purity titanium foils (99.6% purity) of thickness 0.2 mm that was used in this study were purchased from STREM Chemicals. Prior to anodization, Ti foils were degreased by sonicating in acetone for 15 minutes followed by a deionised water rinse and then dried in a nitrogen stream. The anodisation were performed in a two-electrode configuration bath with titanium foil as the anode and platinum electrode as the counter electrode. The distance between cathodic and anodic electrode was 30 mm. The bath consisted of 1 M Na_2SO_4 with different NH_4F concentration (0.05g, 0.1g, 0.3g, 0.7g and 0.9g). The pH of the solution was adjusted to 3 with H_2SO_4 addition. The anodisation was done potensiostatically with constant potential of 20V applied to the foil. During the experiment, the fluorinated electrolyte was stirred using a magnetic stirrer. With magnetic agitation of the electrolyte, it reduces the thickness of the double layer at the metal/electrolyte interface, and ensures uniform local current density and temperature over the Ti electrode surface [1]. The bath was kept at room temperature. The as-anodised titanium foils were cleaned in distilled water and dried in nitrogen stream. The structural and morphological conditions of the titania nanotubes were characterized using a Field Emission Scanning Electron Microscope (FESEM SUPRA 35VP ZEISS) operating at working distances down to 1 mm and extended accelerating voltage range from 30 kV down to 100 V. In order to obtain the thickness of the nanotubes layer, cross-sectional measurements were carried out on mechanically bent samples. Therefore the actual length

CP989, *Neutron and X-ray Scattering in Materials Science and Biology, International Conference on Neutron and X-ray Scattering 2007*, edited by A. Ikram, A. Purwanto, Sutiarso, A. Zulfia, S. Hendrana, and Z. Nurachman

of tubes will be divided by cos 45°. The insert in the FESEM images represent the cross-sectional of the titania nanotubes. Phase identification was carried out by X-ray diffractometer (Philip PW 1729).

RESULTS AND DISCUSSION

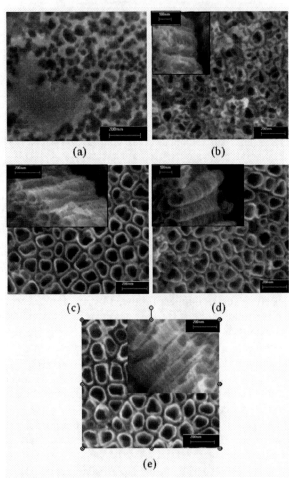

FIGURE 1. FESEM images of nanotubes produced with various fluoride content for sample anodized for 30 minutes (a) 0.05g, (b) 0.1g, (c) 0.3g (d) 0.7g and (e) 0.9g

The first set of experiment was done to monitor the formation of titania nanotube with increasing fluoride for sample anodized for 30 minutes. The experiment was carried out at room temperature with anodization voltage of 20V. The formation of the titania nanotubes was monitored by taking FESEM (Figure 1) images at various fluoride content. Figure 1a shows that titanium anodized in 0.05g F, which results in a non-uniform structure, far from being organized with some small patches of pores between large non-porous areas. Apparently, the morphology is not uniform on the entire surface in the same manner, there are preferentially etched. Figure 1b shows condition which

occurs on titanium anodized in 0.1g fluoride. In this case, the tubes seem to be formed (insert in Figure 1b) but the degree of homogeneity was comparable low. As for 0.3g (Figure 1c), 0.7g (Figure 1d) and 0.9g (Figure 1e), very well defined nanotubes which are homogeneous and clean were formed. In terms of the inner diameter, it was found to increase with fluoride content (Table 1) and more uniform tubes were achieved with electrolyte containing 0.3g. On the other hand, the length of the tube was found to be decreasing with fluoride content. As we know, the formation of titania nanotubes depends on the oxide growth rate and the dissolution rate [3,4]. With increasing fluoride content, the dissolution rate increase and resulted in big pores and short tubes.

TABLE 1. Pore diameter and length of titania nanotubes with various F$^-$ content for sample anodized for 30 minutes

Mass (g)	Diameter (nm)	Length (nm)
0.05	~20-40	-
0.1	~30-70	~540
0.3	~90-100	~690
0.7	~90-120	~540
0.9	~110-130	~500

In the next set of experiments, the time was prolonged to 120 minutes (Table 2) and the nanotubes were successfully formed with electrolyte containing 0.05g fluoride (Figure2a). The entire surface was etched but the diameters of the tube were not uniform; it varies from 50nm to 100nm with the length of 650nm. Figure 2b shows the tube formed in electrolyte consisting 0.1g fluoride for 120 minutes. It results in far more homogeneous tube structure as compared to sample anodized for 30 minutes. The diameter was approximately 80nm to 100nm with length of 760 nm. In the case of electrolyte containing 0.3g fluoride (Figure 2c), bigger tubes were formed but the length become smaller as compared to 30 minutes. As for 0.7g fluoride (Figure 2d), the images do not show any significant difference with 30 minutes. In contrast, sample anodized with 0.9g fluoride (Figure 2e), shows foremost difference because almost the entire area of the pores was covered with precipitate and formed non-uniform tubes. We believe the precipitates hinder the flux of ions to be continuous and uniformly distributed. Therefore, the nanotubes produced are nonuniform and not well organized. In terms of wall thickness, sample anodized with electrolyte containing 0.05g fluoride was large (~30nm) as compared to other samples (0.1g, 0.3g and 0.7g) which were approximately, 15nm.

TABLE 2. Pore diameter and length of titania nanotubes with various F- content for sample anodized for 120 minutes

Mass (g)	Diameter (nm)	Length (nm)
0.05	~50-100	~650
0.1	~70-100	~760
0.3	~110-120	~700
0.7	~90-120	-
0.9	~100	-

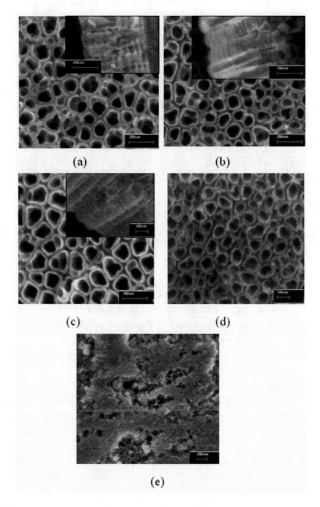

(a)　　　　　(b)

(c)　　　　　(d)

(e)

FIGURE 2. FESEM images of nanotubes produced with various fluoride content for sample anodized for120 minutes (a) 0.05g, (b) 0.1g, (c) 0.3g (d) 0.7g and (e) 0.9g

The XRD measurements of all the as-anodized samples reveal that the self-organized titania nanotubes have amorphous structure as only Ti-peaks were shown. Ti-peaks presented because the information from the substrate are revealed. A representative XRD spectrum is shown in Figure 3. Shortly, the content of fluoride content in the electrolyte has insignificant effect on the crystallography phase of the nanotube formed.

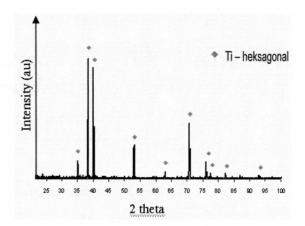

FIGURE 3. XRD pattern of as anodized sample

CONCLUSIONS

Longer anodization time (120 minutes) enhance the formation of tube in the electrolyte containing less fluoride, $\leq$ 0.7g whereas for shorter anodization time (30 minutes), more fluoride content ($\geq$ 0.3g) were consumed in order to form to form well ordered and homogeneous nanotubes. Diameter, length and wall thickness can be controlled by varying the fluoride content and anodization time. The as anodized samples are amorphous.

ACKNOWLEDGMENT

The author would like to thank the Ministry of Higher Education because this work was sponsored through grant FRGS 6070020.

REFERENCES

1. G. K. Mor, O. K. Varghese, M. Paulose, K. Shankar, C. A. Grimes, *Solar Energy Materials & Solar Cells* **90**, 2011–2075 (2006)
2. J. H. Park, S. Kim, A. J. Bard, *Nano Letters* **6**(1), 24-28 (2006)
3. H. Tsuchiya, J. M. Macak, L. Taveira, E. Balaur, A. Ghicov, K. Sirotna and P. Schmuki, *Electrochem. Commun.* **7** (6), 576-580 (2005)
4. J. M. Macak, K. Sirotna and P. Schmuki, *Electrochimica Acta* **50** (18), 10 June 2005, 3679-3684
5. Q. Cai, M. Paulose, O. K. Varghese and C. A. Grimes, *J. Mater. Res.* **20** (1), Jan 2005
6. J. M. Macak, H. Tsuchiya, and P. Schmuki, *Angew. Chem. Intl. Ed.* **44**, 2100 –2102 (2005)

The Influence of Base Concentration on the Surface Particle of Lithium Aluminosilicate System

I. M. Nazri, M. A. Sri Asliza and R. Othman

School of Material and Mineral Resources Engineering, Engineering Campus
Universiti Sains Malaysia, 14300 Nibong Tebal Pulau Pinang,Malaysia

Abstract. The study of base concentration effect toward surface particles of lithium aluminosilicate glass ceramic system has been done by using NaOH solution. The parent glass with composition of 60% SiO_2, 31% Li_2O, 6% Al_2O_3 and 3% TiO_2 in wt% was prepared by melting process at 1250 °C prior to quenching rapidly to room temperature. Sintering and crystallization process on this parent glass were carefully examined by Differential thermal analysis (DTA) and X-Ray Diffraction (XRD). Based on these analyses, the selected crystal has been chosen as a precursor material. There are two controlling parameter involved in this study i.e. NaOH concentration and leaching period. The morphology of the glass ceramic particle was observed by Field Emission Scanning Electron Microscope (FESEM). The result shows that by increasing the basic concentration as well as increasing the soaking leaching period, the tendency of glass ceramic particle to leach out is relatively high.

Keywords: Lithium aluminosilicate, glass ceramic, leaching, porous
PACS: 81.05.Kf

INTRODUCTION

Lithium aluminosilicate (LAS) glass ceramic has been extensively studied and commercialized since 1957 [1]. Some of the commercial examples are enamel in steel (Nucerite®, Pfaudler) and hermetic seal applications in high-voltage connectors, high pressure actuators (S glass, Sandia National laboratory), Nuclear reactor, and turbine engine parts [2]. LAS glass ceramic have gained considerable commercial attention because of their very low thermal expansion, transparency, high chemical durability, and strength[3]. However, their melting temperature is significantly ($\geq$ 1600 °C). In glass ceramic, crystal nucleation is an important theme as well as in glass technology. The crystal nucleation in glass technology is undesirable for producing glass without devitrification but it is an important process for preparing of glass ceramic via controlled crystallization. In studying the nucleation and crystallization behavior of LAS glass ceramic, two techniques are commonly used. First is time-consuming technique, but it is possible to evaluate the nucleation and crystallization effect quantitatively by using microscope. This technique is to observe the crystal growth from nuclei by heat treatment because the dimension of nuclei is too small. The second technique is using differential thermal analysis (DTA).

Since the crystallization peak temperature in DTA curve depends on nucleation treatment of glass, the shift of peak temperature plays a role of indicator showing the influence of nucleation state indirectly[4-6]. In LAS glass ceramic, the main crystalline phases are is β-quartz solid solution ($Li_2OAl_2O_3$-$2SiO_2$), β-spodumene (Li_2O-Al_2O_3-$4SiO_2$) and lithium metasilicate (Li_2O-SiO_2). Most of these crystalline can easily extracted by means of mineral acid or base to form a porous medium. The different in porous glass ceramic with controlled pore sizes and porosities can be prepared by the successive acidic and basic leaching parameter [7]. This work describes a method to obtain porous lithium aluminosilicate glass ceramic by controlling the reagent concentration and period of leaching time.

EXPERIMENTAL METHOD

In order to prepare the parent lithium aluminosilicate glass, the following weight compositions, 60% SiO_2, 31% Li_2O, 6% Al_2O_3 and 3% TiO_2 were used. For the preparation of the glasses, SiO_2, Li_2CO_3, Al_2O_3 and TiO_2 were used as the basic reagents. The reactants were weighted, homogeneously mixed, and melted in a Pt crucible at 1250 °C for 2 hours prior to quenching rapidly to room

CP989, Neutron and X-ray Scattering in Materials Science and Biology, International Conference on Neutron and X-ray Scattering 2007, edited by A. Ikram, A. Purwanto, Sutiarso, A. Zulfia, S. Hendrana, and Z. Nurachman

temperature. The glass was pulverized in an agate planetary mill and shifted. DTA analysis of the glass particles was performed to obtain the values of the glass transition temperature (T_g) and the crystallization temperature (T_c) in order to know the most adequate temperature range for the different thermal treatment. Both T_g and T_c were measured in a Differential Thermal Analyzer at a heating rate of 10 °C/min. The analysis was conducted up to 1000 °C. Glass powder (< 75 μm) was pressed to cylindrical bar of about 16 mm diameter obtained from humidified 3.5% water. The samples were sintered applying different thermal treatment slightly above the T_g of the glass (616 °C) in order to hinder its devitrification. The specimens were heated at 620 °C for an hour before continuously heated at 750 and 1000 °C respectively.

The devitrification or formation of new crystalline phases was evaluated by X-Ray Diffraction (XRD). Leaching process using 0.5M, 1.5M, and 3.0M NaOH for 3, 7 and 14 days soaking period were studied. The morphology of these samples was carefully examined by Field Emission Scanning Electron Microscope (FESEM).

RESULTS AND DISCUSSION

DTA curve of the glass at a heating rate 10₀C/min was shown in Figure 1. The typical thermogram of a glass ceramic point out the glass transition temperature (T_g) inflexion point at 616 °C and the exothermic crystallization peak (T_c) at 965 °C. The result shows that the nucleation rate (or nucleation density) of present glass is a maximum near 965 °C. Based on this data, the nucleation and crystallization studies were carried out on glass samples. The result of initial studies on heating rates and maximum growth temperature pointed out that the temperature of nucleation around 620 °C and the crystallization temperature around 750 °C to 1000 °C along with lower heating rates 5 °C/min were conducive for the development of desired phases.

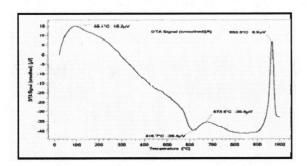

FIGURE 1. DTA analysis of powdered parent glass

In order to evaluate qualitatively the level of devitrification or formation of crystalline phases in this material, XRD analysis was performed. The XRD results on this glass ceramic system suggest consecutive transformation of the parent glass (amorphous) into crystalline phases. The typical XRD spectrum peaks for both 750 °C and 1000 °C were identify as lithium metasilicate phase. The crystalline phase was dispersed in area of micron size and precipitated in different regions of the glass. The evaluation of this lithium metasilicate phase was shown in Figure 2.

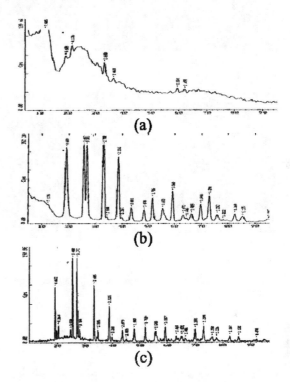

FIGURE 2. XRD spectrum for crystallized glasses treated with different temperatures a) 400 °C; b) 620/750 °C; c) 620/1000 °C

Regarding to the microporosity of the samples, Figure 3 shows the FESEM analysis on lithium metasilicate phase. These samples were leaching out with 0.5M, 1.5M and 3.0M of NaOH for 3, 7 and 14 days respectively. In this work, porous glass ceramic based on lithium aluminosilicate system has been developed. The figure shows that interrelated porous occurred on the leaching surface with range of 2-10 μm for sample 3.0M at 14 days leaching period. However, image analysis results indicated that majority of lithium metasilicate phases is still dominantly presented for sample 0.5M and 1.5M after 7 and 14 days. On the other side, the microstructure of lithium metasilicate was totally been effected with 3.0M NaOH for all these period. Here, it could be observed from extending the concentration of NaOH solution and leaching period, the percentage in getting highest porosity is possible. Figure 3a and Figure 3c proved that totally most of the phases successfully been leaching out in order to be higher degree of interconnectivity. Given that, the most porous of sample in Figure 3a presented a homogeneous range size below 2 μm.

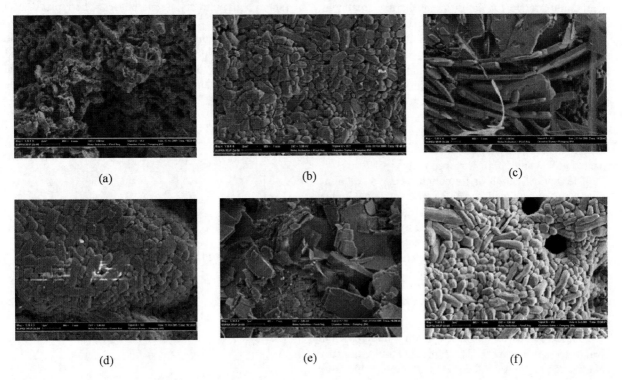

(a)   (b)   (c)

(d)   (e)   (f)

FIGURE 3. FESEM images after leaching process a) 3.0M, 14 days; b) 1.5M, 14 days; c) 3.0M, 7 days; d) 1.5M, 7 days; e) 3.0M, 3 days; f) 1.5M, 3 days.

CONCLUSIONS

In the present work, Leaching process with NaOH is an interesting method to obtain porous lithium aluminosilicate glass ceramic. By controlling the NaOH concentration and leaching period, the percentages of porosity, pore size and percentages of crystallinity can be modulated. Besides, further studies should be perform in order to gain a better understanding of getting porous lithium alumino-silicate glass ceramic.

ACKNOWLEDGEMENTS

The authors acknowledge the research grant (6035111) provided by Universiti Sains Malaysia, Penang that has made this article possible.

REFERENCES

1. Z. Strnad, "Glass-Ceramic Materials, Glass Sci. & Tech." Elsevier, Amsterdam, (1986)
2. M. Bengisu and R. K. Brow. *J. Non-Crys. Solids* **331**, 137 (2003)
3. K.E. Lipinska-Kalita, G. Mariotto, P. E. Kalita and Y. Ohki., *Phys.* **B 365**, 155 (2005)
4. I. W. Donald, B. L. Metcalfe, D. J. Wood, et. al., *J. Mater. Sci.* **24**, 3892 (1989).
5. X. J. Xu, C. S. Ray, D. E. Day, *J. Am. Ceram. Soc.* **80**, 3100 (1997)
6. T. Wakasugi, L. L. Burgner, M. C. Weinberg, *J. Non-Crys. Solids* **244**, 63 (1999)
7. F. Janowski And D. Enki, "Handbook of Porous Solid, Vol 3" (2002)

Effect of Sintering Temperature on the Synthesis of High Purity Cordierite

Y. P. Choo, T. Y. Chow and H. Mohamad

School of Material and Mineral Resources Engineering, Universiti Sains Malaysia,
Engineering Campus 14300 Nibong Tebal, Pulau Pinang, Malaysia

Abstract. Cordierite is silicate material widely used in ceramic industry. The effect of sintering temperature to the properties of cordierite by sol gel method was studied with utilizing magnesium nitrate, aluminum nitrate, ethanol, and tetraethyl orthosilicate (TEOS) as starting materials. Gels are dried and sintered at different temperature (1000°C, 1200°C, 1300°C and 1350°C) then characterized by varies analysis techniques. XRD analysis shows that spinel, μ-cordierite and cristobalite are formed at 1000 °C which spinel as predominant phase. At 1200 °C, μ-cordierite occured as predominant phase, spinel, sapphirine and cristobalite are formed as minor phases. It also confirmed that high purity α-cordierite formed at 1300 °C and clearly observed at 1350 °C. Result of EDX analysis proved that magnesium, aluminum, silicon and oxygen was existed in the cordierite.

Keywords: Cordierite, sol-gel, TEOS, coefficient thermal expansion.
PACS: 81.07.-b

INTRODUCTION

Cordierite ($2Al_2O_3 . 5SiO_2 . 2MgO$) and cordierite based glass ceramics, having a low dielectric constant (5.0 at 1 MHz), high volume resistivity, low thermal expansion coefficient and reasonably high strength, high refractoriness. Cordierite ceramics can be used for structural material, catalyst for exhaust gas control in automobiles, heat exchangers for gas turbine engines, industrial furnaces, packing materials in electronic packing, refractory coatings on metals, etc [1]. Many attempts had been made to produce cordierite. The most common way is via solid state sintering. However, previous study reported that high purity cordierite is very difficult to be obtained through traditional solid-state sintering [3]. This is due to the fact that the optimum sintering temperature range is close to the melting point of cordierite. In addition, calcinations for α-cordierite is difficult without any aid because of the narrow sintering temperature range. Therefore, incorrect calcinations will cause undesired phase appear so that affect its properties [3]. Sol-gel is a prominent method to yield high purity, homogeneous products at lower processing temperature than that required by melting temperature. Sol-gel method had been reported as a useful method to synthesis cordierite by using alkoxides, metal salts as starting materials [4-6]

EXPERIMENTAL METHOD

In this studies, cordierite obtained by sol-gel method using magnesium nitrate hexahydrate, $Mg(NO_3)_2.6H2O$, aluminum nitrate nonahydrate, $Al(NO_3)_3.9H_2O$ and tetraethyl orthosilicate, TEOS as a starting materials. All the starting materials were undergone a mixing process and then stirred at room temperature. Gel obtained after 8 hours stirring and gone through drying and then sintering at different temperature (1000°C, 1200°C, 1300°C and 1350°C). Phase transformation of oxide powders were observed using X-Ray Diffraction (XRD) and composition of elements were determined using Energy Dispersive X-Ray (EDX). The experimental procedure is as shown in Figure 1.

RESULTS AND DISCUSSION

Figure 2 shows XRD results for powder sintered at different temperature; (a) 1000°C, (b) 1200°C, (c) 1300°C and (d) 1350°C. Phase transformations were compiled in Table 1. μ-cordierite (ICDD 14-0249) occurred as predominant phase and spinel (ICDD 21-1152) as a minor phase appeared at 1000°C and it remained up to 1200°C.

CP989, *Neutron and X-ray Scattering in Materials Science and Biology, International Conference on Neutron and X-ray Scattering 2007*, edited by A. Ikram, A. Purwanto, Sutiarso, A. Zulfia, S. Hendrana, and Z. Nurachman
© 2008 American Institute of Physics 978-0-7354-0508-0/08/$23.00

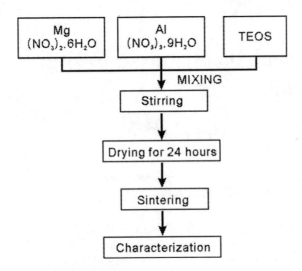

FIGURE 1. Experimental Procedure

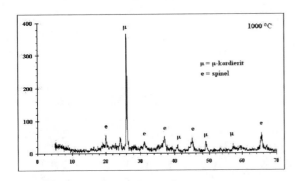

(a)

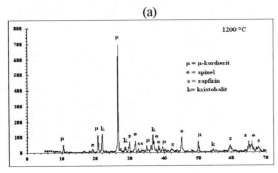

(b)

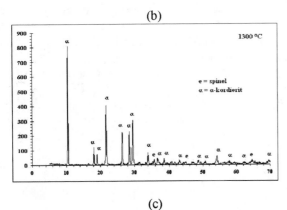

(c)

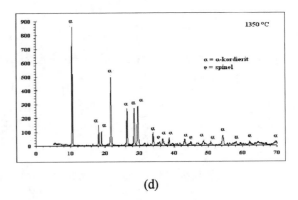

(d)

FIGURE 2. XRD results for powder sintered at different temperature (a) 1000°C, (b) 1200°C, (c) 1300°C and (d) 1350°C.

However, transformation of sapphirine (ICDD 11-0598) and cristobalite (ICDD 39-1425) phase also begins appear at 1200°C. With increase in temperature up to 1300°C, predominant phase has transformed to be α-cordierite (ICDD 13-0293) with small amount of spinel phase. At 1350°C, α-cordierite is clearly observed and spinel phase was decreased. The μ-cordierite is formed due to the reaction between $Si(OAl)_3OSi$ and magnesium octahedral, MgO_6. But the aluminum tetrahedral will react with magnesium tetrahedral to form a spinel phase, $MgAl_2O_4$[1]. At 1200°C, the sapphirine phase is formed by the reaction between intermediate phase and cristobalite or amorphous silica[2].

Transformation of α-cordierite is due to a few reactions: (i) spinel and μ-cordierite (ii) sapphirine and cristobalite (iii) cristobalite and spinel[1,2,7]. However, remaining small amount spinel phase in high purity α-cordierite is possible from the cause of lack cristobalite phase for completely reaction with spinel for forming α-cordierite. This condition means that the lost of quantity for the silica source during the mixing and stirring process.

TABLE 1. Phase transition at different sintering temperature

Sintering Temperature (°C)	Predominant Phase	Secondary Phase
1000	μ- cordierite	Spinel
1200	μ- cordierite	Sapphirine, Cristobalite Spinel
1300	α- cordierite	Spinel
1350	α- cordierite	Spinel (decreased)

EDX analysis in Table 2 shows that the main element in cordierite; magnesium, aluminum, silicon and oxygen were observed where increasing the sintering temperature will increase the atomic percentage of each element closer to the theoretical atomic percentage 8.31(at%)Mg, 18.45(at%)Al, 24.01(at%)Si and 49.23(at%)O.

TABLE 2. Elemental composition from EDX results.

Element	Composition (% atomic weight)			
	1000°C	1200°C	1300°C	1350°C
Mg	6.99	7.19	6.93	7.73
Al	15.15	19.51	17.39	18.59
Si	18	13.24	18.43	20.66
O	59.87	60.07	57.24	53.02

CONCLUSIONS

Phases transformation in synthesized powder were observed where high purity α-cordierite was obtained at 1350 °C. Increasing sintering temperature will increase atomic percentage of four main elements closer to the theoretical value.

ACKNOWLEDGMENTS

The author would like to express their thanks to Malaysian Ministry of Higher Education for funding this project under Fundamental Research Grant Scheme 6070016.

REFERENCES

1. M. K. Naskar and M. Chatterjee, *J. Eur. Ceram. Soc.* **24**, 3499 (2004)
2. S. Kumar, K. K. Singh and P. Ramachandrarao, *J. Mater. Sci. Lett.* **19**, 1263 (2000)
3. A. M. Menchi and A. X. Scian, *Mater. Lett.* **59**, 2664 (2005)
4. S. Drmanic, and D. J. Kostic-Gvozdenovic, *J. Sol-Gel Sci. Technol.* **28**, 111 (2003)
5. S. Mei, J. Yang, and J.M.F. Ferreira, *J. Eur. Ceram. Soc.* **20**, 2191 (2000)
6. U. Selvaraj, S. Komarneni, and R. Roy, *J. Am. Ceram. Soc.* **73**, 3663 (1990)
7. K. Susanta and P. Pramanik *J. Mater. Sci.* **30**, 2855 (1995)

Synthesis and Characterization of Double Perovskite Sr$_2$Mg$_{1-x}$Mn$_x$MoO$_6$ as Anode Materials in Fuel Cell

N. R. Sari, Ismunandar and B. Prijamboedi

Inorganic and Physical Chemistry Research Division, Faculty of Mathematics and Natural Sciences, Institut Teknologi Bandung, Jl. Ganesha 10, Bandung, INDONESIA

Abstract. The anode materials in the solid oxide fuel cell (SOFC) require properties such as high electric conductivity, high catalytic effect for hydrogen decomposition from hydrocarbon and good resistant to sulfur. For these purposes, anode materials based on double perovskite structure, A$_2$BB'O$_6$, were synthesized and studied. In this work, double perovskite compounds, Sr$_2$Mg$_{1-x}$Mn$_x$MoO$_{6-\delta}$ (SMMMO) with $0 \leq x \leq 1$, were synthesized by means of solid state reaction technique at 1350 °C. Crystal structures were determined using X-rays diffractometer and analyzed with the RIETICA program, by employing Le Bail method. The oxides materials showed to adopt monoclinic structure with space group of *P21/n*. For SMMO with higher x, the lattice parameters and the unit-cell volume become smaller. SEM (Scanning Microscope Electron) image, which gave the morphology of SMMMO showed that grains size was increased with the increasing of x. Meanwhile, the DC measurement for electrical conductivity as function of temperature revealed that the conductivity increases with the increasing of x.

Keywords: SMMO, double perovskite, SOFC, X-ray diffraction
PACS: 61.10.Nz, 61.43.Gt, 66.10.Ed

INTRODUCTION

Recently, solid oxide fuel cells (SOFC) gain much attention as efficient and clean energy generating systems. They are capable of generating electricity without Carnot limit, and have high efficiency even in the small size. This is one of reasons why SOFC has potential used in the co-generation and local power generation systems [1].

Many kinds of oxides with perovskite or perovskite related structure have been widely studied to be applied in SOFCs. Tao et al. [2] have reported that (La$_{1-x}$Sr$_x$)Cr$_{0.5}$Mn$_{0.5}$O$_{3-\delta}$ (LSCM) perovskite is an MIEC that has catalytic active properties for CH$_4$ oxidation. However, LSCM has low electronic conductivity in the reducing anodic atmosphere and it is unstable against 10 % sulfur in fuel [3].

The perovskite structure with formula A$_2$BB'O$_6$ is containing transition metals in octahedral (B) sites. The B cations generally determine the physical properties of double perovskites. There are three B-cation sublattice types known for double perovskites, which are random, rock salt, and layered. These sublattices may be differentiated from its powder X-ray diffraction pattern [4]. The ideal double perovskite structures can be viewed as a regular arrangement of corner-sharing BO$_6$ and B'O$_6$ octahedra alternating along the three directions of the crystal, with large A cations occupying the voids in between the octahedra. Depending on the relative size of the B and B' cations with respect to the A cations, the crystal structure can be defined as cubic (*Fm3m*), tetragonal (*I4/m*) or monoclinic (*P21/n*) [5, 6].

Strategy in the selection of Sr$_2$Mg$_{1-x}$Mn$_x$MoO$_{6-\delta}$ (SMMO) was based on four observations by Huang et al. [7].

1) The perovskite structure can support oxide-ion conduction.

2) A perovskite containing a mixed-valence cation from the 4d or 5d block can provide good electronic conduction even where these ions only occupy one sub-array of the double perovskite structure.

3) The ability of Mo(VI) and Mo(V) to form molybdyl ions allows a six-fold coordinated Mo(VI) to accept an electron while losing an oxide ligand.

4) If the two octahedral site cations of the double perovskite are each stable in less than six fold coordination, the perovskite structure can remain stable on the partial removal of oxygen.

The double perovskite Sr$_2$MnMoO$_6$ was first studied in the 1960s and described as cubic with a =

CP989, Neutron and X-ray Scattering in Materials Science and Biology, International Conference on Neutron and X-ray Scattering 2007, edited by A. Ikram, A. Purwanto, Sutiarso, A. Zulfia, S. Hendrana, and Z. Nurachman
© 2008 American Institute of Physics 978-0-7354-0508-0/08/$23.00

7.98 Å [8]. Recently, the compound was reported as cubic, a = 7.9973(1) Å [9] and monoclinic, a = 5.6671(1) Å, b = 5.6537(1) Å, c = 7.9969(2) Å, β = 89.92(2)° [10]. Huang et al. [7] discovered $Sr_2Mg_{1-x}Mn_xMoO_{6-\delta}$ were monoclinic ($P21/n$) characteristic of a well-ordered double perovskite with lattice parameters as well as the unit-cell volume increase with x.

EXPERIMENTAL

The $Sr_2Mg_{1-x}Mn_xMoO_{6-\delta}$ compounds were prepared as polycrystalline powders by the solid state reaction technique. Stoichiometric $SrCO_3$, MgO, Mn_2O_3 and MoO_3 were mixed together with acetone and then were well ground using an agate mortar. MgO were dried in the furnace at 1000 °C overnight before using. The mixed powder was calcinated at 950 °C for 17 h. After cooling down to room temperature, powders were reground and pressed into pellets and heated up again at 1200 °C for 48 h and finally at 1350 °C for 24 h. The X-ray diffraction (XRD) patterns were obtained using the X-ray diffractometer with Cu-K_α radiation. Morphology of SMMO grains was analyzed by scanning electron microscope (SEM). The electrical resistivity (ρ) was measured as a function of temperature at 50-300 °C by using the conventional two-probe technique using electrometers. Before measurement, sample surfaces were covered with silver paste to obtain a good contact with the two electrodes or cables.

RESULTS AND DISCUSSION

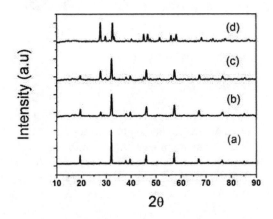

FIGURE 1. X-ray diffraction pattern for double perovskite $Sr_2Mg_{1-x}Mn_xMoO_{6-\delta}$. a) $Sr_2MgMoO_{6-\delta}$, b) $Sr_2Mg_{0.8}Mn_{0.2}MoO_{6-\delta}$, c) $Sr_2Mg_{0.5}Mn_{0.5}MoO_{6-\delta}$, and c) $Sr_2MnMoO_{6-\delta}$.

The powder X-ray diffraction pattern of oxide compounds of $Sr_2Mg_{1-x}Mn_xMoO_{6-\delta}$ ($0 \leq x \leq 1$), which were synthesized by solid state reaction with maximum temperature of 1350 °C, show that the

oxides are double perovskite phase with B-cation sublattice rock salt. The XRD patterns of the $Sr_2Mg_{1-x}Mn_xMoO_{6-\delta}$ oxides are shown in Figure 1.

The refinement results of the powder X-ray diffraction patterns using Le Bail method implemented in the RIETICA program showed that for SMMO with x = 0 and 0.2; pure phases are formed and it can be indexed as monoclinic perovskite structure with a space group of $P21/n$. Meanwhile, in SMMO with x = 0.5 and 1, peaks of $SrMoO_4$ phase were also observed

The refinement result with Le Bail method of $Sr_2MgMoO_{6-\delta}$ diffraction data synthesized by means of solid state reaction technique is shown in Figure 2. The refinement results show that $Sr_2MgMoO_{6-\delta}$ crystallized in monoclinic system with $P21/n$ space group and the refinement and lattice parameters are given as follows: R_p = 4.038, R_{wp} = 4.736, a = 5.5941(3) Å, b = 7.8984(3) Å, c = 5.58862(2) Å, and β = 89.622(4)°.

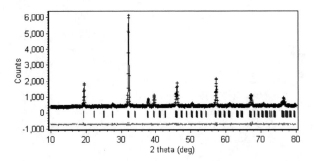

FIGURE 2. Le Bail refinement plot showing the observed (+), calculated, and difference of X-ray diffraction profile for $Sr_2MgMoO_{6-\delta}$ at room temperature fitted with $P21/n$ space group.

Refinement results for $Sr_2Mg_{0.8}Mn_{0.2}MoO_{6-\delta}$ are shown in Figure 3. It shows that $Sr_2Mg_{0.8}Mn_{0.2}MoO_{6-\delta}$ also crystallized in monoclinic system with $P21/n$ space group and with parameters as follows: R_p = 3.992, R_{wp} = 5.133, a = 5.599(2) Å, b = 7.886(3) Å, c = 5.567(1) Å, and β = 89.55(2)°.

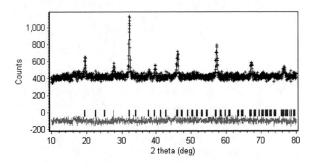

FIGURE 3. Le Bail refinement plot showing the observed (+), calculated, and difference of X-ray diffraction profile for $Sr_2Mg_{0.8}Mn_{0.2}MoO_{6-\delta}$ at room temperature fitted with $P21/n$ space group

For double perovskite of $Sr_2Mg_{0.5}Mn_{0.5}MoO_{6-\delta}$, the refinement results are given in Figure 4. The results show that $Sr_2Mg_{0.5}Mn_{0.5}MoO_{6-\delta}$ also crystallized in monoclinic system with $P21/n$ space group and the parameters are given as follows: R_p = 4.025, R_{wp} = 5.025, a = 5.572(2) Å, b = 7.887(2) Å, c = 5.561(2) Å, and β = 89.59(2)°. Meanwhile, for $Sr_2MnMoO_{6-\delta}$, the refinement result is shown in Figure 5. The refinement results also show that $Sr_2MnMoO_{6-\delta}$ crystallized in monoclinic system with $P21/n$ space group and with the refinement parameters of R_p = 4.974, R_{wp} = 5.931, a = 5.4894(3) Å, b = 7.829(1) Å, c = 5.679(2) Å, and β = 88.19(1)°.

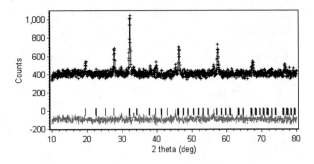

FIGURE 4. Le Bail refinement plot showing the observed (+), calculated and difference of X-ray diffraction profile for $Sr_2Mg_{0.5}Mn_{0.5}MoO_{6-\delta}$ at room temperature fitted with $P21/n$ space group

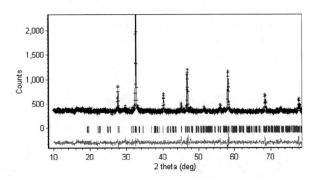

FIGURE 5. Le Bail refinement plot showing the observed (+), calculated and difference of X-ray diffraction profile for $Sr_2MnMoO_{6-\delta}$ at room temperature fitted with $P21/n$ space group.

The cell volume of SMMO decreases linearly with the increasing of Mn concentration as it can be seen in Figure. 6. This observation is in accordance with the smaller ionic size of Mn^{3+} (r_{Mn}^{3+} = 0.58 Å) compared to Mg^{2+} (r_{Mg}^{2+} = 0.72 Å) [11].

The morphology and particle size of the SMMO samples observed by SEM are shown in Figure 7. An increase of particle size with the increasing of Mn content (from an average value of 2.5 µm for x = 0 to about 10 µm for x = 1) was observed. This indicates that the incorporation of Mn into Mg sites facilitates grain growth during the sintering process.

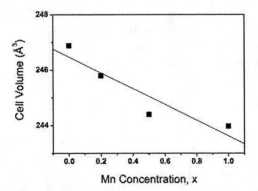

FIGURE 6. Cell volume as a function of Mn concentration in SMMO ($0 \leq x \leq 1$).

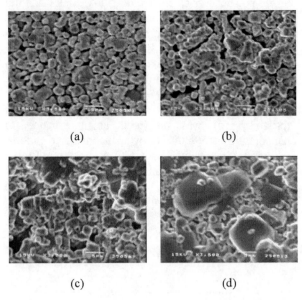

FIGURE 7. SEM images of SMMO ceramics samples at various Mn concentrations. a) Mn = 0, b) Mn = 0.2, c) Mn = 0.5 and d) Mn = 1.

The DC conductivity estimated from the bulk response of the material has been observed as a function of temperature as shown in the Figure 8. DC conductivity behavior appears to be a thermally activated transport of Arrhenius type governed by the relation:

$$\sigma = \sigma_0 \exp\left(\frac{E_a}{k_B T}\right) \tag{1}$$

where σ_0, E_a and k_B represent the pre-exponential factor, the activation energy of the mobile charge carriers and Boltzmann constant, respectively. It was found that doping with Mn gave a positive effect on the electrical conductivity.

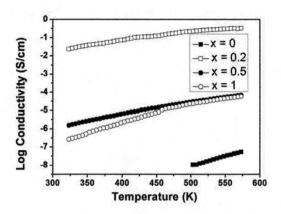

FIGURE 8. Temperature dependence of DC conductivity in SMMO

First, the Mn doping increases conductivity in the SMMO ceramics samples. It is also found that, the activation energy of SMMO decreases with the increasing of x as can be seen in Table 1. Defects in the semiconductor crystal could affect both ionic and electronic conductivity. On $Sr_2MgMoO_{6-\delta}$, Mg has single valence number (2+). Meanwhile, Mn could have multi-valence states of 2+ and 3+. This multi-valence element could provide more defects and act as charge doping in the sample and lead to the higher conductivity and smaller activation energy.

TABLE 1. Conductivity on 300°C and activation energy of SMMO

	σ (Scm^{-1})	E_a (eV)
$Sr_2MgMoO_{6-\delta}$	5.33×10^{-8}	1.469×10^{-5}
$Sr_2Mg_{0.8}Mn_{0.2}MoO_{6-\delta}$	5.73×10^{-5}	1.329×10^{-5}
$Sr_2Mg_{0.5}Mn_{0.5}MoO_{6-\delta}$	6.75×10^{-5}	8.991×10^{-6}
$Sr_2MnMoO_{6-\delta}$	3.19×10^{-1}	6.332×10^{-6}

CONCLUSIONS

Synthesis of double perovskite compounds of $Sr_2Mg_{1-x}Mn_xMoO_{6-\delta}$ by means of solid-state reaction method have been carried out. The structural study was performed by measuring the powder X-ray diffraction data that were refined using Le Bail method in the RIETICA program. The ceramic samples of $Sr_2Mg_{1-x}Mn_xMoO_{6-\delta}$ were crystallized in monoclinic system with space group of *P21/n*. Unit cell parameter values were *a* = 5.5941(3) Å, *b* = 7.8984(3) Å, *c* = 5.58862(2) Å, β = 89.622(4)°, R_p = 4.038, R_{wp} = 4.736 and *V* = 246.88(2) Å^3 for $Sr_2MgMoO_{6-\delta}$. The lattice parameters and the unit-cell volume of SMMO decrease for SMMO with the increasing of Mn concentration.

The morphology of SMMO samples that were studied using SEM showed that grains size increases with the increasing of x. Also, the DC conductivity measurement showed that the electrical conductivity increases with the increasing of Mn concentration and all samples showed Arrhenius type of conductivity.

ACKNOWLEDGMENTS

We would like to thank the Ministry of Research and Technology of Republic of Indonesia, through the insentif program for the financial support.

REFERENCES

1. T. Kobayashi, H. Watanabe, M. Hibino, T. Yao, *J. Solid State Chem.* **176**, 2439-2443 (1996).
2. S. W. Tao, J. S. S. Irvine, *Nat. Mater.* **2**, 320 (2003).
3. S. Zha, P. Tsang, Z. Cheng, M. Liu, *J. Solid State Chem.* **178**, 1884 (2005).
4. M.T. Anderson, G. B. Kevin, A.T. Gregg, R. P. Kenneth, *Prog. Solid State. Chem.* **22,** 197 (1993).
5. A. K. Azad, S. -G. Eriksson, S. A. Ivanov, H. Rundolf, J. Eriksen, R. Mathieu, P. Svedlindh, *Ferroelectrics* **269,** 105 (2002).
6. R. P. Borges, R. M. Thomas, C. Cullinan, J. M. D. Coey, R. Suryanarayanan, L. Bendor, L. P-Gaudart, A. Revcolevski, *J. Phys. Condens. Matter.* **11**, 445 (1999).
7. Y. H. Huang, R. I. Dass, Z. Xing, J. B. Goodenough, *Science* **312**, 254 (2006).
8. G. Blasse, *J. Inorg. Nucl. Chem.* **27**, 993 (1965).
9. M. Itoh, I. Ohta, Y. Inaguma, *Mater. Sci. Eng.* **B41**, 55 (1996)
10. A. Munoz, J. A. Alonso, M. T. Casais, M. J. Martinez-Lope, M. T. Fernandez-Diaz, *J. Phys. Condens. Matter.* **14**, 8817 (2002).
11. R. D. Shannon, *Acta Cryst.* **A32**, 751 (1976).

Phase Compositions of Self Reinforcement Al_2O_3/$CaAl_{12}O_{19}$ Composite using X-ray Diffraction Data and Rietveld Technique

D. Asmi[1], I. M. Low[2], and B. O'Connor[2]

[1]*Department of Physics, Faculty of Mathematics and Natural Sciences, University of Lampung,*
Jl.Sumantri Brojonegoro No. 1, Bandar Lampung 35145, Indonesia
[2]*Materials Research Group, Department of Applied Physics, Curtin University of Technology,*
GPO Box U 1987, Perth, Western Australia

Abstract. The analysis of x-ray diffraction (XRD) patterns by the Rietveld technique was tested to the quantitatively phase compositions of self reinforcement Al_2O_3/$CaAl_{12}O_{19}$ composite. Room-temperature XRD patterns revealed that α-Al_2O_3 was the only phase presence in the CA0 sample, whereas the α-Al_2O_3 and $CaAl_{12}O_{19}$ phases were found for CA5, CA15, CA30, and CA50 samples. The peak intensity of CA_6 in the self reinforcement Al_2O_3/$CaAl_{12}O_{19}$ composites increased in proportion with increase in $CaAl_{12}O_{19}$ content in contrast to α-Al_2O_3. The diffraction patterns for CA100 sample shows minor traces of α-Al_2O_3 even in relatively low peak intensity. It is suggesting that the *in-situ* reaction sintering of raw materials were not react completely to form 100 wt % $CaAl_{12}O_{19}$ at temperature 1650 $^\circ$C. Quantitative phase compositions of self reinforcement Al_2O_3/$CaAl_{12}O_{19}$ composites by Rietveld analysis with XRD data has been well demonstrated. The results showed that the GOF values are relatively low and the fluctuation in the difference plots shows a reasonable fit between the observed and the calculated plot.

Keywords: X-ray diffraction, Rietveld technique, Al_2O_3/$CaAl_{12}O_{19}$
PACS: 61.10.Nz

INTRODUCTION

For more than 40 years, the Rietveld technique[1,2] has been used as a powerful tool for crystal structure refinements based on powder diffraction data, especially in materials science. The atom parameters in the unit cell in this technique are calculated fitting the entire powder diffraction patterns by the least-squares method. The relative weight fractions of crystalline phases in a multiphase sample can be calculated directly from scale factors of the respective calculated intensity as described by Hill and Howard[3]. Using the Rietveld technique, time consuming calibration measurements can be avoided, and the phase abundance can be determined if all phases are identified and also their crystal structure parameters and chemical compositions are known. Beside that, this technique offers other advantages over the traditional integrated-intensity technique of quantitative phase analysis[4] and also have been extensively use for structural studies of many ceramic compound such as a-SiAlON[5], (Pb,La)TiO3[6], and ZrO2[7]. The purpose of this paper was to test the Rietveld technique for the quantitative phase compositions of self reinforcement Al_2O_3/$CaAl_{12}O_{19}$ composites processed by in-situ reaction sintering of Al_2O_3 and $CaAl_{12}O_{19}$ precursor. The X-ray diffraction refinements were performed by the Rietica program.

EXPERIMENTAL METHOD

The self reinforcement Al_2O_3/$CaAl_{12}O_{19}$ composites were prepared from well-mixed of Al_2O_3 and (0, 5, 15, 30, 50, 50, 100 wt%) of $CaAl_{12}O_{19}$ precursor[8]. The samples were labeled as CA0, CA5, CA15, CA30, CA50 and CA100 respectively. The powder was mixed with 150 ml methanol and wet ball-milled for 3h with 3 mm diameter high purity alumina media. The slurry was then dried in oven at 60 $^\circ$C for 48 h and ground with mortar and pestle. The obtained powder was then screened with 150, 75 and 45 μm grid-size sieves. The powder mixture was then uniaxially pressed in a metal-die at 150 MPa to yield cylindrical pellets with dimensions of 5 mm height and 19 mm diameter and bar-shaped with dimensions of 5 mm depth, 10 mm width, and 60 mm length. These

CP989, Neutron and X-ray Scattering in Materials Science and Biology, International Conference on Neutron
and X-ray Scattering 2007, edited by A. Ikram, A. Purwanto, Sutiarso, A. Zulfia, S. Hendrana, and Z. Nurachman

samples were fired in air at 1400 °C for 12 h, followed by 1650 °C for 2 h, and then furnace-cooled Laboratory X-ray diffraction (XRD) patterns for all heat-treated samples were conducted with an automated Siemens D500 Bragg-Brentano instrument using CuKα radiation (λ = 0.15418 nm), produced at 40 Kv and 30 Ma over the 2θ range 5° – 130°, step size 0.04 and counting time 2.4 s/step. Samples were mounted onto aluminium sample holders using a viscous adhesive and adjusted to the correct height with a glass slide. The X-ray diffraction refinements were performed by the Rietica program for windows 95/98/NT version 1.6.5 which derives from Hill-Howard-Hunter LHPM program[9]. The crystal structure models used in the calculation were taken from the Inorganic Crystal Structure Data Base (Fach Informations Zentrum and Gmelin Institut, Germany)– ICSD # 75725[10] for α-Al2O3 and # 34394[11] for CaAl12O19 phases. The final Rietveld scale factors were converted to the phase compositions-by-weight using the *ZMV* expression[3]

$$W_k = \frac{s_k (ZMV)_k}{\sum_{i=1}^{n} s_i (ZMV)_i} \qquad (1)$$

where W_k is the weight fraction of phase k, s is the Rietveld scale factor, Z is the number of formula units per unit cell, M is the mass of the formula unit, and V is the unit-cell volume.

RESULT AND DISCUSSION

Figure 1 shows room-temperature X-ray diffraction (XRD) patterns for the sintered of A/(0-100 wt%) CaAl12O19 samples. The phase α-Al2O3 was the only phase observed for the unmodified samples. For the compositions CA5, CA15, CA30 and CA50, the phases present were α-Al2O3 and CA6. However, for the composition CA100 (100 wt % CA6) the major phase CA6 and a trace of α-Al2O3 were observed. This result can be attributed to the incomplete reaction of the raw materials (Al2O3 and CaCO3) to form CA6 under the test conditions at 1650 °C for 2 h. The *in-situ* formation of CA6 phase in A/CA6 composites is clearly evident for each composition. The peak intensity of CaAl12O19 increased in proportion to the content of CaAl12O19 precursor added. In contrast, for the peak intensity for α-Al2O3 decreased in proportion to the content of CaAl12O19 precursor added.

Selected output difference plots obtained from the XRD patterns Rietveld calculations for CA15 and CA30 samples are shown in Figures 2. The results show no unassigned Bragg reflections, thus indicating that the correct phases were used in the Rietveld

calculations. The fluctuations in the difference plots indicate reasonable fit between the observed and the calculated plot.

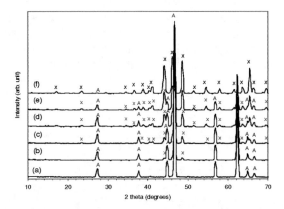

FIGURE 1. Room temperature X-ray diffraction (XRD) patterns of Al2O3/(0-100 wt %) CaAl12O19 composites heat-treated at 1650 °C for 2 hours: (a) CA0, (b) CA5, (c) CA15, (d) CA30, (e) CA50 and (f) CA100, respectively. Legends A = α-Al2O3 and x = CaAl12O19.

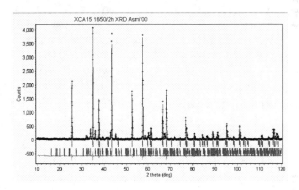

(a)

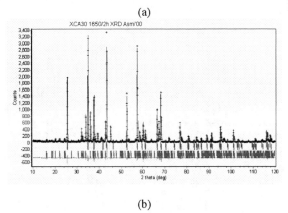

(b)

FIGURE 2. XRD Rietveld refinement plot of Al2O3/ CaAl12O19 composites (a) CA15 and (b) CA30. The observed data are shown by a (+) sign, and the calculated data by solid line. Vertical line represents the positions of diffraction lines of α-Al2O3 and CaAl12O19 respectively. The line below the vertical line is the difference profile.

The figures-of-merit for CA0, CA5, CA15, CA30, CA50, and CA100 samples obtained by Rietveld analysis with the XRD data is presented in Table 1. The RB factors for the individual phases in each composition obtained are 3.3 – 6.9 %. The degree of the refinements is also demonstrated by the small goodness-of-fit (GOF) values, i.e. less than 3 %, indicating that all refinement results are acceptable[12].

TABLE 1. Figures-of-merit from Rietveld refinement with XRD data for CA0, CA5, CA15, CA30, CA50, and CA100 samples.

Sample	R_{exp}	R_{wp}	GOF	R_B α-Al_2O_3	R_B $CaAl_{12}O_{19}$
CA0	10.8	16.4	2.3	5.0	-
CA5	10.5	15.2	2.1	3.5	6.9
CA15	10.7	14.5	1.8	3.4	5.5
CA30	10.9	14.6	1.8	3.8	4.2
CA50	11.1	14.7	1.8	3.6	4.2
CA100	10.9	14.9	1.9	3.3	4.6

The relative phase compositions (wt%) for CA0, CA5, CA15, CA30, CA50, and CA100 samples obtained by Rietveld analysis with the XRD is presented in Tables 2. The relative phase compositions (wt%) of $CaAl_{12}O_{19}$ obtained from the XRD patterns agree well with the as-weighed $CaAl_{12}O_{19}$ compositions, except for the CA15 and CA100 samples for which there are discrepancies of 3.1 and 5.1 wt% of $CaAl_{12}O_{19}$, respectively. This discrepancy indicates incomplete reaction of the raw materials used to form *in-situ* $CaAl_{12}O_{19}$ at 1650 °C for 2 h. However, assuming that the reaction was complete, the agreement is satisfactory in general. The agreements between the weight fraction of $CaAl_{12}O_{19}$ (wt%) obtained from the Rietveld analysis versus the as-weighed $CaAl_{12}O_{19}$ content (wt%) are shown in Figures 3. The correlation obtained was 0.9953.

TABLE 2. Relative phase composition from Rietveld refinement with XRD data for CA0, CA5, CA15, CA30, CA50, and CA100 samples.

Sample	α-Al_2O_3 (wt%)	$CaAl_{12}O_{19}$(wt%)
CA0	100.0 (0.5)	-
CA5	94.6 (0.9)	5.4 (0.3)
CA15	88.1 (0.9)	11.9 (04)
CA30	68.5 (0.8)	31.5 (0.5)
CA50	48.6 (0.6)	51.4 (0.7)
CA100	5.1 (0.3)	94.9 (1.1)

CONCLUSION

The phase relations of the self reinforcement of Al_2O_3/$CaAl_{12}O_{19}$ composite was confirmed by a compositional study using quantitative XRD Rietveld analysis showing the $CaAl_{12}O_{19}$ content in the Al_2O_3/$CaAl_{12}O_{19}$ composites increased in proportion with increase in $CaAl_{12}O_{19}$ precursor added in contrast to α-Al_2O_3. The refinement x-ray diffraction data analysis showed that the goodness of fit values are relatively low and the fluctuations in the difference plots showed a reasonable fit between the observed and the calculated plot.

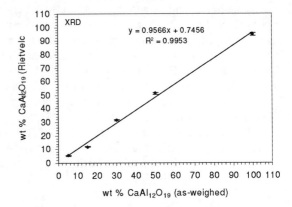

FIGURE 3. Relationship between the $CaAl_{12}O_{19}$ composition (wt%) derived from Rietveld refinement with X-ray diffraction (XRD) patterns versus the $CaAl_{12}O_{19}$ (wt%) as-weighed in Al_2O_3/$CaAl_{12}O_{19}$ composite. Error bars indicate two estimated standard deviations (2σ).

ACKNOWLEDGMENTS

D. Asmi is grateful to the World Bank through the DUE Project University of Lampung, Indonesia for granting a research scholarship.

REFERENCES

1. H. M. Rietveld, *Acta. Cryst.* **22**, 151-152 (1967).
2. H. M. Rietveld, *J. Appl. Cryst.* **3**, 65-71 (1969).
3. R. J. Hill, C.J. Howard. *J. Appl. Cryst.* **20**, 467-474 (1987).
4. R. J. Hill, *Powder Diffraction* **6**, 74 (1991).
5. M. Herman, S. Kurama, H. Mandal. *J. Eur. Ceram. Soc.* **22(16)**, 2997-3005 (2002).
6. M. Mir, C.C. de Paula, D. Garcia, R.H.G.A Kiminami, J.A. Eiras, Y.P. Mascarenhas, *J. Eur. Ceram. Soc.* **27(13-15)**, 3719-3721 (2007).
7. G. Stefanic, S. Music, A. Gajonic, *J. Eur. Ceram. Soc.* **27(23)**, 1001-1016 (2007).
8. D. Asmi, *PhD Thesis*, Curtin University of Technology, Perth, Australia. 2001.
9. R.J. Hill, C.J. Horward, B.A. Hunter, *Computer Program for Rietveld Analysis of Fixed Wavelength X-ray and Neutron Powder Diffraction Patterns*, Australian Atomic Energy Commission (now ANSTO). Rept. No. M112, Lucas Heights Research Laboratories, New South Wales, Australia, 1995.

10. E. N. Maslen, V.A. Streltsov, N. R. Streltsova, N. Ishizawa, Y. Satow, *Acta Cryst*. **49B**, 973-980 (1993).

11. K. Kato, H. Saalfeld, *Neues J. Min.* **109**, 192-200. (1968).

12. E.H. Kisi, *Mat. Forum*, **18**, 135-153 (1994).

Synthesis and Structural Properties of Fe Doped La$_{0.8}$Sr$_{0.2}$Ga$_{0.8}$Mg$_{0.2}$O$_{3-\delta}$ (LSGM) as Solid Electrolyte for Solid Oxide Fuel Cell

Rusmiati, B. Prijamboedi, and Ismunandar

Inorganic and Physical Chemistry Research Division, Faculty of Mathematics and Natural Sciences, Institut Teknologi Bandung, Jl. Ganesha 10, Bandung, INDONESIA

Abstract. La$_{0.8}$Sr$_{0.2}$Ga$_{0.8}$Mg$_{0.2}$O$_{3-\delta}$ (LSGM) is perovskite base oxide material, which exhibits high ion oxygen conductivity, and can it be applied as electrolyte material in solid oxide fuel cell (SOFC). In order to reduce the fuel cell operational temperature, high ionic conductivity must be obtained at lower temperature. High ionic conductivity can be achieved by introducing impurity or defect into material. Doping with Fe for Mg site is expected to increase oxygen ion conductivity in LSGM, since Fe atom has higher valence number (+3) compared with Mg (+2). The LSGM and LSGMF (La$_{0.8}$Sr$_{0.2}$Ga$_{0.8}$Mg$_{0.2-x}$Fe$_x$O$_{3-\delta}$ with x = 0, 0.05, 0.1, 0.15) perovskite structure were synthesized by solid state reaction technique at high temperature and it was sintered at 1350 °C for 24 hours. Crystal structures were analyzed using X-rays diffractometer and refined using Rietica program. The lattice parameters were determined using Le Bail method in cubic structure with space group of *Pm3m*. The cell parameters for La$_{0.8}$Sr$_{0.2}$Ga$_{0.8}$Mg$_{0.2-x}$Fe$_x$O3$_{-\delta}$ (with x = 0, 0.05, 0.1, 0.15) were a = 3.92023(1) Å, a = 3.91056(7) Å, a = 3.89459(9) Å, and a = 3.92463(0) Å. Scanning Electron Microscope (SEM) and Energy Dispersive X-Ray Spectroscopy (EDX) was used to study the grain morphology and elements composition of the LSGMF in order to analyze the effect of Fe substitution.

Keywords: Perovskite, Solid Oxide Fuel Cell, LSGM dope-Fe, Solid State Reaction, X-rays diffraction
PACS: 61.10.Nz, 61.43.Gt, 66.10.Ed

INTRODUCTION

Fuel cell is expected to be an alternative energy source to replace the fossil fuel. There are several benefits of using fuel cell as an energy source since fuel cell exhibits high efficiency, produce water and less pollutant substances at the end of process and uses renewable energy source such as hydrogen. Fuel cell is an electrochemical cell, which consists of two electrodes where reduction and oxidation process occur and the electrolyte placed between electrodes that acts as ionic conductor.

Fuel cells are classified according to the type of electrolyte [1]. One of the fuel cell types is the solid oxide fuel cell (SOFC), which consists of solid oxide material as electrolyte, anode and cathode. For the SOFC application, oxide materials used as electrolyte must have high ionic conductivity in order to make oxygen ion move efficiently from cathode to anode to complete reduction-oxidation reaction and then producing electrical energy.

Lanthanum gallate is an oxide material with perovskite structure and have high ionic conductivity

at high temperature. It has been used as electrolyte for SOFC. Doping of Sr ion at La site and Mg ion at Ga site with La$_{1-x}$Sr$_x$Ga$_{1-y}$Mg$_y$O$_{3-\delta}$ (LSGM) composition have increased ionic conductivity due to defect introduced by doping [2]. In order to reduce the fuel cell operational temperature, high ionic conductivity of electrolyte must be obtained at lower temperature. High ionic conductivity can be achieved by introducing impurity or defect into material. The effects of doping Co on the ion conductivity of LSGM have investigated [3]. Now, doping with Fe for Mg site is expected to increase oxygen ion conductivity in LSGM, since Fe has higher valence number (+3) compared with Mg (+2).

In this work, the LSGM and LSGMF (La$_{0.8}$Sr$_{0.2}$Ga$_{0.8}$Mg$_{0.2-x}$Fe$_x$O$_{3-\delta}$ with x = 0, 0.05, 0.1, 0.15) perovskite structure were synthesized by solid state reaction technique at various temperature, and it was sintered at highest operable temperature of the equipment that available in our laboratory, that is 1450 °C for 12 hours. Crystal structures were analyzed using X-rays diffractometer and refined using Le Bail method in Rietica [4] program.

CP989, Neutron and X-ray Scattering in Materials Science and Biology, International Conference on Neutron and X-ray Scattering 2007, edited by A. Ikram, A. Purwanto, Sutiarso, A. Zulfia, S. Hendrana, and Z. Nurachman
© 2008 American Institute of Physics 978-0-7354-0508-0/08/$23.00

EXPERIMENTAL

All specimens were prepared by a solid state reaction method [5] using metal oxide powders. The specimen with composition of $La_{0.8}Sr_{0.2}Ga_{0.8}Mg_{0.2-x}Fe_xO_{3-\delta}$ and with x = 0, 0.05, 0.10, and 0.15, were prepared by weighting La_2O_3 (99%), $SrCO_3$ (99%), Ga_2O_3 (99%), MgO (99%), and Fe_2O_3 (99%) in the stoichiometric proportion. La_2O_3 and MgO were first dried at 1000 °C and then were weighed and mixed in agate mortar for 1 hour, then calcinated at 1150 °C for 12 hours. After calcination process, the mixture were grinded again for about 1 hour then pressed into pellet with diameter of 10 mm and thickness of 1 mm. Pellets were sintered again at 1300 °C for 4 hour, 1350 °C for 30 hours and finally at 1450 °C for 12 hours with intermediate grinding process.

The crystal structures of the samples were determined by X-rays powder diffraction technique with Cu-K_α illumination. Refinement was carried out to determine the lattice parameters [6]. Grains morphology and elements composition were determined by SEM and EDX measurement.

RESULTS AND DISCUSSION

The cell parameters of XRD pattern for $La_{0.8}Sr_{0.2}Ga_{0.8}Mg_{0.2}O_{3-\delta}$ (LSGM) that was sintered at 1300 °C were found in monoclinic structure with space group I12/a1 and Z = 4, a = 7.637(4) Å, b = 5.540(1) Å, c = 5.515(1) Å, and V = 233.364(2) Å³. R_p (%) = 10.63, R_{wp} (%) = 13.93, and χ value = 0.295. R_p and R_{wp} are the index of fit between calculated and measured patterns [6]. The Le Bail plot for this sample can see in Figure 1. The XRD pattern shows that the perovskite structure was present in the sample. However, other phases could be observed clearly also.

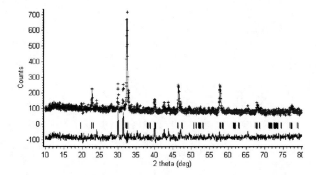

FIGURE 1. Le Bail plot for specimen that sintered at 1300°C for 4 hours.

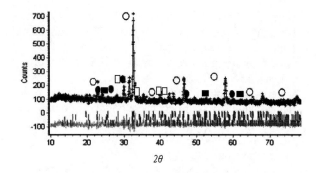

FIGURE 2. Le Bail plot for refinement 4 phases with all impurities phase, with O are peaks of LSGM, ● are peaks of $LaSrGa_3O_7$, □ are peaks of $LaSrGaO_4$, and ■ are peaks of $La_4Sr_3O_9$

Further heating at 1350 °C for 30 hours does not eliminate the impurity peaks. Some peaks, however, decrease after long heating at 1350 °C. The impurities that are presence in the sample can be identified as $LaSrGaO_4$ (PDF, 83-1004), $LaSrGa_3O_7$ (PDF, 45-0637), and $La_4Sr_3O_9$ (PDF, 72-0893) and indicated by peaks with 2θ value of 27.925°, 29.945°, 31.440°, 42.595° and 43.760°[7]. Refinement result, which includes the other 4 phases are shown in Figure 2. With the impurities peaks included in the calculation, the refinement gave better value of R_p (%) = 8.75, R_{wp} (%) = 11.05 and χ = 0.187. These values shows that better refinement parameters could be obtained for multiphase systems by calculating the all of the phases.

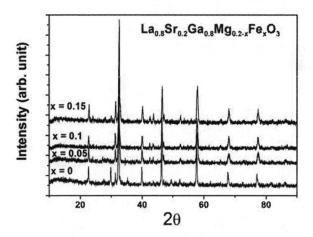

FIGURE 3. XRD pattern of $La_{0.2}Sr_{0.8}Ga_{0.8}Mg_{0.2-x}Fe_xO_{3-\delta}$ with x = 0, 0.05, 0.1, and 0.15 after sintered at 1450 °C.

The next XRD patterns that were studied are LSGM with Fe doped in the composition of $La_{0.2}Sr_{0.8}Ga_{0.8}Mg_{0.2-x}Fe_xO_{3-\delta}$ with x = 0, 0.05, 0.1, 0.15 and all specimens were sintered at 1450 °C for 12 h. The X-rays diffraction patterns are shown in Figure 3. In the XRD pattern the highest impurity peaks are from the $LaSrGa_3O_7$ compound. The $LaSrGaO_4$ peaks have high intensity in the Fe free LSGM sample and tend to decrease with the increasing of Fe concentration in the samples. In the LSGMF with x = 0.1, the $LaSrGaO_4$ peak can not be seen in the XRD pattern. We note also that there is no impurity peaks that contain Fe in their chemical composition observed with the significant intensity. For cell volume comparison purpose, the refinements were carried out in the cubic structure with space group of *Pm3m*. The lattice parameters as a result of refinements with Le Bail method can be seen at Table 1.

TABLE 1. Cell parameters as a result of refinement for $La_{0.8}Sr_{0.2}Ga_{0.8}Mg_{0.2-x}Fe_xO_{3-\delta}$.

Cell Parameters	x = 0	x = 0.05	x = 0.1	x = 0.15
Crystal System	cubic	cubic	cubic	cubic
Space Group	*Pm3m*	*Pm3m*	*Pm3m*	*Pm3m*
a (Å)	3.924 (1)	3.9183 (7)	3.9126 (1)	3.9061 (3)
V (Å^3)	60.246	60.161 (2)	59.859 (8)	59.599 (7)
R_p (%)	8.69	11.133	9.866	9.913
R_{wp} (%)	10.08	12.477	11.172	10.984
χ^2	0.286	0.452	0.172	0.237
Z	2	2	2	2

Plot of crystal cell parameters and unit cell volume for the above data are given in Figure 4. The cell lattice parameter decreases linearly against Fe initial concentration, x. The same trend is also observed for cell volume in the LSGMF sample. It is consistent with the fact that the ionic size of Fe that replaces Mg is smaller than ionic size of Mg [8].

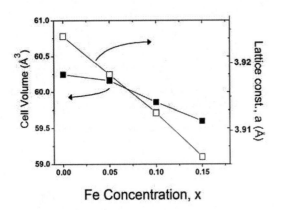

FIGURE 4. Plot of cell parameter, a and cell unit volume V against Fe concentration, x.

Scanning Electron Microscope (SEM) and Energy Dispersive X-Ray Spectroscopy (EDX) is used to study the grain morphology and elements composition of the LSGM and LSGMF in order to analyze the effect of Fe substitution. The SEM images of samples

surface of the LSGMF compound made into pellet form with pressure of 10 N/cm^2 show that it consists of grains with porous distributed between some grains (Figure 5). The LSGM and LSGMF grains diameter were found to be from about 0.5 μm up to 2 μm. With the increasing of Fe concentration, the grain size tends to be smaller and the number of porous also increases.

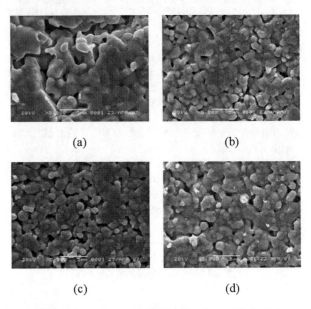

(a)   (b)

(c)   (d)

FIGURE 5. SEM images of LSGMF compounds for (a) x = 0, (b) x = 0.05, (c) x = 0.1 and (d) x = 0.15.

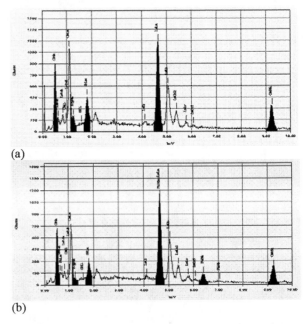

(a)

(b)

FIGURE 6. EDX data which describe elements composition of the specimen. On the top for LSGM, and the bottom for LSGMF which shows the presence of Fe at around 6.4 keV

Meanwhile, the EDX data (see Figure 6 for examples) detect the presence and concentration of Fe in the samples. The Fe peak is located around 6.4 keV. From those data, we can find the actual Fe and other elements in the samples. However for x < 0.15, the obtained Fe concentration was found to be smaller than the calculated in the starting materials (Figure 7). The agreement between the obtained from EDX and the calculated was only observed in the x = 0.15.

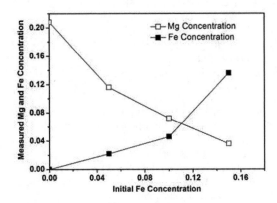

FIGURE 7. A plot of Mg and Fe concentration from the EDX data against Fe initial concentration

CONCLUSIONS

LSGM and LSGMF with various concentration of Fe have been synthesized by means of solid state reaction method. The effect of Fe cations doping in the LSGM crystal at Mg site were investigated. From XRD data, doping of Fe ion into LSGM at Mg site can be effectively carried out. This can seen in the a cell parameter and volume of unit cell data, which decreases with the increasing of Fe concentration from $La_{0.2}Sr_{0.8}Ga_{0.8}Mg_{0.2-x}Fe_xO_{3-\delta}$ and it is consistent with fact that the ionic size of Fe is smaller Mg. Substitution of Fe at Mg site in the LSGM tend to make smaller grain size of LSGM and increases the porous in between the grains.

ACKNOWLEDGMENTS

We would like to thank the Ministry of Research and Technology of Republic of Indonesia, through the insentif program for the financial support.

REFERENCES

1. US DOE, *Fuel Cell Hand Book*, EG & G, West Virginia Technical Service Inc., 2004.
2. T. Ishihara, *J. Am. Chem. Soc.* **119**, 2747-2748 (1994).
3. T. Ishihara, *Chem. Mater.* **11**, 2081-2088 (1999).
4. C. J. Howard, B. J. Kennedy, and B. C. Chakoumakos, *J. Phys. Condens. Matter.* **12**, 349 – 365 (2000).
5. A. R. West, *Basic Solid State Chemistry, 2nd ed.*, New York, Plenum Press, 1998.
6. W. Clegg, *Crystal Structure Determination*, New York Oxford University Press, 1989, p. 45.
7. PCPDFWIN v. 20.1 Database, JCPDS-International Center for Diffraction Data, Newtown Square, 1998.
8. R. D. Shannon, *Acta Cryst.* **A32**, 751 (1976).

Size and Correlation Analysis of Fe_3O_4 Nanoparticles in Magnetic Fluids by Small-Angle Neutron Scattering

S. A. Ani[1], S. Pratapa[1], S. Purwaningsih[1], Triwikantoro[1], Darminto[1], E. Giri Rachman Putra[2], A. Ikram[2]

[1] Department of Physics, Faculty of Mathematics and Sciences, Sepuluh Nopember Institute of Technology, Kampus ITS Sukolilo, Surabaya 60111, Indonesia
[2] Neutron Scattering Laboratory, BATAN, Kawasan Puspiptek Serpong, Tangerang 15314, Indonesia

Abstract. Size and correlation analyses of Fe_3O_4 nanoparticles have been carried out by the measurement of small angle neutron scattering. Particle size of magnetite is analyzed by employing Guinier region approximation in order to obtain gyration radius, which will be used to further determine the average radius of magnetite particles resulting in the value between 17.8 nm and 53.6 nm. Further approximation using a polycore-shell ratio model shows the average particle size of 25 nm with polydispersity of 0.4. The effect of magnetic fluids concentration on the thickness of surfactant layer on the particles surface has not shown a significant change with the value between 6 Å and 8 Å relating to the molar concentration between 0.5 M and 3 M, signifying no correlation among the particles in magnetic fluids.

Keywords: Fe_3O_4 nanoparticle, small angle neutron scattering, magnetic fluids.
PACS: 61.12.Ex, 61.46.Df

INTRODUCTION

Monodisperse magnetic nanoparticles in a self-assembled structure is of technological importance for possible data storage media and is a model system for the study of magnetic interaction in a ferromagnet[1]. The knowledge of magnetic fluids microstructure is very important to understand and control the mechanisms of their stabilization[2]. The neutron scattering methods have been largely used in the last two decades for determination of structural properties of magnetic liquids at microscopic level. Besides, this method can also be developed for examinations of particle size, phenomena of aggregation, magnetic fluid dynamics, particle interaction with surface active agent and magnetic behavior as well[3]. For studies on magnetic phenomena occurring over these length scales of magnetism, small angle neutron scattering (SANS) is exactly appropriate. SANS involves a monochromatic beam of neutrons scattered by the sample and measures the scattered neutrons intensity as a function of the scattering angle. The wave vector transfer extracted from SANS experiment is expressed by

$$Q = \frac{4\pi}{\lambda} \sin 2\theta \tag{1}$$

where 2θ is the scattering angle and λ is the neutron wave length [4]. SANS is a technique that allows us for characterizing structures or objects on the nanometer scale and exploits the big momentum transfer and wavelength reaching out of 3 – 6 Å, which is applicable to observe particle structure in low ordering scattering system being close to the district distance of measurement between 20 – 1250 Å[5].

In this research, we report an analysis concerning the particle size in a low order scattering system in the form of magnetic fluids containing Fe_3O_4 particles using SANS technique. Particle size has been analyzed by Guinier region approximation. Besides, this research also studies the effect of magnetic particle content in magnetic fluids on the layer thickness of surface active agent and their particles correlations.

EXPERIMENT

The Fe_3O_4 magnetic nanoparticles have successfully been synthesized in an aqueous solution by means of the co-precipitation process. We have used iron-sands taken form some rivers in East Jawa (Sungai Brantas, Kali Madiun and Sungai Regoyo), from which the natural magnetite (Fe_3O_4) can be extracted using a magnetic separator. First, 20 g of natural magnetite was dissolved into 38 mL of 12.063

CP989, Neutron and X-ray Scattering in Materials Science and Biology, International Conference on Neutron and X-ray Scattering 2007, edited by A. Ikram, A. Purwanto, Sutiarso, A. Zulfia, S. Hendrana, and Z. Nurachman

M HCl at 70 °C under vigorous stirring for 20 minutes. To this solution, 24 mL of 6.5 M NH₄OH was added and the mixture was continuously heated at 70 °C for 20 minutes to induce precipitating process. The precipitate was then washed using de-mineralized water in step by step to gain magnetite nanoparticles using filter papers. The magnetic fluids were finally prepared by adding tetra-methyl ammonium hydroxide to the wet magnetite nanoparticles with varying ratio between surfactant and nanoparticles.

Characterizations of magnetite nanoparticle were conducted by employing X-ray diffractometer (XRD), magnetic force microscope (MFM) and SANS. The structural properties of the as-prepared nanoparticle were analyzed by X-ray powder diffraction with a JEOL JDX 3530 Diffractometer system. The average crystal size was estimated using Rietica internal program. The size distribution of particle has been measured from the enlarged MFM images. Experiments using non-polarized SANS were carried out at Neutron Scattering Laboratory, BATAN, Kawasan Puspiptek Serpong, Indonesia.

The objective of SANS experiment is to determine the differential cross section, which contains all the information concerning with shape, size and interaction of the scattering bodies in the sample [6]. The mathematical expression related to the SANS patterns scattered by the magnetic fluids in conjunction with differential cross section is given by

$$\frac{d\Sigma}{d\Omega}(Q) = n_p \left(V_p^{\ 2}\right)\left(\Delta\rho\right)^2 \left(P(Q)^2\right)S(Q) \tag{2}$$

where n_p is the number concentration of scattering bodies, V_p is the volume of one scattering body, $\Delta\rho$ is the difference in neutron scattering length density called the contrast for convenience, $P(Q)$ is a function known as the form or shape factor, $S(Q)$ is the interparticle structure factor, and Q is the modulus of the scattering vector[6].

In the dilute particulate system, particles are uniformly dispersed in a matrix. When the particles content is sufficiently small (dilute), the positions of individual particles are apart quite far away from each other to form a uncorrelated system. If the shape of the particles is known, or assumed from the basis of independent information, the intensity of scattering from individual particle can be calculated and compared with the observation. If the particles are of irregular or unknown shape, the data may be analyzed according to the Guinier law to determine the radius of gyration characterizing the size of the particles[7].

The differential scattering cross section for the dilute system is given assuming that the interparticle correlation is small enough and can therefore be neglected [8]. Using extrapolation for $Q = 0$, since $S(Q)=1$, the cross section is given by

$$\frac{d\Sigma}{d\Omega}(0) = n\left(V_p^{\ 2}\right)\left(\Delta\rho^2\right) \tag{3}$$

Guinier [7] has shown that for small values of QR_g, where R_g is the linier dimension of the particle, $P(Q)$ is related to a simple geometrical parameter which is so-called the radius of gyration R_g. Guinier showed further that

$$P(Q) = \exp\left(-\frac{Q^2 R_g^2}{3}\right) \tag{4}$$

scattering function given by equation (4) is called the Guinier law.

$$I(Q) = n\left(V_p^{\ 2}\right)\left(\Delta\rho^2\right)\exp\left(-\frac{Q^2 R_g^2}{3}\right) \tag{5}$$

R_g for a particle is defined by

$$R_g = \left[\frac{\int r^2 \rho(r)dr}{\int \rho(r)dr}\right]^{\frac{1}{2}} \tag{6}$$

which is valid typically for $QR_g < 1$. A plot of scattered intensity in the logarithmic scale versus Q^2 results in a straight line for the low Q, and its slope gives the value of R_g.

$$\ln I(Q) = -\frac{1}{3}R_g^{\ 2}Q^2 \tag{7}$$

The particle size, no matter whether its shape is geometrically well defined or irregular, can be conveniently characterized by its radius of gyration R_g. The concept of R_g is applicable to particles of any shape, for the purpose of determining the radius of particle [7].

RESULTS AND DISCUSSION

The average crystal size of the Fe_3O_4 analyzed by the powder XRD was around 7 nm, while the particle diameter shown by the MFM image, Fig. 1 varied from 10 to 50 nm. This indicates that Fe_3O_4 particle is secondary particle, meaning that each particle contains several grains. Further, the measured scattering data were converted to the total cross section by subtracting the scattering from an empty cell and ambient background, and by normalizing with the sample transmission and thickness. Fig. 2 shows the one-dimensional average of corrected scattering pattern from magnetic fluids sample at 0.5M solution concentration and 0.28% of magnetite concentration.

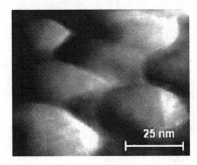

FIGURE 1. MFM images from magnetite particle with average diameter 25 nm

Size and correlation of magnetite particles were analyzed based on the experimental result using two methods. The first method is a fitting program using Guinier region approximation at interval Q-range between 0.8 nm^{-1} and 1.2 nm^{-1} to determine the average particle size. The gradient of ln (I) to Q^2 plot is applicable to determine radius gyration (R_g) which will be used then to determine the radius of magnetite particle. Gradient at Guinier region is proportional to $-^1/_3 R_g^2$. In Fig. 3 there are three characteristics of particle size for the magnetic fluids with the concentration of 0.5 M and magnetite concentration of 0.28% (Fig. 3). They correspond to the length intervals of 17.8 nm, 29.4 nm, and 53.6 nm. We suppose that the reason may probably be connected with the polydispersity of magnetite particle size in magnetic fluids.

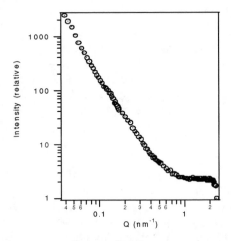

FIGURE 2. Scattering pattern from magnetic fluids with the concentration of 0.5 M and magnetite concentration of 0.28 %.

It is interesting to go further using the approximation of polycore-shell ratio model to determine the polydispersity of magnetite particles. It

has been analyzed in this regards by assuming that the particle is spherical and the fluids contain of magnetite as core, surface active agent as shell and the water as solvent. This curve calculated according to the expected model of polydispered spheres with scattering length density of magnetite gives result $\rho_c = 6.921 \times 10^{-5}$ Å^{-2}. The polydispersity of 0.4 shows that the most particles have average radius of 25 nm, Fig. 4).

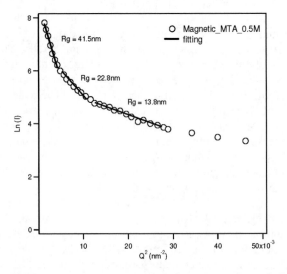

FIGURE 3. Characterize the radius of gyration of the magnetite particle

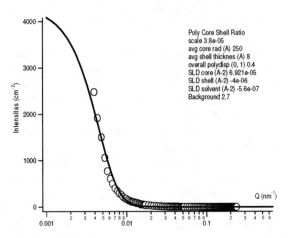

FIGURE 4. Polycore-shell ratio model to determine polydispersity of magnetite particle

The thickness of surfactant shell is the key parameter, which determines the stability of the fluids [11]. It has been shown that in the case of magnetite/tetramethyl ammonium hydroxide/water, the thickness of the surfactant layer will change, when the magnetite concentration is varied. The effect of magnetic fluids concentration on the thickness of surfactant layer on the particles surface has not significantly been observed. A change with the value between 6 Å and 8 Å corresponds to the molar concentration between 0.5 M and 3 M, signifying no correlation among the particles in magnetic fluids. In the case of the magnetite concentration of 0.28 %, the large penetration between sub-layers of the double surfactant was detected.

TABLE 1. The result of pattern analysis of SANS to determine the thickness of surfactant layer

Concentration of magnetic fluids (M)	volume of magnetite particle Fe_3O_4 (%)	Q (nm^{-1})	Intensity (cm^{-2})	The thickness of surfactant layer (Å)
0.5	0.28	0.112	118.65	8
1	0.55	0.112	193.73	7
2	1.11	0.112	344.18	6
3	1.67	0.112	344.18	6

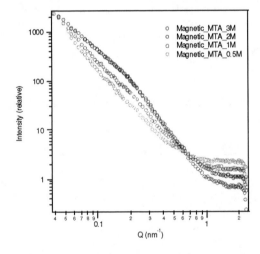

FIGURE 5. Scattering pattern of intensity $I(Q)$ versus Q at different magnetic fluids concentration

The average radius of particle has not been changed significantly with increasing the magnetic concentration. The particle radius ranges at interval between 17.8 nm and 53.6 nm using polycore-shell ratio model with average radius of 25 nm. It shows that there are no correlations of magnetite particle in the concentration range of 0.5 M and 3 M, Fig. 5. A significant decrease in the background of scattering pattern was observed from the sample with increasing magnetite concentration.

CONCLUSIONS

Size and correlation in magnetic fluids structure of magnetite/tetramethyl ammonium hydroxide/water were determined by means of non-polarized SANS. Three characteristics of the average radius of magnetite with the magnetite concentration in the range of 0.28-1.67 %-Volume were obtained and determined using Guinier region approximation and compared with polycore-shell ratio model to result in the thickness of surfactant layer, corresponding to the radius intervals of 17.8 nm, 29.4 nm, and 53.6 nm. We conclude that the reason may be connected with the polydispersity of magnetite particle in the magnetic fluids. The polydispersity of 0.4 shows that the most particles have average radius of 25 nm using polycore-shell ratio model. The effect of magnetic fluids concentration on the thickness of surfactant layer on the particles surface has not been shown significantly, signifying no correlation among the particles in the magnetic fluids.

ACKNOWLEDGMENTS

This work was partially supported by the program of Research Incentive in Nanoscience and Nanotechnology, prepared by the State Ministry of Research and Technology (KMRT), Republic of Indonesia, 2006.

REFERENCES

1. D.F. Farrell, Y. Ijiri, C.V. Kelly, J.A. Borchers, J.J. Rhyne, Y. Ding, S.A. Majetich. *J. Magn. Magn. Mater* **303**, 318-322 (2006).
2. M.V. Avdeev, M. Balasoiu, V. L. Aksenov, V.M. Garamus, J. Kohlbrecher, D. Bica, L.Vekas. *J. Magn. Magn. Mater* **270,** 371-379 (2004).
3. M. Balasoiu, M.V. Avdeev, V. L. Aksenov, V. Ghenescu, M. Ghenescu, Gy. Torok, L. Rosta, D. Bica, L. Vekas, D. Hasegan. *Romanian Rep.Phys.* **56**, 4, 601-607 (2004).
4. V. K. Aswal, P. S. Goyal, *Curr. Sci.* **79**, 7 (2000).
5. J. Kohlbrecher, "Characterization of Nanomagnetic Structures by Polarized Small Angle Neutron Scattering", at http://kur.web.psi.ch/
6. S. M. King. Small Angle Neutron Scattering. Link fixed Sep 2003.
7. Ryong-Joon Roe, "Methods of X-Ray and Neutron Scattering in Polymer Science", Oxford University Press, Inc., New York, 2000
8. P.S. Goyal, "Small Angle Neutron Scattering", IAEA Workshop on SANS, BARC, Mumbai, India, April 1995.

The Use of Rietveld Technique to Study Phase Composition and Developments of Calcium Aluminate

I. Ridwan and D. Asmi

[1]Department of Physics, Faculty of Mathematics and Natural Sciences, University of Lampung, Jl.Sumantri Brojonegoro No. 1, Bandar Lampung 35145, Indonesia

Abstract. The phase composition and development of calcium aluminates (CA, CA_2, and CA_6) processed by *in-situ* reaction sintering of Al_2O_3 and $CaCO_3$ have been studied by Rietveld refinement technique. The formation of calcium aluminates is temperature-dependent. X-ray diffraction result revealed that the CA, CA_2, and CA_6 phases starts to develop at approximately 1000 °C, 1100 °C and 1375 °C, respectively. The relative phase compositions obtained from x-ray diffraction patterns for the α-Al_2O_3 phase decreased markedly with increasing temperature, i.e. from 86.0(1.1) wt % at 1000 °C to 34.7(0.4) wt % at 1400 °C. The wt % of CA decreased from 10.9(0.3) – 1.9(0.2) wt % at 1100 – 1200 °C but disappeared at 1300 °C. The wt % of CA_2 reached 36.0(0.7) wt % at 1300 °C and decreased to 18.5 (0.5) wt % at 1400 °C. The wt % CA_6 increased markedly from 1375 to 1400 °C, i.e. 12.80(0.6) – 47.3(0.9) wt %. The goodness of fit values is relatively low and the fluctuation in the difference plots shows a reasonable fit between the observed and the calculated plot.

Keywords: X-ray diffraction, Rietveld technique, calcium aluminates (CA, CA_2, and CA_6).
PACS: 61.10.Nz

INTRODUCTION

The binary compound of the CaO-Al_2O_3 system has been used in a wide range of applications such as ceramic materials, metallurgical slag, and high alumina cement industry[1]. Some of the amorphous calcium aluminates in this system have also been used as information storage device owing to its high photosensitive properties[2]. The chemical and thermodynamic study of this system has been investigated by a number of researchers[3-5], however differences of opinion still exist on several points such as the presence types and the temperature formation of calcium aluminates compound in the CaO-Al_2O_3 system. Shepherd[3] found the presence of four compounds (C_3A, C_5A_3, CA, and C_3A_5) in this system and Langerquist *et al* [4] in X-ray studies identified C_5A_3 as $C_{12}A_7$ and C_3A_5 as CA_2 and they reported a new high alumina phase C_3A_{16} and this later compound was identified as CA_6 by Filonenko and Lavrov[6] in their microscopic study. This paper present the use of Rietveld technique to study the phase composition and development of calcium aluminates (CA, CA_2, and CA_6) processed by *in-situ* reaction sintering of alumina and calcium carbonate fired at temperature 1000-1400°C. The chemistry notation used in this paper are C for CaO and A for Al_2O_3.

EXPERIMENTAL METHOD

Commercial Al_2O_3 and $CaCO_3$ were used as raw materials. The sample was prepared from a well-mixed 6:1 mixture of alumina and calcium carbonate powders. The powder was mixed with 150 ml methanol and stirred for 7h. The slurry was then dried in oven at 60 °C for 48 h and ground with mortal and pestle. The obtained powder was then screened with 150, 75 and 45 μm grid-size sieves. Finally these samples were fired in air at 1000, 1100, 1200, 1300, 1350, 1375, and 1400 °C for 2 h and then furnace-cooled. Laboratory X-ray diffraction (XRD) patterns for all heat-treated samples were conducted with an automated SHIMADZU XD-600 instrument using CuKα radiation (λ = 0.15418 nm), produced at 30 kV and 30 mA over the 2θ range 5° – 130°. Qualitative X-ray diffraction was performed by comparing the diffraction lines with standard Powder Diffraction File data base using the search match procedure. The parameters refined for all samples were those controlling pattern intensity (scale factors), peak profile (width and shape), peak position (zero point

CP989, Neutron and X-ray Scattering in Materials Science and Biology, International Conference on Neutron and X-ray Scattering 2007, edited by A. Ikram, A. Purwanto, Sutiarso, A. Zulfia, S. Hendrana, and Z. Nurachman
© 2008 American Institute of Physics 978-0-7354-0508-0/08/$23.00

and unit cell), background polynomial parameters, and the individual atom thermal parameters. The crystal structure models used in the calculation were taken from the Inorganic Crystal Structure Data Base (Fach Informations Zentrum and Gmelin Institut, Germany) – ICSD files #75725[7] for α-Al_2O_3, #75785[8] for CaO, 260[9] for $CaO.Al_2O_3$ (CA), # 34487[10] for $CaO.2Al_2O_3$ (CA_2) and # 34394[11] for $CaO.6Al_2O_3$ (CA_6). The refinements were performed by the Rietica program for Windows 95/98/NT version 1.6.5 which derives from the Hill-Horward-Hunter LHPM program[12] and the final Rietveld scale factors were converted to the phase compositions-by-weight using the *ZMV* expression proposed by Hill and Horward [13].

RESULT AND DISCUSSION

Room-temperature XRD patterns for fired samples Al_2O_3:$CaCO_3$ in ratio 6:1 at various temperatures are shown in Figure 1. Various phases were observed for this system. The conventional solid state reaction process for fabrication of the CA_6 compound was based on the following reactions:

$$6\ Al_2O_3 + CaCO_3 \rightarrow CaO.6Al_2O_3 + CO_2 \qquad (1)$$

However, the solid state reaction processes were usually more complex. Fired at 1000 °C (Figure 1a) the phases observed were α-Al_2O_3, CaO and CA. The peak intensity for the α-Al_2O_3 phase decreased with increasing in temperature. The peak intensities for CA was also decreased with increasing in temperature and disappeared by 1300 °C, followed by the appearance of CA_2 phase at 1100 °C, which remained at temperature of 1400 °C. The formation temperatures of CA_6 occurred at approximately 1375 °C. The formation of *in-situ* CA and CA_2, and their transformation to CA_6, are believed to occur through exothermic reactions between Al_2O_3 and CaO, at temperature 1000 – 1400 °C according to the following reaction [14]:

$$Al_2O_3 + CaO \rightarrow CaAl_2O_4\ (CA) \qquad (2)$$

$$CA + 2Al_2O_3 \rightarrow CaAl_4O_7\ (CA_2) \qquad (3)$$

$$CA + 5Al_2O_3 \rightarrow CaAl_{12}O_{19}\ (CA_6) \qquad (4)$$

$$CA_2 + 4Al_2O_3 \rightarrow CaAl_{12}O_{19}\ (CA_6) \qquad (5)$$

Traces of α-Al_2O_3 and CA_2 often accompany CA_6 in the final product as the completion of such reactions depends on the diffusion distances, i.e. on the particle sizes and the degree of mixing of the reactant powders[15]. The formation of CA_6 is believed to

occur via *in-situ* reactions according to Equations 2-5 and also temperature dependent[16]. Therefore, in this case, Equations 4 and 5 are important for understanding the development of elongated CA_6 grains, it being known that CA and CA_2 is the intermediate phase during the solid state formation of CA_6.

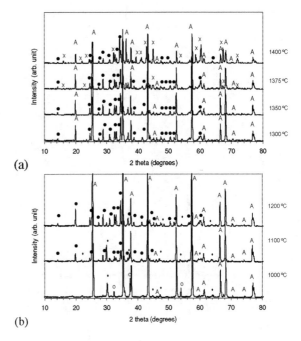

(a)

(b)

FIGURE 1. Room temperature X-ray diffraction (XRD) patterns of simples fired at 1000, 1100, 1200 °C in (a) and 1300, 1350, 1375, 1400°C in (b). Legends A = α-Al_2O_3, o = CaO, + = CA, •=CA_2, and x=CA_6.

TABLE 1. Figures-of-merit from Rietveld refinement with XRD data for samples fired at 1000, 1100, 1200, 1300, 1350, 1375, and 1400°C for 2 h.

Temp. (°C)	R_{exp}	R_{wp}	GOF	R_B Al_2O_3	R_B CaO	R_B CA	R_B CA_2	R_B CA_6
1000	11.0	15.3	1.9	2.9	8.2	5.8	-	-
1100	10.6	15.2	2.0	2.9	-	8.2	5.2	-
1200	10.5	13.7	1.7	2.9	-	7.4	5.6	-
1300	10.3	14.1	1.9	2.8	-	-	5.4	-
1350	10.7	14.9	1.9	2.8	-	-	6.2	-
1375	11.6	15.2	2.0	6.0	-	-	6.0	4.9
1400	10.8	15.1	2.0	5.4	-	-	5.7	4.3

The profile figures-of-merit for samples fired at 1000, 1100, 1200, 1300, 1350, 1375, and 1400°C for 2h as determined from Rietveld refinement with room temperature XRD is shown in Tables 1. The R_B factors for each phase in each refinement with XRD data are approximately 2.8 – 6.0 % for α-Al_2O_3, 8.2 % for CaO, 5.8 – 8.2 for CA, 5.2 – 6.2 % for CA_2, and 4.3 - 4.9 % for CA_6. The GOF values were relatively low, i.e. all approximately less than 2 %, indicating that the qualities of refinements are acceptable.

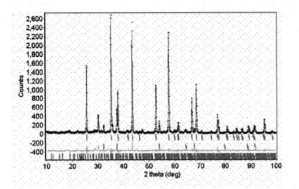

(a)

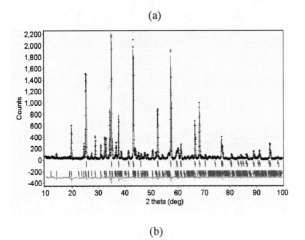

(b)

FIGURE 2. XRD Rietveld difference plots for CA100 sample fired at (a) 1000 °C and (b) 1350 °C. The observed data are shown by a (+) sign, and the calculated data by a solid line. Vertical line represents the positions of diffraction lines for α-Al$_2$O$_3$, CA and CaO in (a), and α-Al$_2$O$_3$ and CA$_2$ in (b), respectively. The line below the vertical lines is the difference profile.

TABLE 2. Relative phase composition from Rietveld refinement with XRD data for samples fired at 1000, 1100, 1200, 1300, 1350, 1375, and 1400°C for 2h.

Temp. ($^{\circ}$C)	α-Al$_2$O$_3$ (wt%)	CaO (wt%)	CA (wt%)	CA$_2$ (wt%)	CA$_6$ (wt%)
1000	86.0 (1.1)	3.1 (0.1)	10.9 (0.3)	-	-
1100	78.8 (1.1)	-	9.0 (0.3)	15.2 (0.4)	-
1200	66.9 (0.9)	-	1.9 (0.2)	31.2 (0.4)	-
1300	64.5 (1.0)	-	-	35.5 (0.6)	-
1350	64.0 (1.0)	-	-	36.0 (0.7)	
1375	55.5 (1.0)	-	-	31.7 (0.7)	12.8 (0.5)
1400	34.7 (0.8)	-	-	18.0 (0.5)	47.3 (0.9)

Table 2 shows the relative phase composition (wt %) of the samples fired at 1000, 1100, 1200, 1300, 1350, 1375, and 1400°C from the XRD data. The relative phase compositions obtained from XRD

patterns for the α-Al$_2$O$_3$ phase decreased markedly with increasing temperature, i.e. from 86.0(1.1) wt % at 1000 °C to 34.7(0.8) wt % at 1400 °C. The wt % of CA decreased from 10.9(0.3) – 1.9(0.2) wt % at 1100 – 1200 °C but disappeared at 1300 °C. The wt % of CA$_2$ increased slightly up to the temperature 1350 °C, and reached 36.0(0.7) wt % at this temperature and decreased to 18.0 (0.5) wt% at 1400 °C. The wt % CA$_6$ increased markedly from 13750 to 1400 °C, i.e. 12.8(0.5) – 47.3(0.9) wt %.

CONCLUSION

Quantitative phase composition and development of calcium aluminate (CA, CA$_2$, and CA$_6$) by Rietveld analysis with XRD data has been well demonstrated. The formation of calcium aluminates is temperature dependent. XRD result revealed that the CA, CA$_2$ and CA$_6$ phases starts to develop at approximately 1000 °C, 1100 °C and 1375 °C and the Rietveld analysis results showed that the GOF values obtained are relatively low and the fluctuation in the difference plots shows a reasonable fit between the observed and the calculated plot.

ACKNOWLEDGMENTS

D. Asmi is grateful to the World Bank through the Directorate General of Higher Education Republic of Indonesia under the Hibah Bersaing XIV research grant program 2nd, 2007.

REFERENCES

1. L.D. Hard (ed), "Alumina Chemical Science and Technology Handbook", American Ceramics Society, Westerville, OH, 1990.
2. E.S. Shepherd, G.A. Rankin, and F.E. Wright, F.E. *Am. J. Sci,* **28**, 293-333 (1909).
3. R.W. Nurse, J.H. Welch, and A.J. Majumdar, A.J. *Trans. Brit. Ceram. Soc.,* **64**, 409-418 (1965).
4. B. Haldstted, B. *J. Am. Ceram. Soc.* **73(1)**, 15-23(1991).
5. K. Lagergvist, S. Wallmark, and A. Westgren, *Z. Anorg. Allg. Chem,* **234**, 1-16 (1937).
6. N.E. Filonenko and I.V. Lavrov, I.V. *Dokl. Akad. Naak SSSR.* **66**. 673-676 (1949).
7. E. N. Maslen, V.A. Streltsov, N. R. Streltsova, N. Ishizawa, and Y. Satow, *Acta Cryst.* **49B**, 973-980 (1993).
8. Q. Huang, O. Chmaissem, J.J. Caponi, C. Challout, M. Marezio, J.L. Tholence and A. Santoro, *Phys.* **C 227**, 1-9 (1994).
9. W. Höerkner and H. Mueller-Buschbaum, *J. Inorg. Nucl. Chem.,* **38**, 983-984 (1976).

10. V.I. Ponomarev, D.M. Kheiker, N.V. and Belov, *Krist*, **15**, 1140-1143 (1970).

11. K. Kato and H. Saalfeld, (1968). *Neues J. Min.*, **109**, 192-200 (1968).

12. R.J. Hill, C.J. Horward, and B.A. Hunter, *Computer Program for Rietveld Analysis of Fixed Wavelength X-ray and Neutron Powder Diffraction Patterns*, Australian Atomic Energy Commission (now ANSTO). Rept. No. M112, Lucas Heights Research Laboratories, New South Wales, Australia, 1995.

13. R. J. Hill and C.J. Howard. *J. Appl. Cryst.* **20** 467-471 (1987).

14. D. Asmi, *PhD Thesis*, Curtin University of Technology, Perth, Australia, 2001.

15. Cinibulk, M.K. *Ceramics Engineering and Science Proceedings,* **16**, 633-640 (1995).

16. C. Dominguez, J. Chevalier, R. Torrecillas, and G. Fantozzi, *J. Eur. Ceram. Soc.* **21(3)** 381-387 (2001).

INSTRUMENTS AND METHODS

Improvements in the Image Quality of Neutron Radiograms of NUR Neutron Radiography Facility by Using Several Exposure Techniques

T. Zergoug[1]., A. Nedjar[1], M.Y. Mokeddem[1], L. Mammou[1]

Centre de Recherche Nucléaire de Draria, (CRND) BP 43 Draria 16000, Algiers, Algeria

Abstract. Since the construction of NUR reactor neutron radiography facility in 1991, only transfer exposure method was used as a non destructive technique. The reason is the excess of gamma rays in the neutron beam. To improve radiation performances of the NR system, a stainless steal hollow conical cylinder is introduced at the bottom of the facility beam port, this filter reduce gamma infiltration through the edges of the NR structure without disturbing neutron beam arriving from the in pool divergent collimator. First results confirm our prediction; a gamma rays diminution and a relatively stable neutron flux at the point object are confirmed, consequently the n/γ ratio reaches a value of 2.104 n/cm² mR. Radiograms obtained by using the direct exposure method reveal the feasibility of the technique in the new NR configuration facility, but a weak resolution and contrast of the image is observed. In this paper, we describe a procedure to improve the image quality obtained by direct exposure technique. The process consists of using digitized images obtained by several exposure techniques (NR, gamma radiography or X radiography) for a comparison study and then better image definition can be attained.

Keywords: neutron radiography, direct exposure technique, image quality, gamma rays
PACS: 07.05.Rm, 81.70.-q

INTRODUCTION

Neutron radiography (NR) is a powerful non-destructive imaging technique for the internal evaluation of materials or components. The method is similar in principle to X or gamma rays radiography, and is complementary in the nature of information supplied. Both techniques use beams of penetrating radiation, that on passing through an object, the beam generates an image.

However, neutrons are attenuated very differently than are X or gamma rays. While X or gamma rays interact with the electron cloud surrounding the nucleus of an atom, neutrons interact with the nucleus itself,

The attenuation coefficient for the X or gamma rays varies gradually with increasing atomic number, however, for the neutron, attenuation is more selective. The probability of an interaction is determined by both the incident neutron energy and the detailed structure of the target nucleus (nuclear cross section). Materials with high atomic numbers preferentially attenuate X-rays, while, materials having low atomic numbers preferentially attenuate neutrons.

Neutron radiography exploits nuclear characteristics of the various elements and their isotopes. Dense materials such as steel, lead, tungsten or uranium are easily penetrated by thermal neutrons, whilst light materials such as hydrogen, boron, are clearly imaged. This also means that fluids or generally hydrogenous materials can be observed through metal walls with neutron radiography (distribution, homogeneity and filling status of pyrotechnics devices).

Neutron radiography facility of NUR reactor in Algeria was installed at the radial channel N°1 and in spite of cautions taken in the design study, the gamma flux at point object of the installation was relatively high which can not allow the use of direct exposure technique.

In order to improve the characteristics of the facility i.e. increasing the neutron flux and reducing the gamma radiation dose, many modifications for this installation are done. In a first step we have introduced a hollow conical gamma filter in the beam port to reduce gamma radiation dose infiltration without disturbing the neutron beam. This experiment shows satisfactory results and a new value n/γ of 2 104 n/cm².mR is reached.

CP989, Neutron and X-ray Scattering in Materials Science and Biology, International Conference on Neutron and X-ray Scattering 2007, edited by A. Ikram, A. Purwanto, Sutiarso, A. Zulfia, S. Hendrana, and Z. Nurachman

As a result, direct method is possible, but not without inconvenient. The weak of the image quality is remarkable enough that was predictable since an excess of gamma which blur the image still remain in the neutron beam.

This paper describes a theoretical approach for removing optical densities due to the excess of gamma from the radiogram obtained by direct exposure technique to get an image with a better contrast. Practically, a combination of two images (neutron + gamma for the direct exposure technique and gamma radiography) is used to reach as possible the pure neutron image.

IMAGE RESTORATION

To develop the procedure of enhancing image contrast using the combination of NR direct exposure image and gamma radiography image, we apply the theory of image reconstruction developed by KOBAYASHI and et. al. [1].

The relation of optical film densities by gamma rays and neutrons are [1]

$$D_n = D_{n0}.e^{-\mu_n d}$$
$$D_y = D_{y0}.e^{-\mu_y d} \tag{1}$$

where D_n and D_γ respectively are Optical Density (OD) obtained by neutron and gamma radiography. D_{n0} and $D_{\gamma 0}$ are OD of film on the locations in the absence of objects (100% Transmission). μ_n and μ_γ respectively neutron and gamma linear attenuation coefficients, and d is a sample thickness.

These equations are based on the following hypothesis:

1. The gamma rays and neutrons incident are uniformly over the field of exposure.
2. The effect of scattering is neglected.
3. The total exposure are not very large and sample size and thickness are both small.

For a film combination of neutron and gamma rays, we can write by analogy of equation (1):

$$D_{(n+y)} = D_{(n+y)0}.e^{-\mu_{(n+y)}.d} \tag{2}$$

$D_{(n+\gamma)}$ is OD by combination of neutron and gamma radiography. $D_{(n+\gamma)0}$ is OD on the locations in the absence of objects (100% Transmission), and $\mu_{(n+\gamma)}$ is theoretical linear attenuation coefficient. If we postulate that:

$$\mu_{(n+y)} = \mu_n + \mu_v \tag{3}$$

we can easily find the relation:

$$D_n = D_{n0}.\left\{(D_{y0})/(D_{(n+y)0}.(D_{(n+y)}/D_y)\right\} \tag{4}$$

The OD of the pure neutron radiograph at a given position is then a function of OD at the same location and background optical densities of the two types of films.

IMAGING AND METHOD OF WORK

We use the neutron radiography facility at NUR reactor [2], as a neutron and gamma mixed source. The characteristics of the NUR reactor neutron radiography installation are mentioned in Table 1.

The neutron radiography direct exposure technique image (direct image) is obtained by using a gadolinium converter in a contact with a D7 Agfa film and the neutron radiography transfer exposure method image (transfer image) is performed by using a dysprosium converter. The same experiments are done for the gamma radiography (gamma image), by exposing only an X-ray film with the object of study (no converter) [3].

TABLE 1. Characteristics of NUR reactor NR facility

Neutron beam	Thermal neutron flux (1 MW) at point object [n/cm².s]	6.10^5
	Cadmium ratio	2.64
	Dose gamma value at point object [R/h]	140
	ratio n/γ [n/mR.cm²]	$\sim 2.10^4$
Geometrical disposition	L/D	113
	Useful area at point object [cm x cm]	34x34

The X-ray films obtained are then digitized with the same conditions, then, OD is correlated to the image pixel intensity for the treatment of the images [4].

EXPERIMENTAL

To obtain images of our experiment, the objects of study were first to gamma, to obtain two X-ray films (gamma image). Then the second film is transferred to put it in contact with a D_y converter (which was exposed before to realize neutron radiography of the same objects at the same position). We took the caution to move the image plane of the neutron in relation to gamma radiography as is shown clearly in Figure 1 (see cadmium disc).

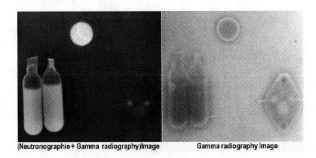

(Neutronographie + Gamma radiography)Image   Gamma radiography Image

FIGURE 1. Experimental (n+γ) and γ images

The images are then treated by computer by using the procedure described above to obtain a new image (see Figure 2).

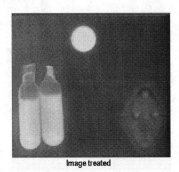

Image treated

FIGURE 2. (n+γ) Image treated

We notice that the contrast and homogeneity of the oil filling in the tubes are better. The penumbra of the cadmium caused by the translation is practically removed. Connectors and image re-composition of the electronic device are clearly visible.

APPLICATION TO DIRECT EXPOSURE METHOD IN NUR NR FACILITY

As mentioned in Table 1, the n/γ ratio of our facility is of about 2.10^{+04} n/cm^2.mR, the direct NR exposure method is possible, but the quality of the image is weak. To improve the image contrast of the direct NR film, a second exposure (of the same objects at the same position with the same conditions) to gamma rays coming from the neutron beam of the NR facility is done. By using our procedure the image quality become better.

The result shows (see Figure 3) best contrast of the image, the powder filling of the bullet is clearly discernible.

Direct Image   Gamma Image   Image treated

FIGURE 3. Example of bullet direct image enhanced

CONCLUSION

The present paper described a practically technique of combining gamma and neutron (polluted by gamma) radiography to reach the pure NR image. The applications described above prove that the technique was effective to enhance image contrast. The results are satisfactory but can be more accurate by using a microdensitometer to digitalize the images.

This method can be useful for NR facilities where only transfer technique is mainly used because of time image processing gain.

ACKNOWLEDGMENT

The authors like to thank the NUR reactor staff, for their helps to do experiments.

REFERENCES

1. H. Kobayashi and K. Tomura, "Enhancement of Image Contrast by Combination of Neutron and X-Ray Radiography", Neutron Radiography (2), Proceedings of the Second World Conference, Paris, France, June 16-20 1986. D. Reidel Publishing Company
2. T. Zergoug, M.Y. Mokeddem, A. Nedjar and S. Aouien, "Neutron Radiography at NUR Reactor: Status and Prospects", the 8th World Conference on Neutron Radiography, Gaithersburg, Washington, USA, 16-19 October 2006
3. K. Kamali Moghadem and M.M. Nasseri, "Recognition of Internal Structure of Unknown Objects with Simultaneous Neutron and Gamma Radiography", *Appl. Radiation and Isotopes. Lett.* **61**, 461(2004)
4. S. Casalta, G.G. Daquino, L. Metten, J. Ouadaert and A. Van De Sande, "Digital Image Analysis of X-Ray and Neutron Radiography for The Inspection and The Monitoring of Nuclear Materials", *Ndt&E International Lett.* **36**, 349 (2003)

Performance of SMARTer at Very Low Scattering Vector q-Range Revealed by Monodisperse Nanoparticles

E. Giri Rachman Putra, A. Ikram, Bharoto, E. Santoso, Sairun

Neutron Scattering Laboratory, BATAN, Kawasan Puspiptek Serpong, Tangerang 15314, Indonesia

Abstract. A monodisperse nanoparticle sample of polystyrene has been employed to determine performance of the 36 meter small-angle neutron scattering (SANS) BATAN spectrometer (SMARTer) at the Neutron Scattering Laboratory (NSL) – Serpong, Indonesia, in a very low scattering vector q-range. Detector position at 18 m from sample position, beam stopper of 50 mm in diameter, neutron wavelength of 5.66 Å as well as 18 m-long collimator had been set up to achieve very low scattering vector q-range of SMARTer. A polydisperse smeared-spherical particle model was applied to fit the corrected small-angle scattering data of monodisperse polystyrene nanoparticle sample. The mean average of particle radius of 610 Å, volume fraction of 0.0026, and polydispersity of 0.1 were obtained from the fitting results. The experiment results from SMARTer are comparable to SANS-J, JAEA - Japan and it is revealed that SMARTer is powerfully able to achieve the lowest scattering vector down to 0.002 Å^{-1}.

Keywords: SANS, monodisperse, nanoparticle, form factor
PACS: 25.40. Dn, 81.07. Wx, 83.85. Hf

INTRODUCTION

The electronic, chemical and biotechnology industries and biomedical field are now poised by the nanotechnologies-development revolution. Preparation and development of nanomaterials, such as polymer or inorganic materials nanoparticles is one of the most important aspects on the nanotechnologies development itself. The nanoparticles can be synthesized on different types and materials and as small as 10 Å[1]. The most interesting nanoparticles beside silica nanoparticles that have been applied for many industrial uses including ceramics, chromatography, catalysis, and chemical mechanical polishing[2] is polystyrene nanoparticles. The needs for well-defined monodisperse nanoparticle such as polystyrene or silica nanoparticles have increased for very specific applications in the high technology industries, e.g. electronic (semiconductor nanocrystal - quantum dots), biotechnology as well as pharmaceuticals (biosensors and immobilization of various biomolecules such as enzymes, proteins, DNA, etc.)[1,3]. Pure spherical monodispered polystyrene nanoparticles generally can be synthesized by conventional microemulsion polymerization which consists of three stages.[4] The particle size and polydipersity are controlled by the comparison of amount of styrene, emulsifier and initiator.

On the other hand, a low or small-angle diffraction is one of the most powerful techniques for studying materials in the nanometer scales, e.g. 10 – 10000 Å like conformation of polymer-chains in the melt or solution, ultra-fine grained crystals (nanocrystalline or nanoparticles), phase separation and precipitates in alloys, porous and fractal structures, self-assembly or aggregates of micelles in microemulsions, gels, colloids, etc.[5] The information about the size, shape or distribution of particles or inhomogeneties can be extracted from scattering data.

The experiment involves scattering of monochromatic beam from the sample and measuring the elastically scattered beam intensity as a function of scattering angle. The scattering vector q is the modulus of the resultant between incident, k_i and scattered k_s, wave vectors and is given by

$$|q| = |k_s - k_i| = \frac{4\pi}{\lambda} \sin \theta \qquad (1)$$

where 2θ is scattering angle and λ is the neutron or X-ray wavelength. By substituting equation (1) into Bragg's Law of diffraction, thus a very useful expression is given as

$$d = \frac{2\pi}{q} \qquad (2)$$

where d is a distance. Here, the intensity $I(q)$ of small-angle scattering defined as the coherent differential

CP989, *Neutron and X-ray Scattering in Materials Science and Biology, International Conference on Neutron and X-ray Scattering 2007,* edited by A. Ikram, A. Purwanto, Sutiarso, A. Zulfia, S. Hendrana, and Z. Nurachman

cross section $(d\Sigma/d\Omega)$ as a function of scattering vector q can be expressed as

$$\frac{d\Sigma}{d\Omega}(q) = n(\rho_p - \rho_s)^2 V^2 P(q) S(q) \quad (3)$$

where n denotes the number density of the particle, ρ_p and ρ_s are the scattering length densities of the particle and the solvent, respectively. The term $(\rho_p - \rho_s)$ is called contrast factor. V is the volume of a particle. $P(q)$ is the intraparticle structure factor and depends on the shape and size of the particles, meanwhile $S(q)$ is the interparticle structure factor and is dictated by the interparticle distance and the particle interaction. For the simple shape of monodisperse spheres and in dilute solution where $S(q) = 1$, the size parameters are obtained by least square fits of the theoretical $P(q)$ for a standard geometries, i.e. spherical. Thus the form factor $P(q)$ can be expressed as

$$P(q) = \left[\frac{3(\sin qR - qR \cos qR)}{(qR)^3} \right]^2 \quad (4)$$

A 36 m small angle neutron scattering (SANS) spectrometer (SMARTer) was installed at the neutron scattering laboratory in Serpong, Indonesia. According to its specification, a theoretical q-range of 0.001 Å^{-1} that relates to the dimension size of 6000 Å (0.6 μm) can be achieved. Therefore a set of experiment using a dilute solution monodisperse polystyrene nanoparticle sample was carried out with the intension of reaching as lowest as possible scattering vector q-range (effective q-range) of SMARTer. The data then analysed by applying some model calculations with a package of data analysis program[6]. This work is very important to exhibit over all performance of SMARTer that has not been accomplished before.

EXPERIMENTAL METHOD

The small angle neutron scattering (SANS) measurements were performed using the 36 meters SANS spectrometer at the neutron scattering laboratory (NSL) – BATAN. The spectrometer composes of a mechanical velocity selector, an 18-meter long collimator tube consists of four sections of movable guide-tube and one section of a fixed collimator (non-reflecting tube), sample stage, and another 18-meter long flight tube accommodating a 128x128 ^{3}He two-dimensional position sensitive detector (2D-PSD). Collimation of the incident thermal neutron beam is made by adjustable apertures or pinholes at discrete distances of 1.5, 4, 8, 13, and 18 m from sample position.

The rotational speed of the mechanical velocity selector was set to 5000 and 3500 rpm with tilting angle of 0° producing neutron wavelengths of 3.9 Å and 5.66 Å, respectively. The pinhole settings are 30 mm and 10 mm at neutron source and sample position respectively. Sample to detector distances of 1.5 and 4 m, and neutron wavelength of 3.9 Å were chosen along with that of 18 m and the longest neutron wavelength $\lambda = 5.66$ Å to achieve the lowest scattering vector possible. A polystyrene nanoparticle sample with 1 mm thick was exposed to thermal neutron beam for 2 hours for each sample-to-detector position. The scattered neutrons were detected using a two-dimensional position sensitive detector (2D-PSD) with a beam stopper of 50 mm in diameter. Meanwhile, measurements of background, electronic noise and detector efficiency were carried out for 2, 12 and 12 hours, respectively and temperature was maintained at room temperature during the experiment.

RESULTS AND DISCUSSION

The log-log plot of the one-dmensional measured neutron scattering intensity $I(q)$ versus scattering vector q coming from three different sample-to-detector position measurements is shown in Figure 1.

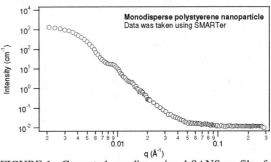

FIGURE 1. Corrected one-dimensional SANS profiles from dilute polystyrene nanoparticle sample in a wide scattering vector q-range using neutron wavelength of 3.9 Å and 5.66 Å

The scattering curve in Figure 1 shows the scattering profiles from a monodisperse system in dilute solution. There are maxima related to the form factor $P(q)$ from the monodisperse spherical system, equation (4), and appeared in a wide scattering vector q-range. However, for further analysis, we are going to analyse the neutron scattering data at very low scattering vector q-range. The best instrument resolution $\Delta\lambda/\lambda$ of 0.11 achieved by applying the longest wavelength $\lambda = 5.66$ Å as well as 18-meter collimation length was chosen to obtain the best performance of SMARTer at very low scattering vector q-range[7].

Corrected data is produced by subtracting the neutron scattering data from the sample with the empty cell as a background, blocked beam as an electronic noise as well as water as a detector efficiency data. One-dimensional scattering pattern like the one shown in Figure 1 is produced by a radial average data calculation from two-dimensional pattern data as shown in Figure 2.

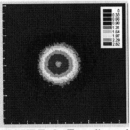

FIGURE 2. Two-dimensional scattering pattern of dilute polystyrene nanoparticle (a) and the background (b) at 18 m of sample-to-detector distance with the neutron wavelength of 5.66 Å.

Corrected data is produced by subtracting the neutron scattering data from the sample with the empty cell as a background, blocked beam as an electronic noise as well as water as a detector efficiency data. One-dimensional scattering pattern like the one shown in Figure 1 is produced by a radial average data calculation from two-dimensional pattern data as shown in Figure 2.

The log-log plot of the one-dimensional measured neutron scattering intensity $I(q)$ versus q in very low scattering vector q-range together with the theoretical calculation and fitting result are presented in Figure 3.

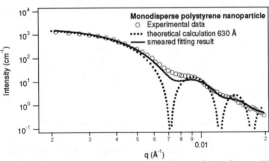

FIGURE 3. The corrected SANS profiles from dilute polystyrene nanoparticle at low scattering vector q range together with the theoretical calculation (dashed line) and fitting result (solid line).

A theoretical calculation of unsmeared monodisperse spherical model with a particle radius of 630 Å has been simply applied to the experimental data and in general showed a good agreement. In spite of this, to have more reasonable fitting on the experiment data, the additional parameter functions

such as polydispersity, volume fraction, scattering length density from particle and solvent, background, smeared data calculation, as well as the instrument resolution values must be applied in the calculation. A data analysis program from National Institute of Standards and Technology (NIST) in United State of America[8] was employed to obtain the best fit from the experiment data and Figure 3 shows the three of them all together.

From the best fitting on the monodisperse polystyrene nanoparticle data we found that the volume fraction of polystyrene particle in the system is 0.0026 with the polydispersity of 0.1 and the particle size of 610 Å as well as 0.003 for the background. To obtain all these values, a known data input such as scattering length of polystyrene particle and the solvent values are definitely required and then fixed during the iteration. Nevertheless, as seen in Figure 3 the best fitting result is also not fitted perfectly to the experimental data of dilute solution monodisperse polystyrene nanoparticle. It suggests that further analysis in term of the structure factor $S(q)$ effect must be deliberated. Solution concentration as well as charge effects in colloidal system of polystyrene nanoparticle must be added into the calculation with more complicated functions to fit the experimental data. Detail of structure factor $S(q)$ effect on data analysis of monodisperse polystyrene nanoparticle system is described elsewhere[9].

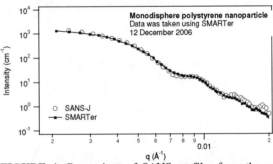

FIGURE 4. Comparison of SANS profiles from the same sample of dilute monodisperse polystyrene nanoparticles using SANS BATAN and SANS-J spectrometers.

Finally, the inter-laboratory comparison on polystyrene nanoparticle sample was conducted between SMARTer and SANS-J spectrometer with the instrument set up of 10 m sample-to-detector distance, 13% of instrument resolution and incident neutron wavelength of 6 Å, Figure 4. It shows that the corrected scattering profile of monodisperse polystyrene nanoparticle produced from SMARTer is comparable to and in a good agreement with the one from SANS-J. The lowest scattering vector q of 0.002 Å^{-1} that corresponds to a size regime of about 3000 Å can be achieved theoretically by SMARTer. However, particle radius of about up to 1000 Å in a dilute solution system can be determined effectively due to the instrument resolution. Figure 5 shows a theoretical calculation from a monodisperse particle model with particle radius of 1000 Å (dashed line) and it also shows that the scattering data profile is just above the lowest scattering vector q-range of SMARTer.

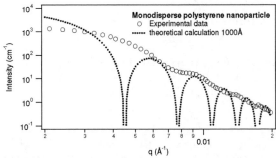

FIGURE 5. SANS profiles from theoretical calculation with the particle radius of 1000 Å compared to experimental data of dilute monodisperse polystyrene nanoparticles with particle radius of 610 Å.

CONCLUSIONS

An excellent performance in revealing the dilute solution of monodisperse polystyrene nanoparticle has been proved by SANS BATAN at very low scattering vector q-range. The experimental data, down to 0.002 $Å^{-1}$ shows a good agreement with the theoretical calculation fitting. Particles with radius of up to 1000 Å in the dilute solution can be determined experimentally using SMARTer.

ACKNOWLEDGMENTS

The authors are indebted to Prof. H. Hasegawa of Kyoto University, Japan for providing the polystyrene nanoparticle sample and Dr. J. Suzuki of JAEA, Japan in providing the experimental data of polystyrene nanoparticles using SANS-J, JAEA, Japan. The authors also acknowledge some advices on data analyses from Dr. J. Kohlbrecher, PSI, Switzerland and Dr. S. Kline, NIST, USA.

REFERENCES

1. M. Qhobosheane, S, Santra, P. Zhang, W. Tan, *Analyst* **126**, 1274-1278 (2001)
2. R. K. Iler, *The Chemistry of Silica*, Wiley, New York, 1979.
3. D. L. Green, J. S. Lin, Y. F. Lam, M. Z. C. Hu, D. W. Schaefer, M. T. Harris, *J. Colloid Interface Sci.* **266**, 346-358 (2003)
4. K. Liu, Z. Wang, *Front. Chem. China* **2**, 17-20 (2007)
5. M. T. Hutchings, C. G. Windsor, *Industrial Applications*, in the *Methods of Experimental Physics*, 23, Part C, edited by K. Skold and David L. Price, Academic Press, Inc., 1987, pp. 405 – 479; *Modern Aspect of Small-Angle Scattering*, edited by H. Brumberger, Kluwer Academic Publishers, Netherlands, 1995.
6. S. R. Kline, *J. Appl. Cryst.* **39**, 895 (2006)
7. E. Giri Rachman Putra, A. Ikram, Bharoto, E. Santoso, "Wavelength Calibration and Instrumental Resolution of 36m SANS BATAN (SMARTer) using Silver Behenate Powder", submitted to *J. Nucl. Related Tech.* 2007
8. http://www.ncnr.nist.gov/programs/sans/ June 2005.
9. E. Giri Rachman Putra, A. Ikram, Bharoto, E. Santoso, *Structure Factor in A Dilute Polystyrene-latex Nanoparticle System*, paper in preparation.

Monochromatic Neutron Tomography Using 1-D PSD Detector at Low Flux Research Reactor

N. Abidin Ashari[1], J. Mohamad Saleh[1], M. Zaid Abdullah[1],
A. Aziz Mohamed[2], A. Azman[2], R. Jamro[2]

[1]School of Electrical and Electronic Engineering, Universiti Sains Malaysia, Engineering Campus,
14300 Nibong Tebal, Pulau Pinang, Malaysia
[2]Malaysian Nuclear Agency (Nuclear Malaysia), 43000 Bangi, Selangor, Malaysia

Abstract. This paper describes the monochromatic neutron tomography experiment using the 1-D Position Sensitive Neutron Detector (PSD) located at Nuclear Malaysia TRIGA MARK II Research reactor. Experimental work was performed using monochromatic neutron source from beryllium filter and HOPG crystal monochromator. The principal main aim of this experiment was to test the detector efficiency, image reconstruction algorithm and the usage of 0.5 nm monochromatic neutrons for the neutron tomography setup. Other objective includes gathering important parameters and features to characterize the system.

Keywords: monochromatic neutron, neutron tomography, Position Sensitive Neutron Detector
PACS: 81.70.Tx, 41.85.Si, 06.70.Dn

INTRODUCTION

Neutron tomography was one of the non-destructive method that begin to receive recognition from research community these days because of its ability to reveal interior structure of certain object which is difficult to image using conventional techniques. Material examples include turbine blade, biological sample or woods. Presently, neutron tomography utilizes polychromatic neutron beam even though the monochromatic beam is still being applied for solving certain specific applications. The later is produced by modulating the neutron beam using monochromatic filter and monochromator. The work reported here used the cooled-type monochromatic neutron beam produced by polycrystalline Beryllium filter. In this way the beam hardening is minimized and the system homogeneity is increased.

Position Sensitive Detector (PSD) is based on gaseous neutron detector. This sensor is chosen due to its good spatial resolution, high dynamic range, improved signal-to-noise ratio and optimal gamma discrimination [1]. The detector is specially designed to detect thermal neutron using neutron absorber gas He^3. For better image quality, a shielding unit is

designed, permitting a 2 mm parallel neutron beam to pass through.

One example in which the monochromatic neutron beam is used for tomography is the Hans-Meitner Institute (HMI). Here, the neutron produced from double crystal diffractometer (DCD) is employed to perform tomographic measurements in both scattering and refraction modes [2]. Similar technique is used at Nuclear Malaysia and reconstruction is based on attenuation measurement.

This paper describes the application of monochromatic neutron beam for neutron tomography. In doing so, the tomographic hardware and image reconstruction software are developed for parallel neutron beam application. Tomographic imaging is performed using data collected using PSD and CCD sensors and resulted compared.

RADIOACTIVE SOURCE

A radioactive source used in this study is capable of producing the monochromatic neutron beam with wavelength up to 0.5 nm. In this case the Beryllium filter together with three layers of highly pyrolitic graphite (HOPG) crystals monochromator is used to modulate the neutron generated by the source. The liquid Nitrogen operating at temperature 190°

CP989, Neutron and X-ray Scattering in Materials Science and Biology, International Conference on Neutron and X-ray Scattering 2007, edited by A. Ikram, A. Purwanto, Sutiarso, A. Zulfia, S. Hendrana, and Z. Nurachman
© 2008 American Institute of Physics 978-0-7354-0508-0/08/$23.00

Celsius is used to cool the Beryllium filter. The monochromator with height at 9 cm and length at 6 cm is mounted on XY goniometer, allowing greater reflectivity and improved alignment. The collimation system for the monochromatic neutron beam is obtained from Small Nagle Neutron Scattering (SANS) facility. The details are shown in Fig. 1.

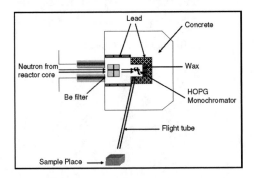

FIGURE 1. SANS collimation system and sample location.

Although the set-up shown in Fig. 1 produced low flux concentration (flux density of approximately 10^3-10^4 ncm^{-2}s^{-1} [3], however, it still produces relatively good image contrast [4]. Good image contrast is one of the most important criteria for successful imaging applications. This monochromatic beam also helps to avoid the cupping effects that normally occur in polychromatic beam due to unwanted low energy neutron beam. These artifacts are amplified in the tomogram due to imperfection in data collection and error in reconstruction. Two important characteristics of the neutron beam produced in the set-up shown in Fig. 1 are that they are homogenous and monochromatic. Thus, the used of PSD-type detector is appropriate.

A neutron beam produced by the generator is projected into a collimator unit of slit size 2.5 mm slit. The collimator is made from 27 mm High Density Polyethylene (HDPE) material. A 2 mm lead is attached in order to filter the unwanted neutron source including gamma-rays. Fig. 2 shows the collimator unit.

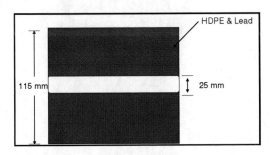

FIGURE 2. Collimator slit designed to allow only parallel neutron beam and to filter gamma-ray.

POSITION SENSITIVE DETECTOR

Position sensitive neutron detector (PSD) used in this research is based on counting detectors and single wire proportional counter (SWPC). When thermal neutron hits the detector, the ^{3}H is produce through the following reaction:

$$^3He + \pi \longrightarrow ^3H + p + 764keV \qquad (1)$$

From this reaction two additional fragments are produced which are proton and triton. The former is about 2/3 and later is 1/3. Electrons are produced as a result of this reaction which is the attracted to the anode. This produces the charge current on the anode strips [5] which is firstly amplified and secondly digitized.

The preferred detectors in this application are those that are capable of locating the interfering patterns and the time-dependence changes occurring inside the samples [1]. It is for this reason that the gaseous-based PSD sensor is a chosen detector. The improved homogeneity demonstrated by this detector with respect to thermal neutron detection is an added advantage. With position resolution better than 2 mm, this detector provides a relatively good resolution for most tomographic applications. However, it lacks pixel resolution compared to CCD based detector. Therefore, the PSD detector is suitable when the size of the imaging target is sufficiently large.

RESULT AND DISCUSSIONS

Based on the experiments that have been performed, we concluded that a fairly good image can be reconstructed using the neutron beams with PSD detector. Comparing real and reconstructed objects, we noticed the images produced by the system accurately characterized the objects. In this case, the sample used is made-up from stainless steel containing air and neutron absorber material in the centre of the block. Fig. 3 below shows the sample phantom.

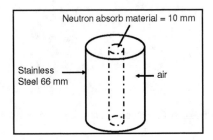

FIGURE 3. Tomography Sample.

As shown in Fig. 3, the stainless steel is shaped into cylindrical rod with outer and inner diameter of 66 mm and 10 mm respectively. The inner region is filled with the neutron absorber material made-up from 10 mm diameter Polyvinyl Chloride (PVC). Fig. 4 shows the open beam profile for this experiment while Fig. 5 shows the sample profile.

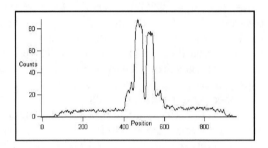

FIGURE 4. Open beam profile for the experiment.

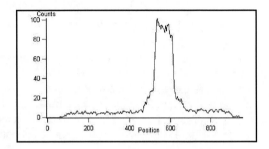

FIGURE 5. Projection data for the sample.

Clearly from Fig. 5, the system produces a fairly good projection profile in spite of low neutron flux.

In this experiment, the numbers of total projection data collected are approximately 108,000 and the sample is rotated at 1.8° intervals. These data are then processed tomographically using three different algorithms. Namely they are the filter back-projection, the convolution technique and the iterative method. Results comparing these different techniques are shown in Fig. 6-8.

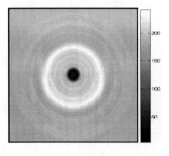

FIGURE 6. Filter back-projection reconstruction.

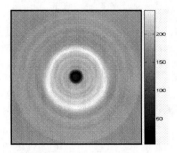

FIGURE 7. Convolution reconstruction.

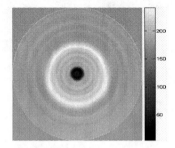

FIGURE 8. Iterative reconstruction.

Careful examination of Fig. 6-8 show that reconstruction images are in good agreement with the real image. The reconstructed diameter of the sample is approximately 60 mm compared to its actual dimension of 66 mm. Similarly, the reconstructed diameter of PVC material is 9.8 mm compared to the actual size of 10 mm. In both cases the spatial location of both the stainless steel and the PVC rods are accurately reconstructed. This experiment was repeated using the commercial spark-plug with CCD sensor as shown in Fig. 9. Fig. 10 shows the image reconstructed at the cross-section line marked 'A'. It can be seen from this image that important features inside the spark-plug, namely the carbon rods are correctly mapped.

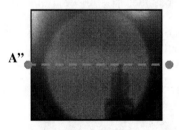

FIGURE 9. Reconstruction of the spark plug Performed at A".

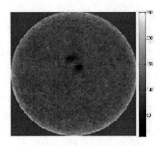

FIGURE 10. Reconstructed image of cross-section A"

CONCLUSIONS

In conclusions, the prototype neutron imaging facility developed at Nuclear Malaysia utilizing monochromatic beam and PSD detection could produce a fairly good reconstruction. More tests need to be carried out in order to improve the accuracy and image-to-noise ratio. Future research includes upgrading the shielding unit so that interference caused by gamma-rays could be minimized.

ACKNOWLEDGEMENTS

The authors would like to thank you Nuclear Malaysia Reactor Interest Group (RIG) for their financial support for this research and allowing access to the neutron facility. The authors also acknowledge support from members of the Universiti Sains Malaysia's School of Electrical and Electronic and Nuclear Malaysia's Material, Reactor and Engineering groups.

REFERENCES

1. A. Oed, "Detector for Thermal Neutron", *Nucl. Instr. and Meth.* **A525**, 62 (2004)
2. W. Treimer, M. Strobl, A. Hilgera, C. Seifert and U. Feye-Treimer, "Tomographic Imaging With a Bonse-Hart camera", 2005.
3. A. Ghafar and A.Fatah, "The Small-Angle Neutron Spectrometer at MINT", *Workshop on the Utilization of Research Reactors,* Jakarta, Indonesia (1994)
4. M. Dierick, "PhD Thesis-Tomographic Imaging Using Cold and Thermal Neutron Beam", Gent University, Belgium, 2004.
5. B. Yu, G.J. Mahler, N.A. Schaknowski and G.C. Smith, "A Position-Sensitive Ionization Chamber for Thermal Neutron", *IEEE Transactions on Nuclear Science* **48**, 336 (2001)

Design of the Mechanical Parts for the Neutron Guide System at HANARO

J. W. Shin, Y. G. Cho, S. J. Cho and J. S Ryu

Korea Atomic Energy Research Institute, 1045 Daedeok-daero, Yuseong-gu, Daejeon 305-353, Korea

Abstract. The research reactor HANARO (High-flux Advanced Neutron Application ReactOr) in Korea will be equipped with a neutron guide system, in order to transport cold neutrons from the neutron source to the neutron scattering instruments in the neutron guide hall near the reactor building. The neutron guide system of HANARO consists of the in-pile plug assembly with in-pile guides, the primary shutter with in-shutter guides, the neutron guides in the guide shielding room with dedicated secondary shutters, and the neutron guides connected to the instruments in the neutron guide hall. Functions of the in-pile plug assembly are to shield the reactor environment from nuclear radiation and to support the neutron guides and maintain them precisely oriented. The primary shutter is a mechanical structure to be installed just after the in-pile plug assembly, which stops neutron flux on demand. This paper describes the design of the in-pile assembly and the primary shutter for the neutron guide system at HANARO. The design of the guide shielding assembly for the primary shutter and the neutron guides is also presented.

Keywords: neutron guide, in-pile plug, primary shutter.
PACS: 61.05.fg

INTRODUCTION

The Cold Neutron Research Facility (CNRF) is a facility to produce and utilize cold neutrons for basic science and nano- and bio-technology research. Since 2003, thanks to a national demand for the cold neutron research, a project called 'the infrastructure construction for cold neutron research and utilization technique development' started at HANARO (High-flux Advanced Neutron Application ReactOr) in KAERI [1]. Developments of the cold neutron source and related systems, the neutron guides and the neutron scattering instruments are to be accomplished by 2010.

The cold neutron research facility will be constructed as shown in Fig. 1. Cold neutrons will be generated from the cold neutron source (CNS) which will be operated in a vertical hole of the reflector tank with liquid hydrogen at 20K [2]. The neutron guide system will be installed to transport cold neutrons from the cold neutron source to the neutron scattering instruments in the neutron guide hall without a significant loss of neutrons and with a low radiation background.

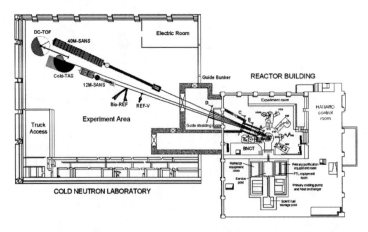

FIGURE 1. The layout of cold neutron research facility at HANARO

CP989, *Neutron and X-ray Scattering in Materials Science and Biology, International Conference on Neutron and X-ray Scattering 2007*, edited by A. Ikram, A. Purwanto, Sutiarso, A. Zulfia, S. Hendrana, and Z. Nurachman
© 2008 American Institute of Physics 978-0-7354-0508-0/08/$23.00

The neutron guide system of HANARO is divided into three different parts. First, the in-pile plug assembly and the primary shutter with in-plug and in-shutter guides, then neutron guides in the guide shielding room with dedicated secondary shutters, and finally neutron guides connected to neutron scattering instruments in the neutron guide hall. The 5 in-pile guides of different cross sections in the in-pile plug assembly start at a distance of 1833mm from the cold neutron source. The guides in the in-pile plug will be incorporated into the primary shutter outside the biological shielding filled with helium. Five neutron guides are named as CG1, CG2, CG3, CG4 and CG5 from north to south and they have incline angles of +2.97°, +1.84°, +0.47°, -1.91° and -2.50° with respect to the beam port axis. After the primary shutter, the guides start to curve in the reactor confinement building in order to remove the gamma rays and fast neutrons, with each providing different curvatures. CG2 and CG5 guides will be separated into two (CG2A, CG2B) and three (CG5A, CG5B, CG5C) guides respectively by using splitters next to the primary shutter. In the end, there are 8 guides in the guide hall, and each instrument will be supplied with high flux neutrons. The overview of the neutron guide system is shown in Fig. 2.

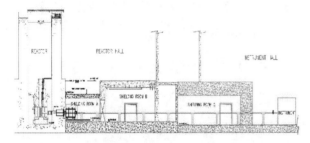

FIGURE 2. The overview of the neutron guide system including reactor, in-pile plug assembly, primary shutter, neutron guides, guide shielding room and instruments

In this paper, design requirements and designs of the in-pile plug assembly and the primary shutter for the neutron guide system were described. The design of the guide shielding assembly for the primary shutter and the neutron guides is also presented.

IN-PILE PLUG ASSEMBLY

An in-pile plug with five in-plug guides will be installed in the CN beam port facing the cold neutron source in the reactor side. It was designed to install in-plug guides at the exact location in a high radiation environment and to enable an easy maintenance and replacement of the guides periodically. The CN beam port is a divergent shape with the nose beam size of 70mm x 150mm and the exit beam size of 150mm x

150mm which is located 630mm away from the nose [3]. The in-pile plug is a two-stepped cylinder type with a 380mm (diameter) x 735mm (length) and a 700mm (diameter) x 1170mm (length) as shown in Figure 3.

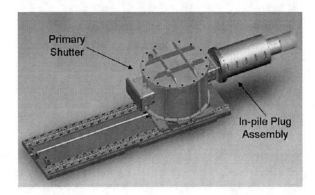

FIGURE 3. The design of the in-pile plug assembly and the primary shutter

The plug (Fig. 4A) is divided into a lower half cylinder and an upper half cylinder. The lower half cylinder is fixed to the base table, which can slide into the beam hole. Once the guide cassette is mounted in the pyramidal center space of the cylinder, the upper half cylinder is solidly fixed to the lower part. These parts are machined from cast carbon steel. All the carbon steel parts are nickel plated, in order to avoid corrosion. The steel shall contain less than 10 ppm cobalt. The base table (Fig. 4B) supports the plug during a mounting, and in its final position. It has an important function as a mechanical reference by being attached to a ring flange centered to the beam tube axis.

The guide cassette (Fig. 4C) is a steel structure protecting and keeping the 5 neutron guides well aligned to the neutron optical path with several mechanics for a high precision vertical and horizontal alignment. Once the neutron guides are mounted inside the cassette, this assembly is called the optical unit. Typical features of an optical unit are reference markings to reproduce the optical path, and leveling jacks for height adjustment. These jacks can be remotely operated by special tools. The new flange (Fig. 4D) is a blind flange, which has the function of keeping the plug in place and confined in Helium atmosphere. Additional features of the new flange are to transfer the neutron optical axis to the reactor face and to support the cold neutron beam window via a window flange (Fig. 4E). The new flange (diameter 830mm, thickness 50mm) is made of stainless steel 304L and it is fixed to the ring flange with M12 screws. The window flange is 30 mm thick and the window is a thin $AlMg_3$ sheet of 2mm thickness.

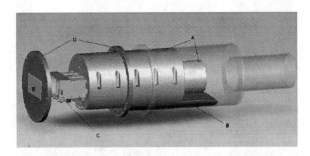

FIGURE 4. The exploded view of the in-pile plug assembly (A: Plug, B: Base Table, C: Cassette, D: Blind Flange & Ring Flange, E : Window Flange)

PRIMARY SHUTTER

The primary shutter is a mechanical structure to block the beam passage, which is horizontally rotated by an electrical system using a stepping motor and positioned at the acceptable tolerance level. When it is closed, no neutron beams will be available. Five neutron guides are incorporated into the primary shutter and establish a continuity of the neutron guides when the shutter is opened. For the installation and the maintenance of the primary shutter, it will be installed on a rail system as shown Fig. 5 and remotely controlled from control panels in the reactor building. An appropriate interlock system will be provided to ensure a safe operation of the primary shutter. It will prevent an access to the guide shielding assembly when the primary shutter is not fully closed.

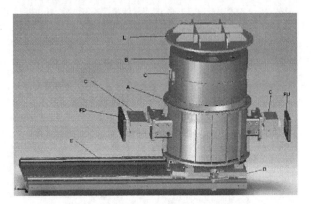

FIGURE 5. The exploded view of the Primary shutter (A: Vessel, B: Drum, C: Cassettes, D: Carriage, E: Rails, FD: Window downstream, FU: Window upstream, L: Vessel Lid)

The primary shutter contains a carbon steel drum (Fig. 5B) that can rotate on its vertical axis to open and close the neutron beam pathway. The drum is housed in a vacuum vessel (Fig. 5A). The neutron guides are located inside the vessel in two static cassettes (upstream and downstream) and one rotating cassette (Fig. 5C). The vessel rests on a support structure with a carriage (Fig. 5D), which can move on rails (Fig. 5E)

for an initial mounting, and for access to the beam tube port. A motor drives the drum through a rack-and-pinion gear.

The vacuum vessel holds the neutron guide components in an alignment and under vacuum. It is a welded structure with an inner diameter of 1450mm and a height of 1000mm, made of steel sheet with 20mm thickness. The vessel lid (Fig. 5L) is welded with a flange that fits to the vessel top flange. The vessel base plate is welded to the shroud. It contains the upper side a ring, which holds a heavy precision bearing. The drum can rotate smoothly on this bearing. The rotating drum is made of lower and upper parts, which envelope the neutron guide rotating cassette. Each part of the drum is a welded cylindrical box made of a 15mm steel sheet, which is filled with heavy concrete and layers of beam catching materials. The density of heavy concrete including steel resin and barite is 6.0 g/cc. Boral, borated polyethylene, boron carbide and lead are used as beam catchers. The vessel support structure consists of a 100mm thick base plate, two rails for their supports, and a 1.6m x 2.0m carriage with rollers. The primary shutter can be installed on the carriage by a crane 2m downstream from its final position. Once the neutron guide cassettes are aligned, the shutter can be moved precisely parallel to the optical axis towards the reactor face. The gap between windows of the in-pile plug assembly and the primary shutter is 5mm at the final location.

GUIDE SHIELDING ASSEMBLY

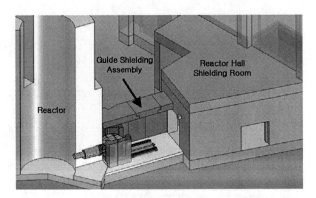

FIGURE 6. The section view of the guide shielding assembly for the primary shutter and neutron guides

The guide shielding room (or assembly) is a biologically shielded area surrounding the neutron guides from the primary shutter up to the secondary shutter at the end of each guide. It is extended from the reactor hall over a part of the neutron guide hall as shown in Fig. 2. It consists of the guide shielding assembly for the primary shutter, the reactor hall shielding room and the guide hall shielding room.

Figure 6 shows the guide shielding assembly which surrounds the primary shutter and neutron guides. The guide shielding assembly will be made of heavy concrete blocks whose densities are 4.5g/cc and thicknesses are 500mm. Ceiling and walls of the shielding assembly consist of three blocks respectively, which have shapes of step at each end. They will be assembled by steel plates and bolts with a gap of 3mm.

CONCLUSION

The design of the in-pile plug assembly and the primary shutter for the neutron guide system at HANARO has been performed by KAERI and MTF GmbH. The in-pile plug assembly and the primary shutter shall be fabricated by domestic companies according to these designs and technical specifications by the end of April 2008.

ACKNOWLEDGMENTS

This work was supported by the Nuclear R&D Program of Ministry of Science and Technology of Korea.

REFERENCES

1. Y. J. Kim, "Conceptual Design of the Cold Neutron Research Facility in HANARO", *HAN-CP-RD-030-04-001, KAERI*, 2004, 170.
2. K. H. Lee, Y. J. Kim, "Current Status of the HANARO Cold Neutron Source Design", *International Symposium on Research Reactor and Neutron Science HANARO*, 2005, 537.
3. B. S. Seong, S. J .Cho, D. G. Whang, C. H, Lee and Y. J. Kim, "Basic Concept of the Neutron Guide System at HANARO," *International Symposium on Research Reactor and Neutron Science HANARO*, 2005, 518.

Neutron Radiographic Inspection of Industrial Components using Kamini Neutron Source Facility

N. Raghu, V. Anandaraj, K.V. Kasiviswanathan and P. Kalyanasundaram

Indira Gandhi Center for Atomic Research, Kalpakkam – 603 102, India

Abstract. Kamini (Kalpakkam Mini) reactor is a U^{233} fuelled, demineralised light water moderated and cooled, beryllium oxide reflected, low power (30 kW) nuclear research reactor. This reactor functions as a neutron source with a flux of 10^{12} n/cm^2 s^{-1} at core centre with facilitates for carrying out neutron radiography, neutron activation analysis and neutron shielding experiments. There are two beam tubes for neutron radiography. The length/diameter ratio of the collimators is about 160 and the aperture size is 220 mm x 70 mm. Flux at the outer end of the beam tube is ~ 10^6 - 10^7 n/cm^2 s. The north end beam tube is for radiography of inactive object while the south side beam tube is for radiography of radioactive objects. The availability of high neutron flux coupled with good collimated beam provides high quality radiographs with short exposure time. The reactor being a unique national facility for neutron radiography has been utilized in the examination of irradiated components, aero engine turbine blades, riveted plates, automobile chain links and for various types of pyro devices used in the space programme. In this paper, an overview of the salient features of this reactor facility for neutron radiography and our experience in the inspection of a variety of industrial components will be given.

Keywords: Kamini, neutron radiography, flux, L/D ratio, thermal neutron, pyro device, turbine blade, roller chain
PACS: 29.25.Dz

INTRODUCTION

Kamini is a unique research reactor and is a national facility for neutron radiography in India. The reactor is water moderated and cooled beryllium oxide reflection, low power 30 kW tank type reactor at Indira Gandhi Centre for Atomic Research (IGCAR), Kalpakkam. The reactor functions as a neutron source with a flux of 10^{12} n/cm^2 s^{-1} at core center and facilitates carrying out neutron radiography of radioactive and non-radioactive objects and neutron activation analysis. Facilities are also available in the reactor for carrying out radiation physics research, irradiation of large samples and calibration and testing of neutron detectors.

The reactor fuel is an alloy of Uranium-233 and aluminium in the form of flat plates and assembled in aluminium casing to form the fuel subassemblies. The reflector is beryllium oxide encased in zircaloy sheath. Demineralised light water is used as moderator, coolant as well as shield. Cooling of the reactor core is by natural convection. Startup and regulation of the reactor is done by adjusting the positions of two safety control plates made of cadmium, which is sandwiched in aluminium. These plates are provided with gravity drop mechanism for rapid shut down of the reactor.

All reactor operations are carried out from a central control panel which uses a hybrid system comprising of hardwired systems and microprocessor based data logging, display and annunciation system. Mimic diagrams, color graphics and alphanumeric displays are provided for guiding the operator and facilitate convenient operation.

EXPERIMENTAL METHODS

Neutron radiographic facilities at KAMINI includes aperture control devices, collimators, beam shutters, film cassette drive mechanism and a radiographic rig for lowering the fuel pins / fuel subassembly in front of the beam tube directly from the hot cells with a indexing mechanism facilitating rotation of the components [1,2].

The neutron radiography rig is made up of three parts viz.; the outer shell, carriage assembly and carriage drive mechanism. The outer shell of the rig is a 7.8 meter long leak tight tube assembly, extending from the cell floor to the reactor vault, with explosively bonded stainless steel/aluminium window at the position corresponding to the south beam port for neutron radiography of irradiation fuel pins and fuel assembly. The outer shell acts as a containment

CP989, *Neutron and X-ray Scattering in Materials Science and Biology, International Conference on Neutron and X-ray Scattering 2007,* edited by A. Ikram, A. Purwanto, Sutiarso, A. Zulfia, S. Hendrana, and Z. Nurachman

vessel for the radioactive objects and is integrated to the cell nitrogen atmosphere with a leak tight joint. The straightness of the inner cylindrical surface is maintained within 0.1mm/meter. Guide rails are screwed to the inner surface of the tube for the movement of the carriage containing the irradiated object. The carriage assembly consists of a rotatable aluminium container which can carry the irradiated fuel subassembly and pins in vertical position. The carriage has a rectangular cut-out on two sides facing the neutron beam to allow neutron to pass through without attenuation. On top of the carriage, a stepper motor with indexing arrangement is positioned along the axis of the aluminium container with the help of dowel pins. With this fuel subassembly can be indexed precisely in steps of ± 0.5 degree rotation for radiography at different orientation.

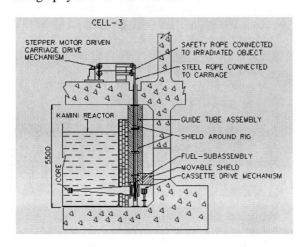

FIGURE 1. Cross-section of the KAMINI facility

The stepper motor controlled carriage drive and indexing mechanism consists of two steel rope drums (main and auxiliary) for moving the carriage up and down and another for lifting the indexing mechanism. The carriage is lowered or raised using a rope and drum mechanism driver by a stepper motor through a worm and worm wheel drive. Two limit switches have been provided in the carriage drive mechanism to indicate the position of the carriage inside the guide tube. At the bottom of the guide tube, a shock absorbing device is provided to bear any shock loads.

For film radiography, a cassette drive mechanism with ten cassettes arranged in a decagonal fashion nad remote indexing has been designed and fabricated (Fig.2). The cassette drive is also stepper motor driver for indexing the converter screens in front of the neutron beam window. A central pneumatic cylinder pushes the cassette holder towards the window to reduce the gap between the object and the screen. The cassettes are of top loading type and can be loaded or

retrieved when the reactor is not in operation and when no radioactive object is present in the rig.

Neutron Radiographic Inspection of Radioactive materials

Initially the beam was characterized using ASTM E545 beam purity indicator and a sensitivity indicator and the quality of the beam determined. The operation was rehearsed with dummy fuel pins using a real time neutron image intensifier system. Paraffin inserts of known dimensions were used as calibration pieces and for the determination of the orientation. The fuel was loaded onto the holder and lowered from the hot cell to the radiography window. Since the neutron beam port window cannot cover the entire length of the pin, radiography is carried out sequentially starting from the spring end of the fuel pin upto the plenum end in three stages.

Evaluation of the radiographs of irradiated fuel pins in general revealed pellet-to-pellet gaps and pellet to clad gaps at the end of the fuel column. Apart from this, chipping of the edges of the insulation pellet, crumbling of the insulation pellet and cracks in fuel pellet were detected in a few cases. Microdensitometric analysis of the radiographs was carried out by profiling the fuel column axially and radially. The density variations were observed to be within the limits of statistics indicating no changes in the fuel region and no actinide redistribution [3].

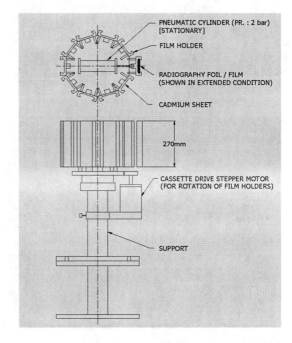

FIGURE 2: Cassette drive mechanism used for multiple exposures

Neutron Radiographic Inspection of Industrial Components

Pyro devices are mission critical devices which contain a small amount of explosive substance in it for performing a desired specific activity. These devices are assembled, self contained and sealed within a metal casing which can be initiated only once. Thermal neutron scattering cross section is high for hydrogen based explosive materials and presence of explosive charge encased within the metal casing is effectively imaged only by neutron radiographic method. In a space mission, hundreds of such pyro devices are used for ignition, stage separation, deployment of satellite solar panels and in the destruct system. Typical devices that are used in the space programme include cable cutters, bolt cutters, explosive manifold, pyro valves, detonating cartridges, pin pushers, pyro thruster and explosive transfer assembly etc. During radiographic inspection, the pyro devices are suitably fastened and integrated with the existing cassette drive mechanism to facilitate multi frame exposure of the pyro devices when the reactor is in service. Figure 3 shows the digitized radiographic images of a few pyro devices. Excellent radiographic contrast and sensitivity enabled clear delineation of the pyro charges, elastomers, filling and integrity of the potting compound, internals, etc.

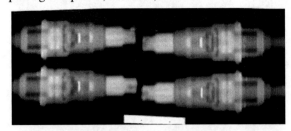

A: Dual Pyro valves

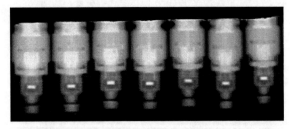

B: Detonating cartridges

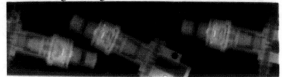

C: Satellite pyro valves

FIGURE 3. Radiographs of a few tested pyro devices

Roller chains are widely used to transmit mechanical power in the industry. The productivity and cost of operations is dependant on the performance and life of roller chains. Adequate lubrication helps to reduce friction and enhance galling resistance and therefore prolongs roller chain wear life. Proper lubrication provides a cushioning effect under impact load and also in heat dissipation of cylindrical rollers which articulate when they enter and leave the sprocket.

The roller chains are coated with petroleum jelly mixed with additives after assembly to ensure adequate lubrication between the pins and rollers within the bushings. The interfacial gap which is available is ~ 25 micrometers and it was necessary to optimize the process parameters. Neutron radiography technique was very helpful to optimize the process parameters to ensure uniform homogeneity of the lubricant inside the bushings, as thermal neutron are effectively scattered in presence of these hydrogen based lubricants. The radiographs of the lubricated and non lubricated chain assemblies are shown in Fig.4. The presence of the lubricant in the bottom chain assembly can be easily identified.

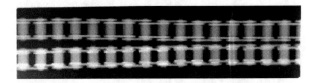

FIGURE 4. Radiograph revealing presence of lubricant in the chain assembly (bottom link)

In the aerospace industry, investment cast gas turbine blades in a ceramic investment shell mould is used. The turbine blades has internal cooling passages extending through the aerofoil and root portions through which compressor bleed air is conducted to cool the aerofoil portion while the engine is in operation. Gadolinium salts are added in a small quantity in the core sand for facilitating radiographic contrast. During casting, the ceramic core is positioned in the investment mould having a configuration corresponding to the internal cooling passages to be formed through the aerofoil and root portion of the turbine blade. The ceramic core is then leached out from the investment cast component in an aqueous caustic bath. The ceramic core can break and remain inside the cooling passage in a cast blade. Neutron radiography helps to reveal the presence of these core materials in a cast blade. The core particles entrapped within the cooling passages on two directionally solidified super alloy turbine blades is shown in Fig.5.

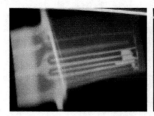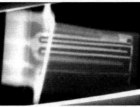

FIGURE 5. Radiographs of HP turbine blades revealing entrapment of core particles within the cooling passage

CONCLUSIONS

The availability of high thermal neutron flux coupled with large L/D ratio of the beam tube has been used to carryout neuron radiographic inspection with high sensitivity and contrast. The Kamini neutron radiographic facility has been well utilized in the post irradiation examination of irradiated fuels pins and control rod assemblies. The neutron radiography activity has also aided in qualification of a number of critical pyro devices and assemblies used in the Indian space mission and aerospace programme.

ACKNOWLEDGMENTS

The authors are thankful to Dr. Baldev Raj, Director, IGCAR for his constant encouragement and support. The authors express their gratefulness to Reactor Operations Division for their kind co-operation in our inspection activities.

REFERENCES

1. K.V. Kasiviswanathan, B. Venkatraman and B. Raj, "Neutron Radiographic Facilities Available with Kamini", Proceeding of the 6th World Conference on Neutron Radiography, Edited by S. Fujine, H. Kobayashi and K. Kanda, Gordon and Breach Science Publishers, Japan, 1999, pp. 117-120.
2. B. Venkatraman, K.R. Sekar, K.V. Kasiviswanathan and P. Kalyanasundaram, "Calibration of KAMINI Neutron Beam Quality for Radiography – Our Experiences", ibid, pp. 145-150.
3. B. Venkatraman, N. Raghu, T. Johny, K.V. Kasiviswanathan and B. Raj, "Radiographic Techniques for Post Irradiation Characterization of Fast Breeder Test Reactor Fuel Pins", International Conference on Quality Control of Nuclear Fuels, (CQCNF), Dec.10-12, 2002, Hyderabad, India.

MODELING AND SIMULATIONS

Dynamical Temperature Study for Spin-Crossover

W. B. Nurdin[1] and K. D. Schotte[2]

[1]*Laboratory of Theoretical and Computer, Department of Physics, FMIPA,
University Hasanuddin, Makassar 90245, Indonesia*
[2]*Fachbereich Physik, Freie Universitaet, Berlin 13353, Germany*

Abstract. Making use of the Rugh's dynamical and micro-canonical approach to temperature, we study the classical model of three dimensional spin-crossover compounds. These compounds are characterized by magnetic ions that can be in a high-spin or low-spin state. We consider the case of diamagnetic low-spin state which are appropriate for Fe-Co compounds. The values of the magnetization average and fraction of high-spin/low-spin are studied over a wide range of values for the system size, temperature, magnetic field, nearest neighbor coupling and exchange interaction. We also address metastability according to the relative values of interaction parameters and the phase diagram of the model.

Keywords: Phase transition, dynamical temperature, spin crossover.
PACS: 05.10.Ln; 05.50.+q; 07.05.Tp; 75.10.Hk; 75.60.Ej

INTRODUCTION

The microscopic dynamic model has been treated so far in the mean field approach. The present study uses the Rugh's micro-canonical approach to temperature [1-6], to be taken into account in the classical model of three dimensional spin-crossover compounds. These compounds are characterized by magnetic ions that can be in a high-spin or low-spin state. We consider the case of diamagnetic low-spin state which is appropriate for Fe-Co compounds. Co-Fe is an attractive material owing to its two-way photo-switching, i.e., magnetic and nonmagnetic. Since the state of Co(II)-Fe(III) which is favorable at high temperature owing to its high degeneracy, we call it a high-temperature (HT) state and the Co(III)-Fe(II) state which is favorable at low temperature owing to its low energy, as a low-temperature (LT) state. In this material, the magnetic coupling between Co and Fe ions is antiferromagnetic. On the dynamics of spin-crossover solids the frontier works have pointed out the role of the interaction on the shape of relaxation curves, i.e. the time dependence of the metastable state fraction after photoexcitation of the system. In some octahedral coordinate iron complexes, a spin crossover (SC) may occur between a high-spin (HS) state and a low-spin (LS) state. An SC transition is induced by external stimulations such as temperature, photo-irradiation, and magnetic field [15–19].

The HS state is favorable at high temperature owing to its high degeneracy. On the other hand, the LS state is favorable at low temperature because it has a low energy with a low degeneracy. Interesting transitions between HS and LS states are provided by the competition between entropy gain and energy gain.

For SC transitions it has been pointed out that cooperative interactions are important. Smooth transition or discontinuous first order phase transition occurs in the SC transition according to the system parameters [15–16]. Control between the HS and LS states has been realized by photoirradiation as light-induced excited spin state trapping (LIESST) [7–12, 19–24] and the structure of the photoinduced metastable state of the systems has become an important topic [11–14]. Wajnflasz-Pick (WP) model gave the theoretical basis of the mechanism of the cooperative transitions using an Ising model with degenerate states [15]. The WP model and its extended models have explained successfully the static and dynamical properties of various cases of SC transitions [17–23]. The metastable state can exist intrinsically at low temperature, which has been supported experimentally [13,14]. The effects of applying a magnetic field have been studied in some SC complexes [25–30]. Since the magnetic moment in the HS state is larger than that in the LS state, the HS state is stabilized by applying a magnetic field. A shift in transition temperature under an applied magnetic field was confirmed experimentally [25–27].

DYNAMICAL TEMPERATURE

Starting from a micro-canonical ensemble, the entropy is a function of energy. It is defined as the logarithm of the number of different states which a physical system has with a given energy, that is $S = ln\ W$ with Boltzmann's constant set to unity. For a classical system with Hamilton function is proportional to the surface of constant energy in the phase space of canonical coordinates. More precisely it

CP989, *Neutron and X-ray Scattering in Materials Science and Biology, International Conference on Neutron and X-ray Scattering 2007*, edited by A. Ikram, A. Purwanto, Sutiarso, A. Zulfia, S. Hendrana, and Z. Nurachman
© 2008 American Institute of Physics 978-0-7354-0508-0/08/$23.00

is the number of points between infinitesimal neighboring surfaces written in the form of a Gaussian surface integral. The direction of the infinitesimal surface element is normal to surface of constant energy coincides with the direction of the gradient. This quantity in general cannot be calculated directly for a mechanical system, but its logarithmic derivative with respect to energy can as has been pointed out by Rugh [6,7]. Using Gauss' theorem the last integral in energy formula can be written as that is as an integral over all phase space where the energy is lower. The derivative with respect to energy is again a constant energy integral as for W. Using a pulsed high magnetic field, the creation of the HS state from the LS branch in the thermal hysteresis loop was observed and these results were interpreted using an Ising-like model [28–30]. However, since both the HS and LS states in the SC complexes are paramagnetic, the study of the effects of applying a magnetic field has been performed only for the paramagnetic region. Both phenomenological mean field model and dynamic Monte Carlo simulation have been reported to described the experimental data. The typical feature of self accelerated relaxation gives a sigmoidal shape was assigned to the effect of interaction. In addition a 'tail' was observed and attributed to the transient onset of correlations. This tail was not obtained in the mean field approach. The purely molecular aspect of relaxation has also received much attention, and the simple idea of thermally activated process in a large temperature interval (for instance 25 K) was developed for the examples of diluted complexes. Accordingly, the dynamic choice which was made to establish the first microscopic dynamic model of spin-crossover solids was Arrhenius-like, rather than widely accepted Glauber choice [7-14]. This choice was crucial to reproduce the experimental non-linear effects, i.e. sigmoidal relaxation curve in the range 30-60 K.

SPIN MODEL

Consider the phenomenon of a bipartite lattice Co-Fe, where one of the sublattices is occupied by Fe ions and the other is occupied by Co ions. The site of the Fe we call sublattice an A-site, and a site of the Co sublattice as B-site. The HT state consists mainly of Fe(III) and Co(II) ions. The LT state consists mainly of Fe(II) and Co(III) ions. Here, we consider the degeneracy of the spin degree of freedom. At an A-site, Fe(II) in the LT state is $S = 0$, whose degeneracy is 1, and Fe(III) in the HT state is $S = 1/2$ with a degeneracy of 2. At a B-site, the degeneracy of $n_B = 0$ is 1 and that of $n_B = 1$ is 4 because Co(III) in the LT state is $S = 0$, and Co(II) in the HT state is $S = 3/2$ with a degeneracy of 4. Besides the state site spin state n

degeneracy spin degeneracy, the system has a degeneracy due to vibrational motions: $g_{phonon}(HS)$ and $g_{phonon}(LS)$. The degeneracy due to phonons is much larger than that of spin. However, in this study, we mainly use spin degeneracy, which is enough to provide the general aspects of the phase structure.

This charge transfer phenomenon is expressed by an electron transfer between Fe and Co atoms. To express this transfer, we introduce the quantity n, which is 1 for Fe in the LT state (i.e., $n_A = 1$) and 0 in the HT state ($n_A = 0$). Correspondingly, it is 0 for Co in the LT state ($n_B = 0$) and 1 in the HT state ($n_B = 1$). Here, i(j) denotes A-(B-)sites. In this system, electrons are transferred between A and B sites. Here, D (> 0) is the difference in on-site energy between A- and B-sites, and J represents the magnetic interaction between A- and B-sites, and s denotes the spin. The external magnetic field is given by h. Since an Fe and Co pair has an electron of the total number of Co and Fe ions of the lattice. The magnetic states of the spins is introduced by s_A and s_B. When $n_A = 1$, $S = 0$ and thus s_A takes only 0, and when $n_A = 0$, $S = 1/2$ and thus s_A takes $-1/2$ or $1/2$. Similarly, when $n_B = 0$, $S = 0$ and s_B takes 0, and when $n_B = 1$, $S = 3/2$ and s_B takes $-3/2, -1/2, 1/2$ or $1/2$. Specifying s_A and s_B, the degeneracy is taken into account naturally. In this model, electrons tend to stay at A-sites energetically, which denotes the LT state. When all the electrons are at A-sites, there is no energy cost due to D. We define this state as the perfect LT state. At high temperature, electrons tend to stay at B-sites which corresponds to the HT state.

Effect of magnetic interaction

The system possesses a HT metastable branch at low temperature in the case without the magnetic interaction (J = 0) as the temperature dependences of n are depicted with the parameters J = 0, 0.02, 0.04, and 0.06,. As the magnetic interaction is included, we find that the metastable branch is enlarged. In the case of J = 0.02, the HT fraction of the metastable HT state is increased. The HT fraction of the unstable solution is also affected. The solution of J = 0 is not changed unless the magnetization has nonzero solution in the MF treatment.

The metastable solution and the unstable solution terminate at different temperatures T = 0.41 and T = 0.39, respectively, and are not connected to each other. This is due to the fact that the present model has three order parameters, i.e., (n, sA, sB). At T = 0.35, there are three solutions that are shown by closed circles for local minimum points and an open circle for a saddle point. When T is increased up to T = 0.4, the saddle point disappears, where two stable solutions correspond to points on the green solid line and black solid line denoting the LT state. At T = 0.42, the metastable point disappear and only one stable point remains, which gives a point on the black line. For large values of J, HT remains locally stable until T = 0, for J = 0.06. A unique dependence of the stability of the ferrimagnetic state depends on J. If J < Jc = D/18, the ferrimagnetic state is metastable and not an equilibrium state at low temperatures. Here, Jc is obtained by the comparison of the energy of the complete ferrimagnetic state and the LT state. We can show an example of J = 0.54 < Jc.

Effect of applied external field

Let us study the case where both thermal hysteresis and the HT metastable exist at low temperature. Consider the effect of the applied magnetic field h. The magnetic field dependence of magnetization at T = 0.1 is presented for various values of J (J = 0.02, 0.03, and 0.05). In the case of J = 0.02, the LT state is a stable state in the absence of a magnetic field. For a weak field, the system is always in the LT phase. The magnetization is suddenly induced up to nearly the saturated value at h = 0.338. The magnetization value of 2 indicates the ferromagnetic state of Fe (S = 1/2) and Co (S = 3/2) ions in the HT state. In the case of J = 0.04, a stable state in the absence of a magnetic field is still the LT state. The magnetization of 1 indicates an antiferromagnetic coupling between Fe and Co ions in the HT state, indicating that the plateau at the magnetization of 1 is the region of the ferrimagnetic state due to the antiferromagnetic interaction J. For the case of J = 0.06, the ground state is ferromagnetic even at h = 0. The magnetization is induced up to 2 in a strong magnetic field. These results indicate that the applied magnetic field can induce transitions between three states: the LT state, antiferromagnetic coupling HT state, and ferromagnetic coupling HT state.

For temperature dependence of magnetization, in the case of T = 0.4, the system is in the LT state in a small field and the magnetization gradually increases until h = 0.175. Then the magnetization jumps to 0.87, and then gradually increases up to 2. The metastable state exists, indicating that the hysteresis loop can be observed. In the case of T = 0.47, the system is in the LT state in a small field and the magnetization increases up to 2 with a jump at h = 0.085.

MONTE CARLO CALCULATION

In this section, we study the model by Monte Carlo simulation and confirm the results obtained in the previous section. Then we study dynamical properties and the metastability of the model.

Equilibrium states

The system size is 16×16×16, which is large enough to study thermal properties. At each step, we performed 7000 MCSs for transient steps and 10000 MCSs to measure the physical quantities. We heat the temperature up to T = 1 and then cool it down to the original temperature. The changes in n(T) and m(T) starting from the initial states of n = 1 and m = 1, respectively, which correspond to the photoinduced HT saturated state at very low temperature. In the case of J = 0, we found a smooth n(T) in Monte Carlo simulation, we find that the HT phase exists down to T = 0 as a metastable state that causes the first-order phase transition at some temperature. This fact indicates that in MC, different types of n(T) appear for the same set of parameters, which is naturally expected. In MC, we find a similar trend of n(T).

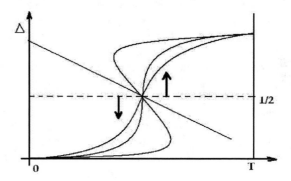

FIGURE 1. First order transition as J increase

Because we sweep the temperature, the result depends on sweeping rate. However, we find that the position of the transition from metastable HT to stable LT changes little with sweeping rate, which suggests that the state is stable or metastable and is rather well defined by the system parameters. We find n(T) at high temperature, where the state is paramagnetic, is also caused by J. When J = 0.025, n(T) changes from that for J = 0, although the system is paramagnetic in both cases. If we increase J further, the system shows a first order phase transition as in Figure 1.

Relaxation of unstable states

We study the relaxation processes from the metastable HT to LT states by MC simulation. The structure of the metastable state e.g., the local minimum of this metastable HT state exists at low temperature. In the field of photoinduced phase transition, the existence or nonexistence of the metastable phase is a significant issue because whether the photoinduced HT phase is a metastable phase or a photogenerated thermodynamically unstable phase.

We adopt here the initial state of n near 1 where the positions of HT sites are chosen randomly. Such a state can be produced by rapid cooling or light irradiation where the HT sites locate uncorrelatedly. We study the system using (D, J, h) = (1, 0.025, 0) at T = 0.1 for different HT fractions. The system size is 64×64×64. When the HT fraction is smaller than 0.9, the relaxation curves show monotonic decreases. In contrast, when n is larger than 0.90, the relaxation curves increase initially. This is a particular phenomenon due to the existence of a local minimum of the HT state. After that, the states relax to the LT state with sigmoidal shapes.

DISCUSSION

A model we take into account is the fact that charge transfer occurs between Co and Fe ions that are ordered antiferromagnetically at low temperatures. The spin states and magnetic properties of antiferro-magnetic Co-Fe were studied. The effects of magnetic interaction and an external magnetic field on the structure of the metastable states. A systematic change in the temperature dependence on the parameters of the system was found as in the previous study. The magnetic interaction induces a magnetic order and enhances the ordering of the metastable states. The low-temperature metastable HT branch is enlarged and even connected to the HT branch. The solution of the metastable states disappears discontinuously, which causes a discontinuous transition. The external magnetic field causes a change in a similar sequence to that found for magnetic interaction. The direction of initial relaxation can be upwards or downwards when the metastability exists depending on the initial. The qualitative features of the effect of magnetic interaction obtained using the MF approximation were confirmed by MC simulation, although in the MC method we found that the magnetic interaction has an effect even in the paramagnetic states. Temperature dependence of nsp, similar to that of the unstable solution.

In the present calculation, we take into account details of microscopic structure changes, such as dynamical temperature, spin values in the HT structure, and thus the formalism will be useful to study the combined phenomena of magnetic ordering, spin-crossover and charge transfer in a wide range of materials. Not only those belonging to Fe-Co but also more general materials in which charge transfer and spin-crossover induce phase transitions, the magnetic properties obtained in this study, particularly the trend of the changes in the temperature dependences of n and m on system parameters will be useful for classifying various materials.

ACKNOWLEDGEMENTS

We thank DFG (Deutsche Forschungs gemeinschaft) for partial financial support. The authors also thank the Laboratory of Theoretical and Computational, Department of Physics Hasanuddin University, Makassar for computational facilities.

REFERENCES

1. W. B. Nurdin, K D. Schotte, *Phys.* **A308**, 209-226 (2002).
2. W. B. Nurdin, K D. Schotte, *Phys. Rev.* **E61**, 3579 (2000).
3. A. I. Khinchin, Mathematical Foundations of Statistical Mechanics (Dover, New York, 1949).
4. K. Huang, Statistical Mechanics, (John Wiley & Sons, New York, 1963).
5. H. H. Rugh, *Phys. Rev. Lett.* **78**, 772 (1997).
6. H. H. Rugh, *J. Phys. A: Math. Gen.* **31**, 7761 (1998).
7. A. Hauser, P. Guetlich, H. Spiering, *Inorg. Chem.* **25**, 4345 (1986).
8. A. Hauser, *J. Chem. Phys.* **94**, 2741 (1991).
9. S. Decurtins, P. Guetlich, Koehler, H. Spiering, A. Hauser, *Chem. Phys. Lett.* **1**, 139 (1984).
10. P. Guetlich, A. Hauser, H. Spiering, *Angew. Chem. Int. Ed. Engl.* **33**, 2024 (1994).
11. A. Hauser, *Comments Inorg.Chem.* **17**, 17 (1995).
12. A. Hauser, *Coord. Chem. Rev.* **11**, 275 (1991).
13. H. Romstedt, H. Spiering, P. Guetlich, *J. Phys. Chem. Solids* **59**, 1353 (1998).
14. A. Desaix, O. Roubeau, J. Jeftic, J.G. Hassnoot, K. Boukheddaden, E. Codjovi, J. Linar_es, M. Nogu_es, F. Varret, *Eur. Phys. J.* **B6**, 183 (1998).
15. H. A. Goodwin, *Coord. Chem. Rev.* **18**, 293 (1976).
16. P. Guetlich, *Struct. Bonding* (Berlin) **44**, 83 (1981).
17. P. Guetlich, A. Hauser and H. Spiering: *Angew. Chem. Int. Ed.* **33**, 2024 (1994).
18. G. A. Renovitch and W. A. Baker: *J. Am. Chem. Soc.* **89**, 6377 (1967).
19. O. Kahn, C.J. Martinez: *Science* **279**, 44 (1998).
20. O. Kahn, Molecular Magnetism (VCH, New York, 1993).
21. S. Decurtins, P. Guetlich, Haselbach, H. Spiering and Hauser, *Inorg. Chem.* **24**, 2174 (1985).
22. A. Hauser, *Coord. Chem. Rev.* **111**, 275 (1991).
23. J. F. L´etard, J. A. Real, N. Moliner, A. B. Gaspar, L. Capes, O. Cador and O. Kahn, *J. Am. Chem. Soc.* **121** 10630 (1999).
23. F. Renz, H. Spiering, H. A. Goodwin and P. G¨utlich, *Hyperfine Interact.* **126**, 155 (2000).
24. K. Nasu, Relaxations of Excited States and Photo-Induced Structural Phase Transitions (Springer - Verlag, Berlin, 1997).
25. T. Tayagaki and K. Tanaka, *Phys. Rev. Lett.* **86**, 2886 (2001).
26. S. Miyashita, Y. Konishi, H. Tokoro, M. Nishino, K. Boukheddaden and F. Varret, *Prog. Theor. Phys.* **114**, 719 (2005).

27. H. Tokoro, S. Miyashita, K. Hashimoto and S. Ohkoshi, *Phys. Rev.* **B 73**, 172415 (2006).

28. J. Wajnflasz and R. Pick, *J. Phys. Colloq. France* **32,** C1 (1971).

Genetic Algorithm Application in Solving Crystallographic Phase Problem

I. Abdurahman[1] and A. Purwanto[2]

[1]*Department of Physics, Faculty of Science and Mathematics, Pelita Harapan University, Tanggerang, Indonesia*
[2]*Center of Technology for Nuclear Industrial Materials, National Nuclear Energy Agency, Serpong, Indonesia*

Abstract. Crystallographic phase problem, the problem of choosing the correct phase set in direct method of crystallographic structure analysis, is a vital step in determining crystal structure. Various methods have been proposed for the problem, with their advantages and disadvantages. Genetic algorithm provides an alternative method of solving phase problem by means of evolution inspired algorithm. Since it tries to solve the phases directly, the method can be seen as a direct method. This paper reports on the computer code based on genetic algorithm to solve the phase problem. It requires Miller index, structure factor, scattering length and space group for input and produces phases along with the resulting scattering length density function as outputs. The genetic algorithm based program is shown to be effective and straightforward in application

Keywords: phase problem, neutron scattering, genetic algorithm.
PACS: 61.05.F-

INTRODUCTION

The surge of neutron sources have created a challenge in the field of crystal structure determination. As data are getting more available to crystallographer, analyzing them fast and accurate enough remains a problem.

One way to solve this is to reduce user intervention in the process hence making the analysis faster with the available resources at hand. However, the approach requires a method that takes some of the work away from crystallographer.

One of the stages where such approach can be done in direct methods crystallography is in solving phase problem. There have been efforts to automate methods like symbolic addition[1] and multisolution[2]. In this paper, genetic algorithm is proposed as a alternative mean to the existing methods. The work done here is a preliminary evaluation on the feasibility of this method.

Miller index, atomic positions[3] from Chromium(III) Oxide (centrosymmetric) and scattering length[4] are used to check whether the new method successfully determine the atomic positions. Structure factors are generated using the positions and indexes to isolate any success or failure only to the performance of the program. Results are validated by comparing the peak of scattering length density relative to the reference atomic positions.

THEORETICAL BACKGROUND

Phase Problem

Structure factor can be separated as magnitude and phase as written below

$$F_H = |F_H| \exp(i\varphi_h) = \sum_1^N f_i \exp\left[2\pi i(hx_j + ky_j + lz_j)\right] \quad (1)$$

which can then Fourier transformed to get the scattering length density

Unfortunately, only structure factor magnitude can be obtained. Phases which contain atomic positions are lost due. Therefore phase problem is about finding the phase information which is irretrievable in a conventional diffraction experiment.

$$\rho(x, y, z) = \sum_{h,k,l} |F_{hkl}| \exp\left[-2\pi i(hx + ky + lz) + \varphi_h\right] \quad (2)$$

Method that tries to do this directly from the magnitudes are called direct methods. The proposed GA-based method can be classified into this.

CP989, Neutron and X-ray Scattering in Materials Science and Biology, International Conference on Neutron and X-ray Scattering 2007, edited by A. Ikram, A. Purwanto, Sutiarso, A. Zulfia, S. Hendrana, and Z. Nurachman
© 2008 American Institute of Physics 978-0-7354-0508-0/08/$23.00

Figure of Merit

Triplets are used to formulate the Figure of Merit. They are formed from sum of three phases with sum of indexes that equals zero. One that is used here is

$$\Phi = \varphi_h - \varphi_k - \varphi_{h-k} \qquad (3)$$

Due to their sole dependency on the structure, triplets are structure invariants and used in determining crystal structure.

The FOM used here is MABS[5]. The optimal value of which is one. Thus, a good phase set that is produced by the program should have

$$\frac{\sum_h \alpha_h}{\sum_h \langle \alpha_h \rangle} \approx 1 \qquad (4)$$

where

$$\alpha_h = \sum_{j=1}^{r} G_j \cos \Phi_j \qquad (5)$$

and G is the concentration parameter which is determined by the product of the normalized structure factor of h, k, and $h\text{-}k$.

Figure of Merit is used in methods like symbolic addition to asses the feasibility of a given set. In genetic algorithm based method, it functions as a fitness for the genetic algorithm. However, it is important to point out that the genetic algorithm based method differs on how a phase set is proposes for evaluation.

METHOD

Genetic algorithm is an evolution inspired algorithm. The algorithm used here is the simple genetic algorithm[6]. It represents solutions in binary form which is analogous to individuals in a population. Genetic operator is performed on selected individuals based on their fitness on every generation it. The selection criteria is determined by fitness which in this case is MABS or any other FOM chosen. The GA based program is only as good as the fitness used. It is very likely that a phase set with the optimal FOM. The GA based program is only as good as the fitness used. It is very likely that a phase set with the optimal FOM (in this case is MABS close to one) value is obtained. However, whether this phase set is correct or not depends on how well the FOM represents the characteristic of a correct phase set.

To obtain MABS, triplet search must be undertaken. Therefore, for a given index h all combination of it with two other indexes that satisfies (3) have to be found. With the data used, a total of 154 triplets are found which consists of triplets that are associated with each Miller index.

Their values must then be evaluated. Once all triplets that involves a Miller index are found, the α values of each index can be calculated. Consequently, MABS can now be used for the fitness function of genetic algorithm. The flowchart of the process is shown below.

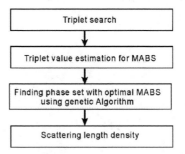

FIGURE 1. Outline of the method

In the GA, the first step is to set up population which is set of individuals. Genetic operator that is applied on the selected individuals either exchange the bits between individuals (crossover) or simply change some of the bits in some individuals (mutation). The process is illustrated on the following flowchart. More detailed discussion on the topic is outside the scope of this paper and available in GA literatures[7].

Determination of the size of the population is given to the user. The population size controls the diversity of the solution. More variation is possible if there are more solutions present in a given population.

There are two important things regarding crossover. The first one is the crossover probability which determines how much bits are exchanged in a generation. The second is crossover method such single point or multiple point[7].

To maintain diversity of solution, mutation is introduced. When population is too uniform, crossover alone will not be effective in generating new solutions. Mutation breaks the uniformity by changing some of the bits. The number of changed bits depends on the mutation probability. By keeping the population diverse, genetic algorithm can proceed generating better solutions.

One more important parameter is the maximum iterations. This is determined by the user. Several run might be necessary to see when convergence happens. If the fitness achieved is still far off from the optimal value, it is very likely that more iterations are needed.

The GA based program The nature of genetic algorithm enables the improvement of fitness in every generation based on schema theorem[8]. One way to use this is to observe the fitness of the best individuals in every generation. The phase set with the corresponding best fitness is used for the scattering length density

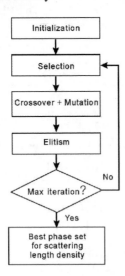

FIGURE 2. The Genetic Algorithm block

RESULTS AND DISCUSSION

The fitness of best individuals is plotted for each generation. An individual with an certain fitness is taken if the plot remains constant after sufficient generation. This simply there is no other individual with better fitness. If such things happen, it is concluded that convergence has happened and the solution represented by the individual is taken as the optimum solution. Using the following parameters: crossover probability = 0.3, mutation probability = 0.01, population size = 30, 1000 generation, multiple point crossover, elitist + roulette wheel selection, the best phase set is ($\pi\pi00\pi0\pi00\pi \pi \pi 0\pi \pi 0\pi 0000\pi 000\pi00\pi 00\pi0\pi0\pi\pi0\pi\pi0\pi 0$). The plot of (1/l1-MABS) is shown on the following figure.

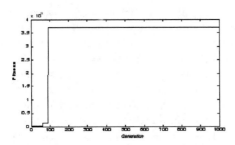

FIGURE 3. Square of Scattering Length Density at z=0. 49

The factor structure in table 2 is generated from atomic positions in table 1 and Miller index from table 2 .Validation is done by checking whether the reference atomic position are located in areas with high scattering length density. The process is described below.

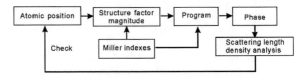

FIGURE 4. Validation process

For practicality, scattering length density is shown in cross sections. It is apparent that the reference atomic positions (shown as black spots) are located at areas with high scattering length density (peaks). Therefore, at several cross section, satisfactory result are produced using the best phase set from genetic algorithm based program. Furthermore, one could argue when sufficient amount of atomic positions are found, the rest can be generated using space group matrices.

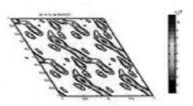

FIGURE 5. Square of Scattering Length Density at z=0.15

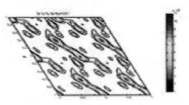

FIGURE 6. Square of Scattering Length Density at z=0. 49

In application, this would mean that the GA based program can generate scattering length density where at least some of atomic positions can be obtained from analyzing the location of the peaks. Better results might be produced by employing other FOM

CONCLUSIONS

Genetic Algorithm-based method can serve as an alternative method in solving phase problem. The main idea is to see phase problem as an optimization problem with FOM as fitness.

TABLE 1. Reference atomic positions

No.	Atom	x	y	z	No.		x	y	z
1	Cr	0	0	0.3458	16	O	0.3058	0	0.7500
2	Cr	0	0	0.1543	17	O	0	0.3058	0.7500
3	Cr	0	0	0.3458	18	O	0.3058	0.3058	0.7500
4	Cr	0	0	0.8458	19	O	0.9724	0.3333	0.5833
5	Cr	0.6667	0.3333	0.6791	20	O	0.6667	0.6391	0.5833
6	Cr	0.6667	0.3333	0.4876	21	O	0.3609	0.0276	0.5833
7	Cr	0.6667	0.3333	0.0124	22	O	0.3609	0.3333	1.0833
8	Cr	0.6667	0.3333	1.1791	23	O	0.6667	0.0276	1.0833
9	Cr	0.3333	0.6667	1.0124	24	O	0.9724	0.6391	1.0833
10	Cr	0.3333	0.6667	0.8209	25	O	0.6391	0.6667	0.9167
11	Cr	0.3333	0.6667	0.3209	26	O	0.3333	0.9724	0.9167
12	Cr	0.3333	0.6667	1.5124	27	O	0.0276	0.3609	0.9167
13	O	0.3058	0	0.2500	28	O	0.0276	0.6667	1.4167
14	O	0	0.3058	0.2500	29	O	0.3333	0.3609	1.4167
15	O	0.3058	0.3058	0.2500	30	O	0.6391	0.9724	1.4167

TABLE 2. Miller index and structure factor used

No.	Miller Index	Structure factor	No.	Miller Index	Structure factor
1	0 1 2	26.5917	23	1 3 1	18.9094
2	1 0 4	21.4377	24	1 2 8	3.9671
3	1 1 0	6.8955	25	3 1 2	39.8141
4	0 0 6	65.5263	26	0 2 10	23.6629
5	1 1 3	87.8548	27	0 0 12	130.3144
6	2 0 2	2.7365	28	1 3 4	8.2153
7	0 2 4	50.7659	29	3 1 5	18.9094
8	1 1 6	89.4432	30	2 2 6	86.2870
9	2 1 1	27.5465	31	0 4 2	62.2323
10	0 1 8	15.2948	32	2 1 10	33.7291
11	1 2 2	7.3298	33	1 1 12	24.6551
12	2 1 4	40.6997	34	4 0 4	14.2029
13	3 0 0	138.8783	35	1 3 7	18.9094
14	1 2 5	27.5465	36	1 2 11	27.5465
15	2 0 8	14.0333	37	3 2 1	33.4755
16	1 0 10	52.9910	38	3 1 8	28.5173
17	1 1 9	87.8548	39	2 3 2	14.9693
18	2 1 7	27.5465	40	2 2 9	79.2177
19	2 2 0	3.7393	41	0 1 14	12.5713
20	0 3 6	56.3305	42	3 2 4	62.9987
21	3 0 6	56.3305	43	4 1 0	3.2898
22	2 2 3	79.2177			

REFERENCES

1. C. Giacovazzo et al, Fundamentals of Crystallography, New York: Oxford University Press, 1992, 352.
2. C. Giacovazzo et al, Fundamentals of Crystallography, New York: Oxford University Press, 1992, 360.
3. A. Purwanto, Data Kromium (III) Oksida, Private Communication., 1999, pp. 212-213.
4. National Institute of Standards and Technology Center for Neutron Research: Neutron scattering lengths and Cross Sections Coh b. http://www.ncnr.nist.gov /nsources.html, [accessed 4 June 2007].
5. C. Giacovazzo et al, Fundamentals of Crystallography, New York: Oxford University Press, 1992, 355.
6. M. Zbigniew, Genetic Algorithms + Data Structures = Evolution Programs, New York: Springer, 1996, 13.
7. M. Zbigniew, Genetic Algorithms + Data Structures = Evolution Programs, New York: Springer, 1996, 33.
8. M. Zbigniew, Genetic Algorithms + Data Structures = Evolution Programs, New York: Springer, 1996, 45.

MATERIALS CHEMISTRY

Structure of Biologically Active Organotin(IV) Dithiocarbamates

Y. Farina, M. Sanuddin and B. M. Yamin

School of Chemical Sciences and Food Technology, Faculty of Science and Technology, Universiti Kebangsaan Malaysia, 43600 UKM Bangi, Selangor, Malaysia

Abstract The diorganotin(IV) complexes of dithiocarbamates derived from from *N*-ethyl-*n*-propylamine (EtPrdtc), 2-dimethylaminoethylamine (Me$_2$Etdtc), 3-dimethylamino-1-propylamine (Me$_2$Prdtc), *p*-tolylmethanamine (TylMetdtc) and *N*-methyl-1-phenylmethanamine (MePhMetdtc) have been synthesized and characterized. Single crystal X-ray diffraction studies on Ph$_3$Sn(EtPrdtc), Me$_2$Sn(MePhMetdtc)$_2$ and Bu$_2$Sn(MePhMetdtc)$_2$ showed that the complexes adopted a monoclinic system with space group $P(2)/n$, $P2_1/n$ and $C2/c$, respectively. The Ph$_3$Sn(EtPrdtc) complex adopted a trigonal pyramidal structure while the Me$_2$Sn(MePhMetdtc)$_2$ and Bu$_2$Sn(MePhMetdtc)$_2$ complexes displayed structures which may be described as distorted octahedrons. Cytotoxicity test using HL60 cells (human promyelocytic leukemic) showed that only Me$_2$Sn(Me$_2$Etdtc), Me$_2$Sn(MePhMetdtc)$_2$ and Bu$_2$Sn(MePhMetdtc)$_2$ complexes were active. The rest of the complexes did not show cytotoxicity behaviour towards HL60 cells.

Keywords: Organotin dithiocarbamates, Biologically active, Structure.
PACS: 61.05.cp

INTRODUCTION

Organotin compounds are known to be biologically active as they are widely used as wood preservatives and fungicides, marine anti-fouling agents, agrochemical fungicides and miticides, and in disinfectants. The potential of such compounds as anti-tumour agents have been studied by several workers [1-4]. The results of the testing showed that these oxygen containing organotins were even more effective than *cis-platin* [4]. The structures of many diorganotin(IV) bis(*N,N*-dithiocarbamates) have been reported [5-10]. The dithiocarbamate ligand in six co-ordinated organotin(IV) complexes may be generally classed into three distinct categories depending on the Sn-S distances [6]; (i) monodentate in which one of the Sn-S bond length is about 2.5 Å equivalent to a single Sn-S covalent bond and the second is greater than the sum of the van der Waals radii; (ii) bidentate in which both Sn-S bond lengths are between 2.5 and 2.8 Å; (iii) intermediate in which one of the Sn-S bond length is about 2.5 Å and the second is about 3.0 Å.

Previous work on hydroxylated dithiocarbamates derived from NH(EtOH)$_2$, NHMeEtOH, NH(^{i}PrOH)$_2$, NH(^{i}Pr)$_2$ have been reported [11-14]. This study was performed utilising dithiocarbamates derived from *N*-ethyl-*n*-propylamine, 2-dimethylaminoethylamine, 3-

dimethlyamino-1-propylamine, *p*-tolylmethanamine and *N*-methyl-1-phenylmethanamine (Figure 1).

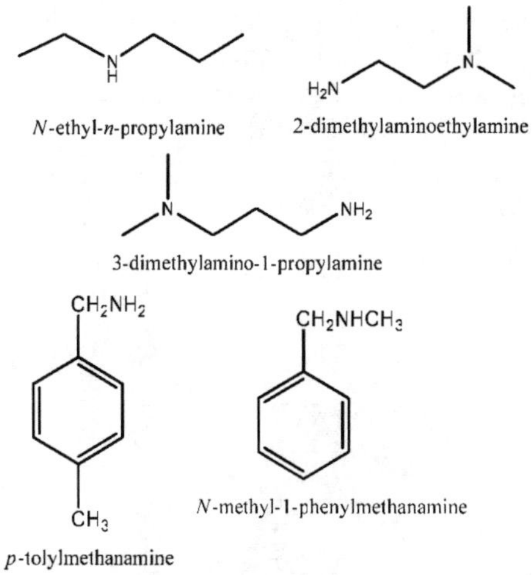

FIGURE 1. Amines used.

METHODOLOGY

A solution of carbon disulphide in methanol was added to a methanolic solution of a mixture of the organotin(IV) chloride and the respective amines. Stoichiometric ratios of 1:2 and 1:1 were used for the

CP989, Neutron and X-ray Scattering in Materials Science and Biology, International Conference on Neutron and X-ray Scattering 2007, edited by A. Ikram, A. Purwanto, Sutiarso, A. Zulfia, S. Hendrana, and Z. Nurachman
© 2008 American Institute of Physics 978-0-7354-0508-0/08/$23.00

diorganotin(IV) and triphenyltin(IV) salts respectively. Both solutions were initially cooled to 273 K and the methanolic solution of carbon disulphide was added at such a rate that the reaction temperature did not rise above 278 K. The reaction mixture was stirred at 275 K for 2 hours. Solid product was filtered and washed with cold methanol.

Single crystals of Ph₃Sn(EtPrdtc), Me₂Sn(MePhMetdtc)₂ and Bu₂Sn(MePhMetdtc)₂ suitable for crystallographic investigation were mounted on a Bruker Smart APEX CCD area detector diffractometer equipped with graphite monochromatized Mo-K$_\alpha$ 0.71073 Å radiation. X-ray diffraction measurements, were performed at room temperature. Structures under study were solved by direct methods with SHELXL-90 [15] and refined using the full matrix least-square techniques on F^2 with SHELXL-97 [16] programs. Non-hydrogen atoms were refined anisotropically.

The HL-60 (human promylocetic leukemia) cells line was obtained from the National Cancer Institute, USA. The cells were cultured in RPMI-1640 medium supplemented with 10% fetal calf serum. Cytotoxicity was determined using the microtitration of 3-(4,5-dimethylthiazol-2-yl)-2,5-diphenyltetrazolium bromide (MTT) assay.

RESULTS AND DISCUSSION

The reaction of diorganotin(IV) chloride with dithiocarbamates derived from the various amines used in this study afforded the 1:2 complexes with the general formula R₂Sn(R'dtc)₂ while the reaction of triphenyltin(IV) chloride with the dithiocarbamates gave the 1:1 R₃Sn(R'dtc) type of complexes.

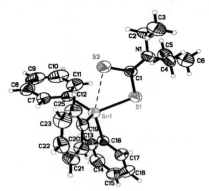

FIGURE 2. ORTEP plot of Ph₃Sn(EtPrdtc)

The crystal data and for the three complexes are:
Ph₃Sn(EtPrdtc): crystal system, monoclinic; space group, $P2_1/n$; a = 9.4395(18) Å, b = 26.225(5) Å, c = 9.8059(19) Å, β = 98.954(3)°, Z = 4, R = 2.84 % for 4928 I > 2σ(I) independent reflections.

Me₂Sn(MePhMetdtc): crystal system, monoclinic; space group, $P2_1/n$; a = 12.189(3) Å, b = 9.174(2) Å, c = 21.115(5) Å, β = 92.5251(4)°, Z = 4, R = 3.26 % for 4626 I > 2σ(I) independent reflections.

Bu₂Sn(MePhMetdtc): crystal system, monoclinic; space group, $C2/c$; a = 19.678(10) Å, b = 6.972(3) Å, c = 22.722(10) Å, β = 108.922(5)°, Z = 4, R = 3.26 % for 2900 > 2σ(I) independent reflections.

The ORTEP plot and numbering system of Ph₃Sn(EtPrdtc) is depicted in Figure 2. The molecular structure of Ph₃Sn(EtPrdtc) shows that the tin atom is bounded to three phenyl groups and two sulphur atoms from one chelating bidentate dithiocarbamate. The Sn-S distances can be classed as the short [Sn1-S1 = 2.4600(7) Å] and long bonds [Sn1–S2 = 3.0950(10) Å]. The long Sn-S distances are however, significantly less than the sum of the van der Waals radii which is 4.0 Å [17] and as such the co-ordination number of the tin atom may be assigned as five [8,10]. The geometry of the Sn atom is close to a trigonal bipyramid, with C2-Sn1-C19, C19-Sn1-S1 and C12-Sn1-S1 angles of 116.49 (3)°, 120.29 (7)° and 110.95 (7)° respectively, in the equatorial positions.

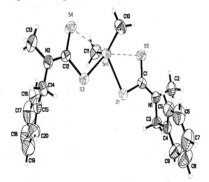

FIGURE 3. ORTEP plot of Me₂Sn(MePhMetdtc)

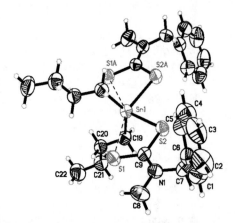

FIGURE 4. ORTEP plot of Bu₂Sn(MePhMetdtc)

The ORTEP plot and numbering system of $Me_2Sn(MePhMetdtc)$ and $Bu_2Sn(MePhMetdtc)$ are depicted in Figures 3 and 4 respectively.

The molecular structures of $Me_2Sn(MePhMetdtc)$ and $Bu_2Sn(MePhMetdtc)$ are similar with the tin atom being six coordinated bounded to two alkyl groups (Me and Bu) and four sulphur atoms from two chelating bidentate dithiocarbamates. The Sn-S bond in both complexes are of the intermediate type where the first Sn-S bond length is about 2.5 Å and the second by the same chelating dithiocarbamate anion is about 3 Å. The geometry around the central tin atom may be described as distorted octahedrons.

The anisobidentate nature of the dithiocarbamate ligand in the three structures may be ascribed to steric requirements of the ligand [6]. A truly isobidentate ligand would show almost equal C-S bond lengths while unequal bond lengths would signify a greater contribution from the canonical form, S=C—S system. The carbon-nitrogen distances in three complexes were found to be around ~1.3 Å. When compared to the values for a C=N double bond (1.29 Å), and a C-N single bond (1.47 Å), it suggested that the carbon-nitrogen bond is intermediate between a single and a double bond. The trends shown by the C-S and C-N distances thus indicate that the π-electron is extensively delocalised in the S_2CN fragment and that the delocalised canonical form has a significant contribution to the overall structure.

The results of the cytoxicity bioassay against HL-60 (human promyelocytic leukemic) cells showed only three of the synthesized complexes gave CD_{50} values of less than 0.6 μg/mL (value for etopside the positive control). The complexes $Me_2Sn(Me_2Etdtc)$, $Me_2Sn(MePhMetdtc)_2$ and $Bu_2Sn(MePhMetdtc)_2$ had CD_{50} values of 0.288, 0.181 and 0.205 μg/mL respectively which was significantly less than the value for etopside. Thus the three complexes $Me_2Sn(Me_2Etdtc)$, $Me_2Sn(MePhMetdtc)_2$ and $Bu_2Sn(MePhMetdtc)_2$ possess potent cytotoxic properties against human promyelocytic leukemic cells.

ACKNOWLEDGMENTS

We would like to thank the Malaysian government for IRPA grant 09-02-02-0133. We would also like to thank all the faculty members that have contributed to this research.

REFERENCES

1. A.J. Crowe, *Metal-Based Antitumour Drugs* **1**, 103-149 (1989).
2. A.K. Saxena and F. Huber, *Coord. Chem. Rev.*, **95**, 109-123 (1989).
3. M.Gielen, M. Mélotte, G. Atassi, G. and R. Willem, *Tetrahedron*, **45**, 1219-1229 (1989).
4. M. Gielen, P. Lelieveld, D. ve Dos and R. Willem, *Meta-Based Antitumour Drugs*, **2**, 29-54 (1992).
5. K. Furue, T. Kimura,, N. Yasuoka, N. Kasai, M. and Kakudo, *Bull. Chem. Soc. Japan*, **43**, 1661-1667 (1970).
6. P.F. Lindley and P. Carr, *J. Cryst. Mol. Struct.*, **4**, 173-185 (1974).
7. T.P. Lockhart, W.F. Manders, E.O. and Schlemper, *J. Am. Chem. Soc.*, **107**, 7451-7453 (1985).
8. T.P. Lockhart, W.F. Manders, E.O. Schlemper and J.J. Zuckerman, *J. Am. Chem. Soc.*, **108**, 7451-7453 (1986).
9. D. Dakternieks, H. Zhu, D. Masi and C. Mealli, *Inorg. Chem.*, **31**, 3601-3606 (1992).
10. J. Sharma, Y. Singh, R. Bohra, A.K. and Rai, *Polyhedron*, **7**, 1097-1102 (1996).
11. Y. Farina, A.H. Othman, I.A. Razak, H.K. Fun, S.W. Ng and I. Baba, *Acta Cryst.*, **E57**, m46-m47 (2001).
12. Y. Farina, I. Baba, A.H. Othman, I.A. Razak, H.K. Fun and S.W. Ng, *Acta Cryst.*, **E57**, m41-m42 (2001).
13. Y. Farina, I. Baba, A.H. Othman and S.W. Ng, *Main Group Metal Chemistry*, **23**, 795-796 (2000).
14. Y. Farina, A.H. Othman, I. Baba, K. Sivakumar, H.K. Fun & S.W. Ng, *Acta Cryst.*, **C56**, e84-85 (2000).
15. G.M. Sheldrick, SHELXS-90. (1990).
16. G.M. Sheldrick, SHELXS-97. (1997).
17. A.J. Bondi, *J. Phys. Chem.*, **68**, 441-451 (1964).

Study of Mg-doped GaN Thin Films Grown on c-Plane Sapphire Substrate by Plasma Assisted Metalorganic Chemical Vapor Deposition Method

A. Subagio[1,2], H. Sutanto[1,2], E. Supriyanto[1,3], M. Budiman[1],
P. Arifin[1], Sukirno[1], M. Barmawi[1]

[1]Laboratory for Electronic Material Physics, Dept. of Physics, ITB, Bandung 40132, Indonesia
[2]Dept. of Physics, Diponegoro University, Semarang, Indonesia
[3]Dept. of Physics, Jember University, Jember, Indonesia

Abstract. Mechanism of doping of Mg in GaN thin film grown on c-plane sapphire substrate by plasma assisted-metalorganic chemical vapor deposition (PA-MOCVD) method have been investigated. The growth was carried out at temperature of 680 °C, which flow rate of Cp_2Mg as dopant source was varied of 0.003; 0.006; 0.009 and 0.012 sccm. The Van der Pauw technique, X-ray diffraction (XRD) and scanning electron microscope (SEM) were used to characterize their electric, crystallographic and morphology properties. Hole concentrations of up to 10^{19} cm^{-3} have been observed for Mg-doped GaN films without any post-growth annealing. The XRD analysis showed change of the FWHMs for the (0002) orientation plane from Mg-doped GaN films. The difference of the atomic radius between Ga and Mg atoms causes lattice distortion in crystal. This lattice distortion generates dislocation and causes growth of the layer to be three-dimensional as showed on SEM images.

Keywords: Mg-doped GaN, PA-MOCVD, Van der pauw technique, XRD, SEM
PACS: PACS category 61.05

INTRODUCTION

The group III-nitrides materials have been the focus of much attention in recent years owing to their applications in optoelectronic devices[1]. Laser diodes (LDs) and light emitting diodes (LEDs) operating in the short wavelength visible and ultraviolet (UV) spectral regions have been successfully fabricated [2-3]. Furthermore, remarkable progress has been made in the development of optoelectronic devices since the success of p-doping in GaN[4,5]. There are still serious problems to be solved such as the high resistivity and low hole concentration of p-doping in GaN if the growth of Mg-doped GaN film using thermal MOCVD method [6-12]. It is known that the Mg-H complex synthesized from decomposed NH_3 as nitrogen source and Cp_2Mg during growth plays an important role in high resistivity and low hole concentration of the Mg-doped p-type GaN film. So, its thermal annealing treatment after growth is absolutely necessary for cracking the Mg-H complex[5].

Plasma assisted MOCVD method has been used as an alternative for growth of GaN and its alloys, which

the growth temperature relatively lower compared to thermal MOCVD. In PA-MOCVD, nitrogen precursor in the form of the N_2 gas was cracked by the plasma and nitrogen radicals were formed. The nitrogen radicals were used as source of nitrogen for growth of Mg-doped GaN film. So, after growth thermal annealing treatment is not necessary for cracking the Mg-H complex, because it is not using NH_3 as nitrogen source.

In this work, we attempt to grow Mg-doped GaN film on (0001) sapphire substrate by PA-MOCVD. Without after growth thermal annealing treatment, we have obtained p-type GaN film with the high hole concentration.

EXPERIMENTAL METHOD

The Mg-doped GaN films were grown on (0001) sapphire substrates with 25 nm thick GaN buffer layer by PA-MOCVD. Trimethylgallium (TMGa), bis-cyclopentadienyl magnesium (Cp_2Mg) and N_2 were used as the source precursors for Ga, Mg dan N, respectively. A low power downstream plasma cavity

CP989, Neutron and X-ray Scattering in Materials Science and Biology, International Conference on Neutron and X-ray Scattering 2007, edited by A. Ikram, A. Purwanto, Sutiarso, A. Zulfia, S. Hendrana, and Z. Nurachman

supplied the reactive N-plasma from nitrogen gas and reactive H-plasma from hydrogen gas. The plasma is generated by 2.45 GHz microwave of 200 Watt. The H_2 carrier gas was purified by passing it through a heated palladium cell.

The (0001) sapphire substrates were rinsed with acetone and methanol in an ultrasonic bath. Substrates subsequently washed with de-ionized water (DI-water) then etched in solution of DI-water:H_3PO_4:H_2SO_4 = 1:1:3 at 70 °C for 10 min. Finally the substrates were input into running de-ionized water and then blown with dry nitrogen. The substrate was immediately introduced into the growth reactor and heated up to 650 °C for thermal cleaning in the H_2 ambient. *In-situ* hydrogen plasma cleaning on substrate was carried out for 10 min using 200 watt plasma power with the H_2 flow rate of 50 sccm. The GaN buffer layer was grown with TMGa of 0.17 sccm and N_2 flow 90 sccm at 500 °C, which produces the thickness of about 25 nm. After the deposition of buffer layer, the substrate temperature was raised to the growth temperature of 680 °C. Different the flow rate of Cp_2Mg was used for depositing a series of Mg-doped GaN films: 0.003; 0.006; 0.009 and 0.012 sccm with the flow rate of TMGa precursors and N_2 were kept as 0.16 and 90 sccm, respectively. Furthermore, the Mg-doped GaN films are treated with *in-situ* plasma nitrogen plasma for cooling-down step.

The X-ray diffraction (XRD), scanning electron microscope (SEM), and Van der Pauw technique, were used to characterize structural, morphology and electrical properties of each Mg-doped GaN film.

RESULTS AND DISCUSSION

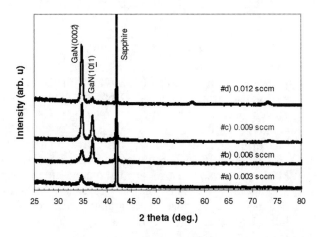

FIGURE 1. X-ray diffraction of Mg-doped GaN films grown with the variation of the Cp_2Mg flow rates.

Figure 1 shows the XRD patterns of Mg-doped GaN films grown with the Cp_2Mg flow rate of 0.003; 0.006; 0.009 and 0.012 sccm. The Mg-doped GaN film

with the Cp_2Mg flow rate of 0.003 has single crystal orientation on (0002) GaN plane, but the (10$\underline{1}$1) GaN plane appear in the other films. The low doping of Mg atom can not change the structure plane of GaN film yet. Increasing of Mg doping caused the (10$\underline{1}$1) GaN plane to appear and this is showed on GaN films with Cp_2Mg flow rate of 0.006; 0.009 and 0.012 sccm.

Figures 2 and 3 show the FWHMs and morphology of Mg-doped GaN films grown with the Cp_2Mg flow rate of 0.003; 0.006; 0.009 and 0.012 sccm. In general, the radius size of Mg atom which is the substitusional element for Ga in GaN, is much longer than that of Ga atom. This difference causes lattice distortion in the crystal. This distortion generates dislocations and causes the growth of the layer to be three-dimensional[13]. As the flow rate of Cp_2Mg increases, the roughness of morphology increases. So, it is possible to conclude that the dislocation density, which is resulted from difference of radius size between Ga and Mg, increases with increased Cp_2Mg flow rate.

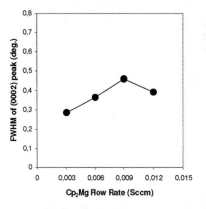

FIGURE 2. FWHM of (0002) peak of Mg-doped GaN films grown with the variation of the Cp_2Mg flow rate.

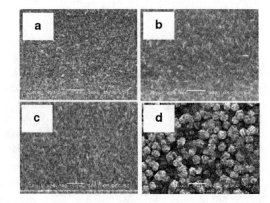

FIGURE 3. Morphology of Mg-doped GaN films grown with the Cp_2Mg flow rate of a) 0.003; b) 0.006; c) 0.009 and d) 0.012 sccm.

Figure 4 shows the resistivities of Mg-doped GaN films. The Au layer was deposited as metal contact on each Mg-doped GaN films for the electrical measurements using the Van der Pauw technique. The contacts were also thermally annealed at temperature of 500 °C for 60 second to decrease the contact resistance of the interface between metal/ semiconductor. The Mg-doped GaN films grown by PA-MOCVD method in this research have low resistivity (<0.3 Ω.cm) than resulted by thermal MOCVD method (>0.3 Ω.cm). In the thermal MOCVD method, the Mg acceptors are passivated by hydrogen with the resulting formation of electrically inactive Mg-H complexes as a result of synthesis from decomposed NH_3 as nitrogen source and Cp_2Mg during growth. As consequence, the Mg-doped GaN films have high resistivity and low hole concentration. However, the problem in this limiting is not only hydrogen passivation, but also self-compensation caused by the formation of a deep donor, $Mg_{Ga}-V_N$, namely a nearest-neighbor associate of the Mg acceptor with a nitrogen vacancy [14-17].

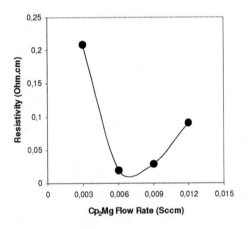

FIGURE 4. Resistivity of Mg-doped GaN films grown with the variation of the Cp_2Mg flow rates.

In the PA-MOCVD method, the problem may be dominated by formation of a deep donor ($Mg_{Ga}-V_N$), which V_N is nitrogen vacancy as the nature of native defect in the nitrides that contribute to n-type conductivity. In the other hand, the Mg-doped GaN films grown by PA-MOCVD method are treated with *in-situ* nitrogen plasma annealing for cooling-down step. The nitrogen plasma is believed to furnish sufficient nitrogen to annihilate the vacancy by reaction of the V_N with the N in the near surface region. The findings herein also suggest that the deep compensating donor of $Mg_{Ga}-V_N$ in Mg-doped GaN is effectively removed by *in-situ* nitrogen plasma annealing treatment, leading to a markedly reduced the resistivity of Mg-doped GaN film. Therefore, without after growth thermal annealing treatment, we have

been obtained Mg-doped GaN film with the low resistivity.

Figure 5 shows the carrier concentration and mobility were measured at room temperature by Hall measurement on Mg-doped GaN films with a Cp_2Mg flow rate of 0.003; 0.006; 0.009 and 0.012 sccm. The sample of 0.003 sccm of Cp_2Mg shows p-type conduction and its hole concentration is about 6.7 x 10^{18} /cm^3. The film of 0.006 sccm of Cp_2Mg also shows p-type conduction with the hole concentration of 4.8 x 10^{19} /cm^3, which is the largest among them. The film of 0.009 sccm of Cp_2Mg also shows p-type conduction having a hole concentration of 3.5 x 10^{19} /cm^3. Judging from this result, it can be understood that the activation of Mg by in-situ nitrogen plasma annealing treatment does not increase proportionally with increased Mg incorporation and there is a saturation point in the activation of Mg at a certain incorporation level.

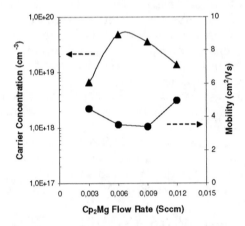

FIGURE 5. Carrier concentration and mobility of Mg-doped GaN films grown with the variation of the Cp_2Mg flow rates.

CONCLUSIONS

We have observed that PA-MOCVD method could be used to grow Mg-doped GaN film without after growth thermal annealing treatment. It has been demonstrated that the Mg-doped GaN films grown by this method have lower resistivity and higher hole concentration than the one produced by thermal MOCVD method. There is a limitation in the Mg activation by *in-situ* nitrogen plasma annealing treatment and the hole concentration decreases from this limit even though the incorporation of Mg increases.

ACKNOWLEDGMENTS

The authors would like to thank The Ministry of Research and Technology, The Republic of Indonesia for financially supporting this research under Program of RUT-IX Project. This work was also supported by Program of ITB Research Grant 2006 and 2007.

REFERENCES

1. N.N. Morgan and Y. Zhizhen, *J. Microwave Optoelectronics* **2** (5), 52-59 (2002).

2. S. Nakamura, *Semic. Sci. Technol.* **14**, R27 (1999)

3. P. Kung and M. Razegui, *Opto-electronic Rev.* **8**, 20 (2000)

4. H. Amano, M. Kito, K. Himatsu, I. Akasaki, *Jpn. J. Appl. Phys.* **28**, L2112-L2114 (1989)

5. S. Nakamura, T. Mukai, M. Senoh, N. Iwasa, *Jpn. J. Appl. Phys.* **31**, L139-L142 (1992)

6. M.G. Cheong, K.S. Kim, N.W. Namgung, M.S. Han, G.M. Yang, C-H. Hong, E-K. Suh, K.Y. Lim, H.J. Lee, A. Yoshikawa, *J. Cryst. Growth* **221**, 734-738 (2000)

7. K.S. Kim, M.G. Cheong, C-H. Hong, G.M. Yang, K.Y. Lim, E.-K. Suh, H.J. Lee, *Appl. Phys. Lett.* **76(9)**, 1149-1151 (2000)

8. C-F. Chu, C.C. Yu, Y.K. Wang, J.Y. Tsai, F.I. Lai, S.C. Wang, *Appl. Phys. Lett.* **77(21)**, 3423-3425 (2000)

9. C-R. Lee, K-W. Seol, J-M. Yeon, D-K. Choi, H-K. Ahn, *J. Cryst. Growth* **222**, 459-464 (2001)

10. R.H. Horng, *Appl. Phys. Lett.* **79(18)**, 2925-2927, (2001)

11. E.R. Glaser, W.E. Carlos, G.C.B. Braga, J.A. Freitas, Jr., W.J. Moore, B.V. Shanabrook, R.L. Henry, A.E. Wickenden And D.D. Koleske, *Phys. Rev.* **B65**, 085312, (2002)

12. H-W. Huang, C.C. Kao, J.T. Chu, H.C. Kuo, S.C. Wang, C.C. Yu, C.F. Lin, *Mat. Sci. Eng.* **B 113**, 19-23 (2004)

13. A. Cros, R. Dimitrov, H. Angerer, O. Ambacher, M. Stutzmann, S. Christiansen, M. Albrecht, H.P. Atrunk, *J. Cryst. Growth* **181**, 197 (1997)

14. J. Neugebauer and C.G. Van De Walle, *Phys. Rev.* **B50**, 8067 (1994)

15. J. Neugebauer and C.G. Van De Walle, *Appl. Phys. Lett.* **68**, 1829 (1996)

16. U. Kaufmann, M. Kunzer, M. Maier, H. Obloh, A. Ramakrishnan, B. Santic, P. Schlotter, *Appl. Phys. Lett.* **72**, 1326 (1998)

17. S.W. Kim, J.M. Lee, C. Huh, N.M. Park, H.S. Kim, I.H. Lee, S.J. Park, *Appl. Phys. Lett.* **76**, 3079 (2000)

18. S. Nakamura, *J. Vacuum Sci. Techn.* **A 13**, 3 (1995)

Characterization of a Water-based Surfactant Stabilized Ferrofluid by Different Scattering Techniques

V.K. Aswal[1], S. N. Chodankar[1], P.U. Sastry[1], P.A. Hassan[2] and R.V. Upadhyay[3]

[1]Solid State Physics Division, Bhabha Atomic Research Centre, Mumbai 400 085, India
[2]Chemistry Division, Bhabha Atomic Research Centre, Mumbai 400 085, India
[3]Department of Physics, Bhavnagar University, Bhavnagar 364 002, India

Abstract. Different scattering techniques dynamic light scattering (DLS), small-angle X-ray scattering (SAXS) and small-angle neutron scattering (SANS) have been used to characterize a water-based surfactant stabilized ferrofluid having ferrite (Fe_3O_4) particles coated with oleic acid. DLS gives the overall size of the particle along with the thickness of the surfactant (oleic acid) coating and water of hydration attached to the particle. SAXS only measures the size of the ferrite particle due to poor contrast of surfactant coating for X-rays. SANS with the possibility to vary the contrast provides both the size of the ferrite particle and the thickness of the surfactant coating on the particle.

Keywords: Ferrofluid, dynamic light scattering, small-angle scattering
PACS: 61.12.Ex, 47.65.Cb, 82.70.Dd

INTRODUCTION

Ferrofluids are the suspension of the nanosized magnetic particles in appropriate carrier liquid [1]. One of the ways to prepare the stable ferrofluids is by the surfactant coating on the magnetic fine particles, which prevent them from agglomeration even when the strong magnetic field gradient is applied to the ferrofluids. The different scattering techniques can be used to get the different information on the magnetic particles. The scattering techniques for such studies depending on the radiation used are: dynamic light scattering (DLS) for light, small-angle X-ray scattering (SAXS) for X-rays and small-angle neutron scattering (SANS) for neutrons. Due to differences in the interaction of these radiations with matter, the techniques DLS, SAXS and SANS together give complementary information on the nanomaterials [2]. In each of these techniques the radiation (light, X-ray or neutron) is scattered by a sample and the resulting scattering pattern is analyzed to provide information about the structure (shape and size), interaction and the order of the components of the samples. Most of the studies on ferrofluids have been using either one of these techniques. In this work, we have used above scattering techniques together to show the usefulness of these techniques to characterize a

ferrofluid. Measurements have been performed on a ferrofluid having ferrite (Fe_3O_4) particles coated with oleic acid and prepared in water.

EXPERIMENT

The magnetic particles of Fe_3O_4 were prepared using co-precipitation method [3]. These ferrite particles were doubly coated with oleic acid and then dispersed in water. DLS measurements on a dilute ferrofluid sample were carried out using a Malvern 4800 Autosizer employing 7132 digital correlator. The light source was an argon ion laser operated at 514 nm. SAXS experiments were performed with wavelength of 0.154 nm using a Rigaku small-angle goniometer mounted on rotating anode X-ray generator. The samples were kept in a capillary of diameter 1.5 mm and scattered X-rays were measured using a scintillator detector. SANS experiments were performed using a small-angle neutron scattering instrument at the Dhruva reactor, BARC [4]. This instrument makes use of a BeO filter as a monochromator, which provides mean incident wavelength of 0.52 nm. The angular distribution of the scattered neutrons was recorded using one-dimensional He^3 position sensitive detector.

CP989, Neutron and X-ray Scattering in Materials Science and Biology, International Conference on Neutron
and X-ray Scattering 2007, edited by A. Ikram, A. Purwanto, Sutiarso, A. Zulfia, S. Hendrana, and Z. Nurachman

ANALYSIS OF SCATTERING DATA

The signal generated by the light scattering from diffusing particles can be analyzed by its electric field autocorrelation function $g^E(\tau)$, which follows a simple exponential decay with decay constant Γ $(=DQ^2)$ as [5]

$$g^E(\tau) = \exp[-\Gamma\tau] \qquad (1)$$

The effective hydrodynamic size (R_h) is calculated from diffusion coefficient (D) by the Stoke-Einstein relations as given by

$$R_h = kT / 6\pi\eta D \qquad (2)$$

where k is Boltzmann's constant, T is the temperature and η is the solvent viscosity.

Small-angle scattering (SAS) experiment with X-rays or neutrons measures coherent elastic scattering intensity $I(Q)$ as a function of wave vector transfer Q. In the case of dilute solution $I(Q)$ is given as [6]

$$I(Q) = nP(Q) \qquad (3)$$

where n is the number density of the particles. $P(Q)$ is the intraparticle structure factor and depends on the shape and size of the particle. $P(Q)$ is given by the intergral [7]

$$P(Q) = \left| \int \left(\rho_p(r) - \rho_s \right) \exp\,(iQ.r)dr \,\right|^2 \qquad (4)$$

For spherical particles with a radius R, $P(Q)$ is given by

$$P(Q) = (\rho_p - \rho_s)^2 V^2 \left[\frac{3\,j_1(QR)}{QR} \right]^2 \qquad (5)$$

where $V = (4/3)\pi R^3$, ρ_s is the scattering length density of the solvent, ρ_p is the mean scattering length density of the particle and $j_1(QR)$ is spherical Bessel function.

$P(Q)$ for a spherical particle having core-shell structure is given by [8]

$$P(Q) = \left[\Delta\rho_1 V \frac{3\,j_1(QR)}{QR} + \Delta\rho_2 V_1 \frac{3\,j_i(QR_i)}{R_i} \right]^2 \qquad (6)$$

where R is the radius and V is the volume of the core of the particle. R_t and V_t are the overall (core + shell) radius and the volume of the particle, respectively. The thickness of shell around the core is given by $t = R_t - R$. $\Delta\rho_1 = \rho_p - \rho_{shell}$ and $\Delta\rho_2 = \rho_{shell} - \rho_s$, where the scattering length densities ρ_p, ρ_{shell} and ρ_s stand for the core, shell and solvent, respectively.

In the case of polydisperse particles Eq. (1) can be written as [9]

$$I(Q) = \int I(Q,R) f(R)dR + B \qquad (7)$$

The polydispersity in the particle size has been accounted by a Schultz distribution as given by

$$f(R) = \left(\frac{Z+1}{R_m} \right)^{Z+1} R_m{}^Z \exp\left[-\left(\frac{Z+1}{R_m} \right)R \right] \frac{1}{\Gamma(Z+1)} \qquad (8)$$

where R_m is the mean value of the distribution and Z is the width parameter. The polydispersity of this distribution is given by $\Delta R/R_m = 1/(Z+1)^{1/2}$.

RESULTS AND DISCUSSION

Dynamic light scattering

In DLS, time-dependent fluctuations in the intensity of scattered light are measured [5]. These fluctuations happen as a result of the Brownian motion, arising out of the particles undergoing random collisions. Small rapidly diffusing particles yield fast fluctuations, whereas larger particles and aggregates generate relatively slow fluctuations. The rate of fluctuations can be determined through the technique of autocorrelation analysis. Analysis of intensity fluctuations enables the determination of diffusion coefficient [Eq. (1)] and it is converted to a size using Stoke-Einstein relationship [Eq. (2)]. Fig. 1 shows the measured intensity autocorrelation function using DLS. The value of diffusion coefficient as obtained using is 4.5×10^{-8} cm^2/sec and this corresponds to the effective hydrodynamic radius of the particle $R_h = 10.5$ nm.

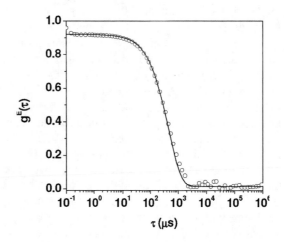

FIGURE 1. DLS data from a ferrofluid.

Small-angle scattering

Small-angle scattering (SAXS or SANS) are the techniques for obtaining structural information of a material on length scale of 1-100 nm [10]. The experiment involves scattering of a monochromatic beam of X-rays (or neutrons) from the sample and measuring the elastically scattered photon (or neutron) intensity as a function of the scattering angle. The magnitude of wave vector transfer Q in SAXS/SANS experiments is typically in the range of ~ 0.01 to 1 nm^{-1}. The experimental details and the data analysis methods used in the two techniques are similar and the only difference is in the radiation used, which can lead to different contrast of the particles for two techniques.

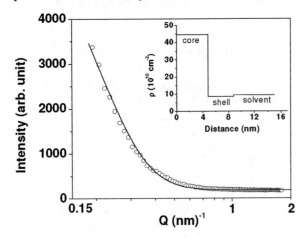

FIGURE 2. SAXS data from a ferrofluid. Inset shows the variation of X-ray scattering length density for different components of the ferrofluid.

Scattered X-ray (or neutron) intensity in SAXS or SANS experiment depends on $(\rho_p - \rho_s)^2$ – the square of the difference between the average scattering length of the particle and the average scattering length density of the solvent. The term $(\rho_p - \rho_s)^2$ is referred to as the contrast factor. The values of ρ_p and ρ_s depend on the chemical composition of the particle and the solvent and are different for X-rays and neutrons. The differences in ρ_p values for X-rays and neutrons arise from the fact that while X-rays are scattered by the electron clouds around the nucleus, neutrons are scattered by the nucleus of an atom. It is seen that as one goes across the periodic table, the X-ray scattering lengths increase linearly with the atomic number of the atom and the neutron scattering lengths vary in a random way. The value of ρ_p or ρ_s for X-ray will be small for hydrogenous materials (organic compounds) as compared to that for heavier elements such as Au, Br, etc. This is not the case with neutrons. Further, the fact that scattering length

of hydrogen is negative (= -0.3742×10^{-12} cm) and that for deuterium is positive (= 0.6674×10^{-12} cm), the value of ρ for neutrons changes significantly on deuteration of a hydrogenous sample. In fact, it is possible to mix the hydrogenous and the deuterated components and thus vary the contrast in a continuous way in a SANS experiment. The SAXS data from ferrofluid are shown in Fig. 2. Due to low contrast of surfactant for X-rays (inset of Fig. 2), there is negligible scattering from the surfactant shell. The scattering is therefore mainly from the ferrite particles. The analysis using Eq. (5) gives mean radius of particle R_m = 4.8 nm and polydispersity σ = 50% as accounted by the Schultz distribution.

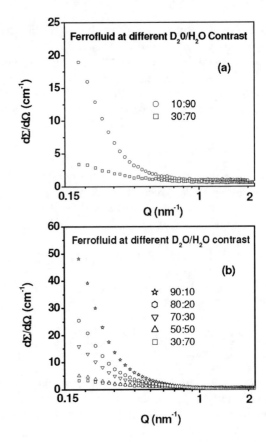

FIGURE 3. SANS data from a ferrofluid as the fraction of D2O in the solvent of mixture of D_2O and H_2O is varied.

Fig. 3 shows the SANS data on ferrofluid with the variation in D_2O composition in a mixture of D_2O and H_2O solvent. It is observed that the scattering intensity first decreases and then increases as the D_2O composition in the mixed D_2O/H_2O solvent is increased. This behavior of scattering intensity can be explained in terms of the variation of scattering length density of solvent (ρ_s) with the change in the composition of the solvent as shown in Fig. 4. The scattering at low D_2O compositions is dominated by the decrease in the contrast for magnetic particles and hence the intensity decreases [Fig. 3 (a)]. On the other hand, at high D_2O compositions the scattering intensity is dominated by the contrast of the surfactant shell and hence scattering intensity increases with the increase in the contrast for shell as the D_2O concentration is increased [Fig. 3 (b)].

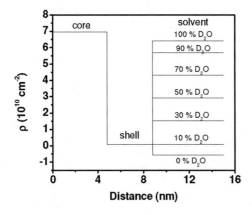

FIGURE 4. The variation of neutron scattering length density as the fraction of D_2O in the solvent of mixture of D_2O and H_2O is varied.

Fig. 4 shows the fitted scattering data from ferrofluid prepared in D_2O to H_2O ratios of 10:90 and 90:10. The data of ferrofluid in D_2O to H_2O ratio of 10:90 have been used to get the size of the ferrite particle as for this sample there is no contrast for the shell (Fig. 4). The analysis gives the similar values of mean radius and polydispersity as obtained using SAXS. The size distribution of the particles is given in the inset of Fig. 5.

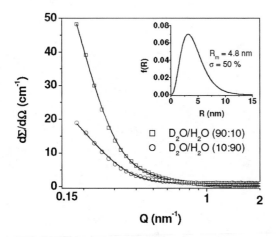

FIGURE 5. Fitted SANS data from a ferrofluid when the surfactant is contrast-matched for D_2O to H_2O ratio of 10:90 and the core-shell structure of the ferrofluid as determined for D_2O to H_2O ratio of 90:10.

In the case of ferrofluid prepared in solvent of higher fraction of D_2O, there exists strong contrast for the shell of surfactant layer around the ferrite particle. The core-shell structure of the particle [Eq. (6)] has been used to fit the data. It is found that double coating of oleic acid has a thickness t = 4.0 nm around the magnetic particles.

Comparison of results of different scattering techniques

Table 1 summarizes the results of different scattering techniques as obtained from the ferrofluid samples. DLS gives the overall size of the magnetic particle with surfactant coating. The value obtained (R_h = 10.5 nm) is larger than the combined value of radius of the particle and thickness of the surfactant coating as obtained using SANS. The larger value as obtained using DLS is expected because of the particle diffuse with water of hydration. The value thus obtained in DLS is the hydrodynamic size and is larger than the bare size of the total particle. SAXS only gives the information on the size of the ferrite particles, as scattering from the surfactant is negligible for the X-rays. SANS with the possibility of contrast variation provides the core-shell structure of the ferrofluid. It may be added further that SANS with polarized neutrons can be used to separate the nuclear and magnetic nanostructures of the ferrofluid [11].

TABLE 1. The various structural parameters of the ferrofluid as obtained by combining different scattering techniques.

Technique	Information obtained	Ferrofluid parameters
DLS	Overall size along with hydration.	R_h = 10.5 nm
SAXS	Only size of the ferrite particle.	$R_m(Core)$ = 4.8 nm σ = 50 %
SANS	Particle size and thickness of surfactant coating	$R_m(Core)$ = 4.8 nm, σ = 50 % $t(Shell)$ = 4.0 nm

REFERENCES

1. S. Odenbach, *Ferrrofluids: Magnetically Controllable Fluids and their Applications*, Springer Verlag, Berlin, 2002.
2. P. Lindner and T. Zemb, *Neutron, X-ray and Light Scattering: Introduction to an Investigative Tool for Colloids and Polymeric Systems*, North Holland, Amsterdam, 1991.

3. R.V. Mehta, R.V. Upadhyay, and B.A. Dasannacharya, P.S. Goyal and K.S. Rao, *J. Magn. Magn. Mater.* **132** (1994) 153.

4. V.K. Aswal and P.S. Goyal, *Curr. Sci. India* **79** (2000) 947.

5. R. Pecora, *Dynamic Light Scattering*, Plenum, New York, 1985.

6. S.H. Chen and T.L. Lin, *in: Methods of Experimental Physics* **23** B (Academic Press, New York, 1987) p. 489.

7. P.S. Goyal and V.K. Aswal, *Int. J. NanoSci.* **4**, 987 (2005)

8. V.K. Aswal and J. Kohlbrecher, *Chem. Phys. Lett.* **424** 91 (2006)

9. V.K. Aswal and P.S. Goyal and P. Thiyagarajan, *J. Phys. Chem.* **B102**, 2469 (1998)

10. L.A. Feigin and D.I. Svergun, *Structure Analysis by Small-angle X-ray and Neutron Scattering*, Plenum Press, New York, 1987.

11. M. Kammel, A. Wiedenmann and A. Hoell, *J. Magn. Magn. Mater.* **252**, 89 (2002)

The Electronic Structure of Wurtzite MnS

O. Murat Özkendir, A. Türker Tüzemen, Y. Ufuktepe

Physics Department, University of Cukurova, 01330 Adana, Turkey

Abstract. Manganese sulfide thin films have been investigated by X-ray Absorption Fine Structure spectroscopy (XAFS). XAFS provides a description of the structure of the films. The paper also presents a structural characterisation of the wurtzite MnS thin films and their crystallisation behaviour by annealing at increasing temperatures. The x-ray absorption fine structure (XAFS) of Mn K-edge and S K-edge in wurtzite MnS have been investigated. The full multiple scattering approach has been applied to the calculation of Mn K edge XANES spectra of MnS. The calculations are based on different choices of one electron potentials according to Mn coordinations by using the real space multiple scattering method FEFF 8.0 code. The crystallographic and electronic structure of the MnS are tested at various temperature ranges from 300 to 573 K. We have found prominent changes in the XANES spectra of Mangeanese sulfide thin films by the change of the temperature. Such observed changes are explained by considering the structural, electronic and spectroscopic properties. The results are consistent with experimental spectra.

Keywords: MnS Thin Film, Electronic Properties, Chalcogenide, XANES
PACS: 61.72.Cc, 64.70.kg, 71.15.Ap, 72.80.Ga, 78.40.Fy

INTRODUCTION

Interesting electronic properties of Manganese attracts a major attention in technological applications. Manganese and its compounds have important semiconducting properties. Manganese compounds are members of Dilute Magnetic Semiconductors (DMS). In DMS, the band electrons and holes strongly interact with the localized magnetic moments and cause a variety of interesting phenomena. Manganese sulfide (MnS) is a dilute magnetic semiconductor material (E_g = 3.1 eV) that is of potential interest in short wavelength optoelectronic applications such as in solar selective coatings, solar cells, sensors, photoconductors, optical mass memories [1-3]. MnS thin films or powders can be found in several polymorphic forms: the rock salt type structure (αMnS) which is the most common form, by low temperature growing techniques it crystallizes into the zincblende (βMnS) or wurzite (γ-MnS) structure [4, 5].

One of the most useful methods serving the electronic structure of materials is the x-ray absorption fine structure (XAFS) spectroscopy [6]. The method XAFS, especially over the last two decades, became an important way for electronic structure calculations, bondings, valency properties, catalytic properties, orbital and atomic configurations. One of the main field of XAFS (X-ray Absorption Fine Structure Spectroscopy) is XANES (X-Ray Absorption Near Edge Structure).

The XANES region yields informations on the lower part of the conduction band. Near-edge structure, observed in X-ray absorption of inner-shell ionization process, can be sensitive to interatomic distances, local coordination and local electronic structure. The shape of absorption spectrum in this region is related to the electronic structure of the material, while the oscillation observed in the high energy part of the spectrum, called as the extended X-ray absorption fine structure (EXAFS) region, is associated with the arrangement of the atoms. It contains signals of interference of outgoing electron wave functions and scattered electron wave functions, gives information about the neighboring atoms, their regional configurations, atomic distances, coordination numbers etc. The amplitudes of the fine structure is related to the number of atoms that surround the absorbing atom. It is known that these amplitudes exhibit a dependence on temperature which is caused by thermal vibrations of absorbing and scattering atoms. To understand the origins of various features it is necessary to calculate the observed fine structures from a suitable theoretical model.

The valence band electronic structure of MnS has been investigated by X-ray and ultraviolet photoemission spectroscopy [7-9]. However, K-edge electronic structure of γ-MnS has not been revealed sufficiently [10-15].

CP989, *Neutron and X-ray Scattering in Materials Science and Biology, International Conference on Neutron and X-ray Scattering 2007,* edited by A. Ikram, A. Purwanto, Sutiarso, A. Zulfia, S. Hendrana, and Z. Nurachman
© 2008 American Institute of Physics 978-0-7354-0508-0/08/$23.00

The purpose of this study is to make a comparison of bonding effects on the electronic states of the S and Mn atoms in γ-MnS and with eachother. Our attention was mainly focused on the bonding effects on K-edge of the atoms. With this purpose we have performed full multiple scattering calculations by using the real-space multiple scattering approach FEFF 8.0 code. Theoretical calculations has been compared with the experimental results which are consistent.

MATERIALS AND METHODS

In order to extract the necessary information of MnS, Mn K-edge absorption fine structure was calculated for the γ-MnS thin film. The calculations are based on different choices of one electron potentials according to Manganese coordinations. Mainly, we made calculations in three sections. Firstly, the bonding effects on electronic structure of Mn K-edge in γ-MnS structure are calculated. Secondly, the same calculations for S K-edge in the same compound are made and finally the results of heating process in wurtzite MnS are investigated. Calculations were performed for 10 Å (Angstroms) thick MnS cluster in wurtzite structure, containing 189 atoms (Mn, S). In γ-MnS, to investigate the bonding effects on Manganese electronic structure, one Mn atom and in another calculation for the same investigation of Sulfur, one S atom was selected as an X-ray absorber also a photoelectron emitter. When an electron leaves from the absorbing atom with enough kinetic energy, it will scatter from a neighboring atom. Its outgoing and scattered wave functions interfere. We have calculated backscattering and phase shift functions for MnS thin film with single and multiple scattering paths using FEFF 8 codes to obtain X-ray absorption spectrum [16]. We applied wurtzite (distorted hexagonal) γ-MnS structure as a model for Manganese sulphide.

RESULTS AND DISCUSSION

Wurtzite hexagonal MnS structure can be crystallized by low temperature growing techniques. Wurtzite structure has asymmetry in its nature. In wurtzite type of (γ-)MnS, some S anions make shorter bonds with Mn atoms than others. Two different structure of MnS compounds, wurtzite and cubic zinc sulfide type is drawn in figure 1.

The structural disorder in wurtzite MnS cause asymmetry in hexagonal form. Due to asymmetry in the wurtzite structure, short distances between Mn and S atoms cause a strong coupling and overlapping between wave functions of outer levels. Strongly overlapping wavefunctions create new hybrid d- and p- like states occuring in the upper part of the valence

band and in the bottom of the conduction state. In γ-MnS structure, this overlapping can be established between Mn 3d state and S 3p state. The main aim of this study is to point out the reflections of hybrid state effects on both sides of the bonds. Mainly, thermal effects and bonding process on γ-MnS structure are investigated.

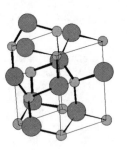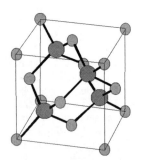

Wurtzite Hexagonal   Cubic Unit Cell

FIGURE 1. Crystal structure of wurtzite and cubic zinc sulfide type

Figure 2 shows Mn K edge XANES structure calculation of α-MnS, γ-MnS, and metallic Mn spectra as a reference material. Energy calibration was performed using metallic Mn foil. The absorption energy difference of Mn K-edge in γ-MnS from metallic Mn K-edge can easily be seen. Bonding with sulfur atoms cause interaction between the outer states of Mn and S which increase the levels of these states. The main absorption peak energies corresponds to transitions both from 1s level to different energy levels of 4s empty state. Mn K-edge absorption peak begins to increase at photon energy 6538.3 eV which is the binding energy of Mn K-shell that corresponds to the transition of 1s core electrons to an unoccupied state above the Fermi level. This transition energy is known as $K_{\beta 5}$ in literature. The pre-edge peak A observed only from wurtzite MnS structure and it has been associated with direct quadrupole transition from 1s to empty Mn 3d states [17]. Pre-edge structure of XANES region has an important role to investigate the electronic and bonding properties of the interested material. Previous studies have shown that structural properties of materials influence the pre edge region [18]. The higher the symmetry of the Mn nearest neighborhood, the weaker is the pre-edge structure observed in absorption spectrum of Mn compounds [11]. Due to asymmetry in the wurtzite structure, short distances between Mn 3d and S 3p orbitals cause to a strong coupling and overlapping. This effect corresponds to a shoulder like shape A (pre-edge) region in Fig. 2.

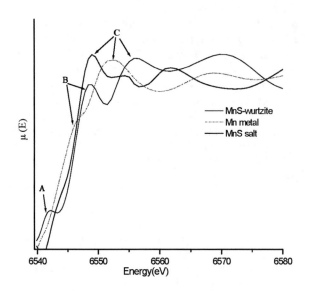

FIGURE 2. Mn K-edge XANES calculated spectra for Mn foil, α-MnS and γ-MnS (wurtzite) structure at 300 K.

The intensity of A is weak because it is due to the 1s → 3d transition which is forbidden by the atomic selection rules in the electric dipole approximation. Peak B observed just after the pre edge peak A is due to atomic-like transitions from 1s core to empty lower level of 4p states. Atomiclike transitions from 1s core to empty 4p states produce multi peaks in the near-edge energy region [19]. The main absorption edge C also corresponds to the transition from 1s state to 4p state and exhibits large chemical shift to higher energies. Strongly overlapped, shorter bonded, Mn 3d and S 3p states create p-like 3d hybridized level in this state. This level support the transitions from 1s core level to 3d hybrid level which is the sign of asymmetry in structure. As a result of bonding with sulphur, main absorption edges of Mn, α-MnS and γ-MnS are calculated as 6538.3 eV, 6543.7, 6551.6 eV, respectively. The energy shifts in α-MnS and γ-MnS spectra are correlated with the length of Mn-S bond. The shorter it is, the greater is the overlap between electron wave functions originating from Mn and S. The energy shifts at Mn K-edge in γ-MnS than the metallic Mn main absorption edge energy for the first and second peaks are 5.4 and 13.3 eV, respectively. Also the hybrid state caused by overlapping outer states of Mn 3d and S 3p increases the transitions to this state in low temperatures.

In figure 3, we have a chance to look at the effects as told above from Sulfur side. In the figure, sulfur K-edge in pure sulfur structure and in γ-MnS is given. Sulfur K-edge absorption spectrum in pure sulfur bulk which has orthorhombic structure begin to increase at photon energy of 2466.75 eV, however S K-edge

absorption spectrum in γ-MnS begins to increase at 2467.78 eV with a pre-edge structure and the main edge at 2472.24 eV. These transitions are called as $K_{\beta3}$ in literature. Energy shift in main absorption edges are obviously seen, like at Mn K-edge energy in γ-MnS. This shift is a result of bonding with outer states of manganese atoms in γ-MnS.

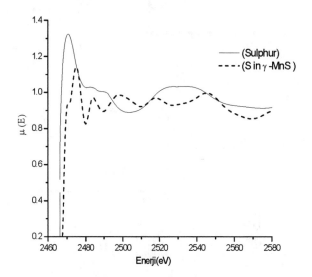

FIGURE 3. Comparison of S K-edge in pure sulfur and in γ-MnS

In figure 3, S atom K-edge in γ-MnS structure has a pre edge shoulder which is the reflection of the hybrid state structure at K-edge. This weak peak shows the transitions from 1s to the hybridized 3p σ^* antibonding state which is caused by overlapping wave functions of shorter bonded S and Mn atoms. The main peak beyond the weak peak corresponds to the transition of 1s electrons to 3.1 eV higher energy levels in 3p σ^* antibonding state. The peaks beyond the main absorption edge are caused by multiple scattering resonances. The facts observed in figure 3 is a reflection of bonding process between Mn-S which is told for Mn atoms in γ-MnS structure.

Heating may cause changes in crystal geometry and electronic structure of atoms. In finite temperatures, atoms oscillates around their point of equilibrium. Raising temperature increase the oscillation frequency and force the atoms to vary their displacements. This process may cause distortions in their crystal structure. Distortions as a result of increasing temperature weaken the bonds and make changes both in electronic properties and geometry of the crystal.

Figure 4 shows Mn K-edge XANES calculated spectra of γ-MnS (wurtzite) structure at 300, 373, 473 and 573 K. Heating of γ-MnS from 300 to 573 K cause thermal distortions. According to the raising temperature, the atoms begin to vibrate strongly around their equilibrium lattice sites and they slightly alter the interference pattern for a given path [20].

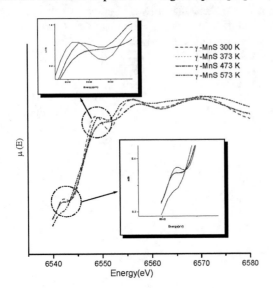

FIGURE 4. Thermal disorder effects on α-MnS wurtzite structure.

Heating of γ-MnS cause thermal distortions which are calculated by XAFS Debye-waller factor, σ^2. The XAFS Debye-Waller factor arises as a natural consequence of fluctuations in interatomic distances and can be viewed as a result of averaging the contribution to the XAFS signal from a given scattering path of a photoelectron over all thermally accessible configurations of the atoms in that path. Short and highly overlapped Mn-S bonds prevents the structure from distortions. As seen in the pre-edge peak window in figure 4, as heat increases, the shape of the peak which is forbidden by the atomic selection rules in the electric dipole approximation, shows a slight change and the peak becomes weaker.

However, Mn K edge spectra calculated at 573 K in Fig. 4, indicate that high temperature begins to distort the wurtzite structure which means that the asymmetric structure (wurtzite) tends to change to a symmetric form (rock salt) which is more clear in the inset windows. This result is consistent with the previous reports [21].

DISCUSSION

In this study, we focused on the bonding effects on both Mn K-edge and S K-edge. We used XANES calculations to observe these effects. Bonding with different type of atoms cause energy shifts both on Mn and S K-edges. Also as a result of asymmetry in γ-MnS structure, short bonded S and Mn atoms wave functions overlap and create a hybridized state. On both atoms K-edge absorption spectra, transitions from 1s core level to this state reveal a pre edge structure. Bonding mechanism whilst the establishment of a compound affect the both side of the atoms exactly with the same magnitude. Calculations are agree with the previous studies. The XAFS calculation for MnS thin film is in good agreement with the experimental results.

REFERENCES

1. D. Fan, X. Yang, H. Wang, Y.Zhang, H. Yan, *Phys.* **B 337**, 165 (2003)
2. B. Piriou, J.D. Ghys, S. Mochizuki, *J. Phys. Condens. Matter.* **6**, 7317 (1994)
3. R. Tappero, P.D'arco, A. Lichanot, *Chem. Phys. Lett.* **273**, 83 (1997)
4. C.D. Lokhande et al., *Thin Solid Films* **330**, 70 (1998)
5. R.L. Clendenen, H.G. Drickamer, *J. Chem. Phys.* **44**, 4223 (1966)
6. J. Stöhr, NEXAFS Spectroscopy, Springer-Verlag, New York (1992)
7. H. Sato et al., *Phys. Rev.* **B 56**, 7222 (1997)
8. H. Franzen, C. Sterner, *J. Solid State Chem.* **25**, 277 (1978)
9. A.B. Mandale, S. Badrinarayanan, S.D. Date, A.P.B. Sihna, *J. Electron Spectrosc. Relat. Phenom.* **33**, 61 (1984)
10. A.N. Kravtsova, I.E. Stekhin, A.V. Saldatov, *Phys. Rev.* **B 69**, 134109 (2004)
11. R. Nietubyc, E. Sobczak, K. E. Attenkofer, *J. Alloys and Compounds* **328**, 126 (2001).
12. P. Zajdel et al., *J. Alloys and Compounds* **286**, 66 (1999).
13. J. Garacia et al., *J. Synchrotron Rad.* **8**, 892 (2001).
14. L. Hozoi, A. H. De Vires, R. Broer, *Phys. Rev.* **B 64**, 165104 (2001)
15. H.P. Myers, Introductory Solid State Physics, Taylor & Francis (1997)
16. A.L. Ankudinov, B. Ravel, J.J. Rehr, S.D. Conradson, *Phys. Rev.* **B 56**, R1712 (1997)
17. F. Bridges et al., *Phys. Rev.* **B 61**, R9237 (2000)
18. F. Bridges et al., *Phys. Rev.* **B 63**, 214405 (2001)
19. K. Tsutsumi, H. Nakamori, K. Ihikawa, *Phys. Rev.* **B 13**, 929 (1976)
20. E. D. Crozier, J. J. Rehr, R. Ingalls, X-Ray Absorption Principles, Konningsberger and R. Prins, John Wiley&Sons pages 375-384 (1988)
21. O. Goede, W. Heimbrdt, V. Weinhold, *Phys. Stat. Solid* **B 143**, 511 (1987)

Effect of Co-doping on Microstructural, Crystal Structure and Optical Properties of $Ti_{1-x}Co_xO_2$ Thin films Deposited on Si Substrate by MOCVD Method

E. Supriyanto[1,2], H. Sutanto[1,3], A. Subagio[1,3], H. Saragih[1], M. Budiman[1], P. Arifin[1], Sukirno[1] and M. Barmawi[1]

[1]*Laboratory for Electronic Material Physics, Department. of Physics, Intitute of Technology Bandung, Jl. Ganesha 10, Bandung 40132, Indonesia*
[2]*Department of Physics, Jember University, Jl. Kalimantan III/25, Jember (68121), Indonesia*
[3]*Department of Physics, Diponegoro University, Jl. Tembalang Semarang, Indonesia*

Abstract. $Ti_{1-x}Co_xO_2$ thin films have been grown on n-type Si(100) substrates by metal organic vapor deposition (MOCVD) using titanium (IV) isopropoxide (TTIP) and tris (2,2,6,6-tetramethyl-3, 5-heptanedionato) cobalt (III) as metal organic precursors. The parameter deposition, such as: bubbler temperature of TTIP T_b(Ti) = 50°C; substrate temperature T_s = 450°C; bubbler pressure P_b(Ti) = 260 Torr; flow rate of Ar gas through TTIP precursor Ar(Ti) = 100 sccm (standard cubic centimeters per minute) and flow rate of oxygen gas O_2 = 60 sccm were found as optimal deposition parameters. The thin films deposited were have rutile (002) crystal plane, whereas those deposited at other parameter were mixing of anatase and rutile phases. Co dopant with concentration of up to 5.77% was not changes the structure of TiO_2. Increase of Co incorporated in thin films was decreasing of band-gap energy.

Keywords: Crystalline orientation, TiO_2, $Ti_{1-x}Co_xO_2$, MOCVD.
PACS: 78.66.Hf

INTRODUCTION

Spintronics has emerged as an interesting technology where both charge and spin are harnessed for new functionalities in devices combining standard microelectronics with spin-dependent effect. This field has been put forth as a promising field in the semiconductor electronics[1,2]. For the realization of spintronic devices, it is essential to hybridize existing semiconductors with highly spin-polarized magnetic materials. Up to now, heterostructures based on metallic-ferromagnet/semiconductor have been extensively studied[3,4]. However, it has been proven difficult to transfer electron spins across the interface still keeping high spin-polarization, mainly due to the large mismatch in electrical conductivity between the two materials[5]. Therefore, much research efforts recently have been focusing on the development of ferromagnetic semiconductors which should not only be easily integrable with existing semiconductors but also be highly spin-polarized[6,7]. Among these, Mn-doped GaAs, successfully grown for the first time

by Ohno *et al.*[6] has shown promising performance in the coherent transfer of electron spins[8]. However, highest Curie temperature (T_C) of this system is limited to 160 K. The most important breakthrough in this field is to elevate T_C well beyond room temperature in order to realize practical devices operable at room temperature. Therefore, various host semiconductors and dopant transition metal elements have been tested for realizing room temperature ferromagnetic semiconductors.

Co-doped TiO_2 anatase, grown by pulsed laser deposition, has been demonstrated to be ferromagnetic up to 400 K by Matsumoto *et al.*[9] using a combinatorial molecular beam epitaxy technique with doping level up to around 8 atomic-%. Titanium dioxide is wide gap oxide semiconductor and has three different crystalline structure phases, namely: rutile, anatase and brookite. Usually rutile is the only stable phase, whereas anatase is metastable and transform to rutile on heating.

It is interesting to note that Matsumoto *et al.* observed a magnetic structure only in Co-doped TiO_2

CP989, *Neutron and X-ray Scattering in Materials Science and Biology, International Conference on Neutron and X-ray Scattering 2007*, edited by A. Ikram, A. Purwanto, Sutiarso, A. Zulfia, S. Hendrana, and Z. Nurachman
© 2008 American Institute of Physics 978-0-7354-0508-0/08/$23.00

anatase but not in the rutile phase, seemingly suggesting that the magnetization of the impurity atoms depend sensitively on the local environment surrounding them. More recently, however, Park et al. have successfully grown ferromagnetic $Ti_{1-x}Co_xO_2$ rutile thin films as DMS materials by sputtering method [10]. The Co atoms were incorporated up to 12 atomic % and the temperature Curie was estimated to be above 400 K.

Since the rutile phase is thermodynamically more stable than the anatase phase, it may have a higher potential for technical applications. The dissimilar behaviors of films grown by different methods indicate that the local environment of Co could have a drastic effect on the magnetic structure of this material. Further more, Soo et al.[11] has found that the local structure around Co could remain virtually anatase like while the nanomaterials underwent an anatase-to-rutile phase transition by annealing the $Ti_{1-x}Co_xO_2$ thin films deposited by sol-gel method at different temperatures.

The above description seems that the characteristics of Co-doped TiO_2 thin films depends on both technique and deposition condition. Therefore, it is evident that the deposition technique of the Co-doped TiO_2 is a centaral issue. Metal organic chemical vapor deposition (MOCVD) is currently among the most important technique for growing thin films, high purity epitaxial films with applications in electronic and optoelectronics[12]. In the technique, metalorganic materials are used as precursors and has many advantages, such as easy to control in-situ of the stoichiometry films, high deposition rates at low temperature, uniform thickness over large areas, conformal coverage of irregular surfaces, and a smooth surfaces. This technique for growth of oxides is low cost and it is easy to control the deposition growth parameters. On the other hand, metal organic precursors have excellent properties since they are completely vaporized and decomposed into oxides at relatively low temperature, commercially available with high purities and at low price[13].

In this letter, $Ti_{1-x}Co_xO_2$ thin films are prepared onto Si(100) substrates using metal organic chemical vapor deposition (MOCVD) method. Effect of Co doping on crystal structure, microstructural and optical properties of $Ti_{1-x}Co_xO_2$ thin films were investigated

EXPERIMENTAL METHOD

A vertical MOCVD reactor was used to growi Co-doped TiO_2 thin films on Si(100) substrates. Titanium (IV) isopropoxide (TTIP) [Ti{OCH(CH$_3$)$_2$}$_4$] 99.99% (Sigma Aldrich Chemical Co., Inc.) and tris (2,2,6,6-tetramethyl-3,5-heptanedionato) cobalt (III), 99%, Co(TMHD)$_3$ powder (Strem Chemical, Inc.) were

used as precursors for Ti and Co, respectively. Oxygen (O_2) gas was used as O source. The Co(TMHD)$_3$ powder was dissolved in a tetrahydrofuran (THF) solvent (Aldrich Chemical Co., Inc.) to form a source solution of 0,1 mol concentration of Co(TMHD)$_3$. Solutions of precursors ([TTIP+ Co(TMHD)$_3$] stored in one bubblers were vaporized and transported into reactor using argon (Ar) carrier gas. Temperature of 50°C was used to vaporized of [TTIP+ Co(TMHD)$_3$] solutions. Flow rate of Ar carrier gas through precursor was maintained at 100 sccm. O_2 gas with flow rate of 60 sccm was supplied through a separate gas line into the chamber. Heating tape was wrapped around the [TTIP+ Co(TMHD)] vapor-transport lines to prevent condensation from the sources to the growth reactor. Base pressure of reactor system was 3×10^{-3} Torr. Substrate temperatures from 400°C to 550°C were used to growing thin films. Si substrates were cleaned by aceton for 10 minutes and then by methanol for 10 minutes. After cleaned, the Si substrate was etched by mixing of de-ionized water and 10% HF for 2 minutes. As soon as the cleaning and etching processes of Si wafers were finished, the substrates were mounted onto a molybdenum susceptor in reactor. A typical growth was carried out for 120 minutes and was following by a slow cooling down to room temperature at rate of 200°C/hour to prevent strain-induced microcracks. After growth, the crystalline structure and morphology of thin films were characterized by X-ray diffraction (XRD) employing Cu-K_α radiation (λ=1.5406 Å) (Philips PAN Analytical 'X'Pert Pro PW3071) and scanning electron microscope (SEM) (Jeol JSM 6360LA), respectively. Chemical composition of thin films was determined by energy dispersive X-ray analysis (EDAXS) (Jeol JSM 6360LA).

RESULTS AND DISCUSSION

As a host material, in this experiment the TiO_2 thin films without Co-dopant were deposited on Si(100) substrates. Substrate temperatures were ranged from 300°C to 550°C. The parameter deposition, such as: bubbler temperature of TTIP $T_{b(Ti)}$ = 50°C; substrate temperature T_s = 450°C; bubbler pressure $P_{b(Ti)}$ = 260 Torr; flow rate of Ar gas through TTIP precursor $Ar_{(Ti)}$ = 100 sccm and flow rate of oxygen gas O_2 = 60 sccm were used and found as optimal deposition parameters. Figure 1 shows the XRD pattern of thin film. XRD pattern showed that the thin films deposited were a single crystal having a rutile structure with (002) orientation (R(002)), whereas those deposited at other parameters were mixing of anatase and rutile structures[14].

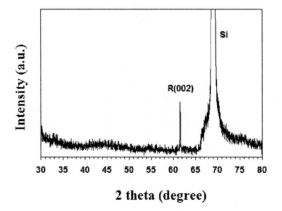

FIGURE 1. XRD pattern of TiO$_2$ thin film grown on Si (100) substrate with parameter deposition of $T_{b(Ti)}$ = 50°C, T_s = 450°C, $P_{b(Ti)}$ = 260 Torr, flow rate of Ar$_{(Ti)}$ = 100 sccm and O$_2$ = 60 sccm

Co-doped TiO$_2$ thin films were grown with a Co dopant. Deposition parameters as mentioned above and the new metal organic compound Co(TMHD)$_3$ were used. Cobalt oxide thin films were usually deposited by using Co(acac)$_2$ (acac=acetylacetonate) and Co(OAc)$_2$ (OAc=acetate) precursors[15]. These two precursors have been used for MOCVD of pure CoO thin films. The deposition processes were must be strictly controlled to avoid oxygen excess into thin films and therefore must be done at low oxygen atmosphere[16]. Co(TMHD)$_3$ has better stability than two others above as a monomer, because each β-diketonate ligand with bulky tert-butyl substituents occupied two coordination sites of the metal and also has an appreciable volatility (vaporization already at 90°C). With this compound, the stoichiometry of Co$_x$O$_y$ could be controlled easily[15], so one could be permitted high solubility of Co into TiO$_2$ matrix. Besides, this compound was relatively stable to air and moisture and could be easily prepared and purified[17].

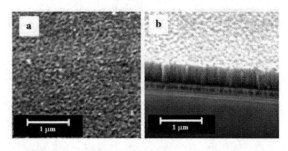

FIGURE 2. SEM images of surface (a) and cross-section (b) of Ti$_{1-x}$Co$_x$O$_2$ (x=0) thin film with parameter depositions as described in figure 1.

Figure 2 shows SEM images of surface morphology (a) and cross-section (b) of Ti$_{1-x}$Co$_x$O$_2$ (x=0) thin film. Thin films were growing with thickness of about 0.7 μm. These micrographs reveal a columnar structure of grains. Grains have similar form. It is correspond to XRD pattern showed in Figure 1 in which the film is has single crystal plane structure of (002) rutile.

Ti$_{1-x}$Co$_x$O$_2$ thin films with x>0 were grown with a Co(TMHD) precursor dopant. Adding of Co(TMHD) precursor vapor was carried out with flowing of Ar gas through Co(TMHD) bubbler. The Ar gas flow rates of (a) 20 sccm, (b) 30 sccm, (c) 40 sccm, (d) 50 sccm, (e) 60 sccm, (f) 70 sccm and (g) 90 sccm were used. The Co contents in each films were determined from EDS data to be: (a) 0.41%, (b) 1.83%, (c) 2.97%, (d) 5.77%, (e) 10.41%, (f) 10.65% and (g) 11.01%, respectively. It appears that different Ar gas flow rate give different numbers of atoms incorporated in the films, and in fact, with increase of Ar gas flow rate, the number of Co atoms which could get into TiO$_2$ host matrix is larger.

Figure 3 shows SEM images cross-sections of selected thin films in which have Co content of (A) 0.97%, (B) 3.84%, (C) 6.21% and (d) 6.5%. Compared to Figure 3a, incorporation of Co atoms could be smoothed-out of TiO$_2$ thin film surfaces. The surface and cross-section images of Ti$_{1-x}$Co$_x$O$_2$ thin films as shown in Figure 3 revealed that there are no significant differences of surface roughness and grain shapes of films. The surfaces are relatively smooth and the grain sizes are mostly the same with columnar structures. The abnormal grains and the clustering Co on thin film surface due to Co atoms are not found.

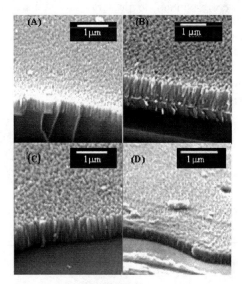

FIGURE 3. XRD pattern of Co doped TiO$_2$ thin film with Co concentration of 5.77% grown on Si (100) substrate.

Crystal structure of films were investigated by X-ray diffraction (Cu $K_{\alpha 1}$ λ=1.5406 Å) measurement. X-ray diffraction patterns show that Co-dopants do not change the structure for a level up to 5.77 %. All films were grown in a single plane (002) rutile structure. Epitaxial growth of $Ti_{1-x}Co_xO_2$ thin films may due to similarity of crystal lattice of both materials. TiO_2 (002) rutile has lattice mismatch about 8 % to Si(100) substrate. XRD pattern of $Ti_{1-x}Co_xO_2$ thin film with 5.77 % of Co content is shown in Figure 4.

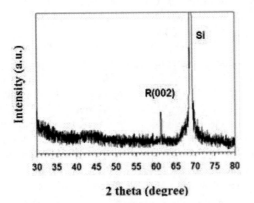

Figure 4. XRD pattern of Co doped TiO_2 thin film with Co concentration of 5.77% grown on Si (100) substrate.

Relationship between optical properties and Co-doping level of $Ti_{1-x}Co_xO_2$ thin films measured at room temperature were investigated by a NIR-UV visible optical reflectance spectroscope. Figure 5 shows the room temperature optical reflectance spectra of $Ti_{1-x}Co_xO_2$ thin films. The last position of the peak oscillation related to the band-to-band transition of the films corresponding to the value of bandgap energy (E_g)[18]. Band gap energy (E_g) value of $Ti_{1-x}Co_xO_2$ thin films with Co content of (a) 0.97%, (B) 3.84%, (C) 6.21% and (d) 6.5% were 3.62 eV, 3.49 eV, 3.39 eV and 3.29 eV, respectively. Increase of Co incorporated in thin films was decreasing of band-gap energy.

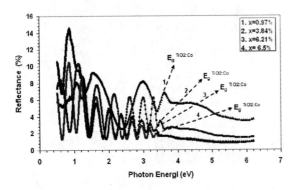

Figure 5. The optical reflectance spectra of $Ti_{1-x}Co_xO_2$ thin films samples grown on Si(100) substrate

CONCLUSIONS

The $Ti_{1-x}Co_xO_2$ thin films have been grown on n-type Si(100) substrate by *metal organic chemical vapor deposition* (MOCVD) method and characterized for structure crystal, microstructure and optical properties as a function of Co doping concentration. The crystal structure of $Ti_{1-x}Co_xO_2$ thin films grown on Si(100) substrates has single crystal orientation growing in the (002) plane. This epitaxial growth occurs in Co concentration up to 5.77%, which indicates a shorter lattice mismatch constant along *c*-axis after Co-doping. Increase of Co incorporated in thin films decreases the band-gap energy

REFERENCES

1. B. T. Jonker, Y. D. Park, B. R. Bennett, H. D. Cheong, G. Kioseoglou, A. Petrou, *Phys. Rev.* **B 62**, 8180 (2000).
2. Malajovich, J. J. Berry, N. Sarmath, D. D. Awschalom, *Nature* **411**, 770 (2001).
3. M. Johnson, *Phys.* **E 10**, 472 (2001).
4. J. A. C. Bland, A. Hirohata, C. M. Guertler, Y. B. Xu, M. Tselepi, *J. Appl. Phys.* **89**, 6740 (2001).
5. G. Schmidt L. W. Molenkamp, *J. Appl. Phys.* **89**, 7443 (2001).
6. H. Ohno, A. Shen, F. Matsukura, A. Oiwa, A. Endo, S. Katsumoto, and Y. Iye, *Appl. Phys. Lett.* **69**, 363 (1996).
7. H. Akinaga, S. Ne´meth, J. De Boeck, L. Nistor, H. Bender, G. Borghs, H. Ofuchi, and M. Oshima, *Appl. Phys. Lett.* **77**, 4378 (2000).
8. Y. Ohno, D. K. Young, B. Beschoten, F. Matsukura, H. Ohno, D. D. Awschalom, *Nature* **402**, 790 (1999).
9. Y. Matsumoto, M. Murakami, T. Shono, T. Hasegawa, T. Fukumura, M. Kawasaki, P. Ahmet, T. Chikyow, S. Koshihara, H. Koinuma, *Science* **291**, 854 (2001).
10. W.K. Park, R.J. Ortega-Hertogs, J.S. Moodera, A. Punnoose, M.S. Seehra, *J. Appl. Phys.* **91**, 8093 (2002).
11. N.J. Seong, S.G. Yoon, C.R. Cho, *Appl. Phys. Lett.* **81**, 4209 (2002).
12. D.M. Dobkin, M.K. Zuraw, Priciples of Chemical Vapor Deposition, Kluwer Academic Publishers, 2003.
13. A. Sandell, M.P. Anderson, Y. Alfedsson, M.K.J. Johansson, J. Schnedt, H. Rensmo, H. Siegbahn, P. Uvdal, *J. Appl. Phys.* **92**, 3381 (2002)
14. H. Saragih, P. Arifin, M. Barmawi, *Jurnal Matematika dan Sains* **9** (3), 263 (2004)
15. S. Pasko, A.Abrutis, L.G.H. Pfalzgraf, V. Kubilius, *J. Cryst. Growth* **262**, 653 (2004)
16. E. Fujii, H. Torii, A. Tomazawa, R. Takayama, T. Hirao, *J. Mater. Sci.* **30**, 6013 (1995)
17. D. Barrreca, C. Massignan, S. Daolio, M,Fabrizio, C. Piccirillo, L. Armelao, E. Tondello, *Chem. Mater.* **13**, 588 (2001)
18. R.Kudrawiec, M. Syperek, J. Misiewicz, R. Paszkiewicz, B. Paszkiewicz, M. Tlaczala, *Superlattices Microstruct.* **36**, 633-641 (2004)

Synthesis and Characterization of In-situ Copper-Niobium Carbide Composite

H. Zuhailawati[1], R.Othman[1], D.L. Bui[1], M. Umemoto[2]

[1]*School of Materials and Mineral Resources Engineering, Engineering Campus, Universiti Sains Malaysia, 14300 Nibong Tebal, Penang, Malaysia*
[2]*Department of Production Systems Engineering, Toyohashi University of Technology, Japan*

Abstract. In this work, synthesis of copper matrix composite powder reinforced by in situ niobium carbide particle was prepared by mechanical alloying of elemental powder and subsequent heat treatment. Elemental powders of Cu-Nb-C correspond to Cu-40wt%Nb-10%wtC composition was milled for 54 hours at room temperature in a planetary ball mill. The effect of heat treatment temperature on the formation of niobium carbide was analyzed. Characterization by X-ray diffraction was done on the milled powder and heat-treated powder in order to investigate NbC formation. Results indicate that NbC began to precipitate after mechanical alloying for about 54h with heat treatment temperature of 900°C and 1000°C.

Keywords: X-ray diffraction, in situ composite, mechanical alloying, copper matrix composite
PACS: 81.05.Ni; 81.07.Bc; 81.20.Ev

INTRODUCTION

Carbides-dispersion-strengthened-copper alloys are potential engineering materials for applications which required good mechanical properties. A copper based alloy with high strength can be achieved by the presence of very rigid particles such as carbides in metal matrix. The benefits of dispersion strengthening coppers offer number of advantageous; the presence of finely dispersed stable particles can increased the strength of materials, which can be explained by the Orowan's mechanism in Equation 1[1].

$$T = \mu b/L \qquad (1)$$

where T is the external stress, μ is the stress modulus of rigidity of the metal matrix metal, b is the magnitude of Burger Vectors and L is the means particle size.

From equation 1, it is shown that the properties of the metal matrix composites are controlled by the refinement of reinforcement at constant volume fractions. As the mean interparticle distance L decreases, the external stress increases, leading to strength of the alloy. The properties of the MMCs are also controlled by the nature of matrix-reinforcement interfaces. An optimum set of mechanical properties can be achieved when fine and thermally stable ceramic particulates are dispersed uniformly in the metal matrix. Efforts have been made to meet such requirements. The work has led to the development of novel composite in-situ MMCs in which the reinforcement are synthesized in a metallic matrix by chemical reactions between elements or between element and compound during composite fabrications [2].

Compared to the conventional MMCs produced by ex situ methods, the in situ MMCs exhibit the following advantages: (1) the in situ formed reinforcement are thermodynamically stable at the matrix, leading to a less degradation in elevated-temperature services; (b) the reinforcement-matrix interface are clean, resulting in a strong interfacial bonding; (c) the in situ formed reinforcing articles are finer in size and their distribution in the matrix is more uniform, yielding better mechanical properties.

In recent years, mechanical alloying technique has been extensively used to fabricate in situ ceramic particle reinforced metal matrix composites. Ma et al. [3] have fabricated in situ Al_4C_3 dispersed and SiC particulate mixture-reinforcement Al composites via the mechanical alloying process. In this process, C, Al and SiC powder mixture was subjected to ball milling. The as-milled powder was then annealed at 873K. Finally, the as-annealed powder compact was extruded to rods at 673K. Biselli et al. [4] have fabricated the in situ TiB_2 particles reinforced copper composite by

CP989, Neutron and X-ray Scattering in Materials Science and Biology, International Conference on Neutron and X-ray Scattering 2007, edited by A. Ikram, A. Purwanto, Sutiarso, A. Zulfia, S. Hendrana, and Z. Nurachman

means of mechanical milling techniques, followed by suitable heat treatments. They reported that the reaction between Ti and B does not occur during milling, but instead takes places during annealing for short time periods at temperature of 873-1073K.

In the present work, NbC is chosen as reinforcement because of its high melting temperature (3783K) and its low solubility in copper. Furthermore, NbC is an extremely hard material, providing wear resistance to the composite. Particular attention is paid on the effect of the heat treatment temperature on the formation of in situ niobium carbide during mechanical alloying and subsequent heat treatment.

EXPERIMENTAL METHOD

Copper, niobium and graphite powder were used as starting material. The initial powder of Cu, Nb and C were used with the composition of Cu-40wt%Nb-10%wtC. The powder was charged together with stainless steel ball of 20 mm with powder to ball weight ratio of 1.10. In order to prevent oxidation the container was filled with argon. The mill was operated at a rotating speed of 400 rpm for 54 hours. A proportion of the powder was heated for 1 hour under argon flow at difference temperature, i.e. 700°C, 800°C, 900°C and 1000°C. The milled and heat treated powder was characterized by X-ray diffraction analysis to determine the formation of the NbC phase.

RESULTS AND DISCUSSION

An XRD pattern for the as-milled mixture powder is shown in Figure 1. The XRD pattern displays all the three strong peaks of copper. These included the peaks at 43.29°, 50.496°, and 74.187°. No peaks of NbC can be detected in the as-milled powders. The XRD pattern also shows that a Cu_2O peak appear at approximately 37° angle.

The as-milled Cu-40wt%Nb-10wt%C powder milled for 54h was heated at different temperature, i.e. 700°C, 800°C, 900°C, 1000°C, in order to study the effect of temperature on the formation of NbC. The XRD results of the heat-treated powder are shown in Figure 2.

According to Figure 2, there are no niobium peaks in the XRD patterns for all samples after heat-treated at temperature ranging from 700 °C up to 1000 °C. However, Nb_2O_5 is found in both of the XRD patterns for samples heat-treated at 700 °C and 800 °C. However, NbO_2 peaks are only detected in samples heat-treated at 900 °C and 1000 °C, whilst Nb_2O_5 XRD peaks completely disappeared at both of these temperatures, as indicated by Figures 2c and 2d. Moreover, NbC XRD peaks only appear in samples

heat-treated at 900 °C and 1000 °C whilst Cu_2O peaks appear in all samples.

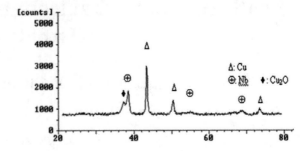

FIGURE 1. XRD patterns for Cu-Nb-C powder milled for 54h

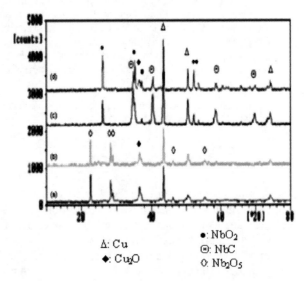

FIGURE 2. XRD patterns for Cu-Nb-C powder milled for 54h with heat treatment temperature: a) 700°C, b) 800°C, c) 900°C and d) 1000°C

Figure 2(a and b) show the presence of Nb_2O_5 peaks with high intensity. However, there is no detection of NbC for these samples which heat-treated at 700 °C and 800 °C. This means that a large amount of Nb has converted to Nb_2O_5 during heat-treatment and this leads to a decrease in Nb content in the samples. In addition, Cu_2O peak also appeared at 36.604 and 36.668 in these samples which were heat-treated at 700 C and 800 C.

The NbC phase starts to be detected in the sample heat-treatment at 900 °C, as shown in Figure 2c. The XRD patterns display strong NbC peaks at approximately 34.881° and 40.380° whilst two other weak peaks are observed at 58.394° and 69.876°. The NbC peaks are strong with high intensity whilst NbO_2 peaks are weak with low intensity. This indicates that the Nb has converted to NbC in larger amounts compared to a smaller amount being converted to NbO_2.

The present research showed that NbC can be formed by mechanical alloying with subsequent heat-treatment at a lower temperature compared to a conventional method. According to [5], synthesis of NbC by conventional method took place at a very high temperature (above 1000°C). Moreover, according to [6], NbC can be formed directly from niobium and graphite at 1300 °C to 1400°C via powder metallurgy method, according to the following reaction:

$$C + Nb \rightarrow NbC \qquad (2)$$

In the conventional method, NbC was produced by coatings of Nb on graphite substrates followed by heat-treatment. The result showed that the NbC phase only formed after heat-treatment at a temperature of 1500°C.

The result in present study is in contrast to a report from [7] who obtained NbC after a short milling time, 16h, without the need of a subsequent heat treatment. However, the present results are in agreement with [8] who reported that NbC began to precipitate after mechanical alloying for about 20h with heat treatment temperature at 600°C. The reason why NbC did not form during the milling process is that the amount of energy transferred to the powders as a result of collision between the milling media was not high enough to overcome the activation energy required for the reaction to form NbC. Furthermore, the ball milled used has been set to a mill and pause cycle.

The formation of NbC phase from Cu-Nb-C powder system by mechanical alloying with subsequent heat-treatment at a much lower temperature can be explained by a decrease in activation energy. Activation energy for diffusion is equal to the sum of the activation energy to form vacancy and the activation energy to move the vacancy[9]. The activation energy, Q, is therefore,

$$Q = Q_f + Q_m \qquad (3)$$

where Q_f is the activation energy for creating vacancies, and Q_m is the activation energy for moving vacancies. According to Lu, the temperature generated by collision is far from reaching the diffusion temperature; this is the reason why NbC cannot be formed during mechanical alloying only. Therefore, the mechanically alloyed powder requires a heat-treatment step to activate the NbC formation.

As shown in Figure 2c, at 900°C, NbO_2 was formed instead of Nb_2O_5. This is because Nb_2O_5 had decomposed to a lower valence to form NbO_2 and oxygen at higher temperature (900°C) according to following decomposition equation:

$$Nb_2O_5 = 2NbO_2 + \tfrac{1}{2}O_2 \qquad (4)$$

In Figure 2d NbO_2 peaks become stronger in the as-milled Cu-Nb-C powder after heat-treatment at 1000 °C. This means that much more Nb_2O_5 had decomposed to a lower valence to form NbO_2 and oxygen due to the higher heat-treatment temperature. In addition, niobium has also reacted with oxygen to convert to NbO_2 according to in the reaction:

$$Nb + O_2 = NbO_2 \qquad (5)$$

Besides, Cu_2O XRD peaks also reflected in 1000 °C heated sample with higher intensity suggesting that the content of Cu_2O in this sample is higher than the Cu_2O formed in powder heat-treated at 900°C.

CONCLUSIONS

No NbC phase was detected in all the as milled powders. Nb_2O_5 was observer at 700°C and 800°C whilst NbO_2 has been formed at 900°C and above. This observation suggests that Nb_2O_5 and decomposes to form NbO_2 and oxygen at 900 °C and above. Cu_2O was found in all samples heat-treated at temperature ranging from 700°C to 1000°C. No NbC phase was formed at temperatures at 800°C and below. However, the NbC phase was formed at a temperature of 900 °C as well as at a temperature higher than 900°C. Relatively, the amount of NbC phase decreased whilst the amount of NbO_2 phase increased with increasing heat-treatment temperatures.

ACKNOWLEDGEMENT

This work was supported by the Japan International Cooperation Agency (JICA) under University Network/Southeast Asia Engineering Education Development Network Program (AUNSEED-Net).

REFERENCES

1. F. Thummler and R. Oberacker, *Introduction to Powder Metallurgy*, Great Britain: The Institute of Materials, (1993) p128-206
2. B.S.S. Daniel, V.S.R. Murthy and G.S. Murty, *J. Mater. Proc. Technol.* **68**, 132-155 (1997)
3. Z.Y. Ma, X.G. Ning, Y.X. Lu, J. Bi, L.S. Wen, S.J. Wu, G. Jangg and H. Daninger, *Scripta Metall. Mater.*, **31**, 31 (1994)
4. C. Biselli and D.G. Morris, N. Randall, *Scripta Metall. Mater.*, **30**, 1327 (1994)
5. A. Raveh, S. Barzalilai and N. Frage, *Thin Solid Films* **496**, 450-546 (2006)

6. M. S. E. Eskandarany. *Mechanical Alloying for Fabrication of Advanced Engineering Materials*, New York: William Andrew, 2001, pp65

7. M.T. Marques, V. Livramento, J. B. Correia. A. Almeida and R. Vilar., *Mater. Sci. Eng.* **A399**, 1-2, 282-286 (2005)

8. T. Takahashi and Y. Hashimoto, "Preparation of carbide-dispersion strengthened coppers by mechanical alloying" in *Material Science Forum*, edited by J.S. Benjamin, Switzerland: Trans Tech Publications, 1992, pp175-182

9. L. Lu, M. O. Lai, *Mechanical Alloying*, USA: Kluwer Academic, 1998, pp48

Mechanical Alloying of Mg$_2$Ni Hydrogen Storage Alloy

A. Azmin Mohamad

School of Materials and Mineral Resources Engineering, Universiti Sains Malaysia
14300 Nibong Tebal, Penang, Malaysia

Abstract. After milling for 20 days, Mg$_2$Ni has produced about 36 μm particles as measured by particle size distribution and scanning electron microscopy. At this stage Mg$_2$Ni alloy phases was formed as proven by X-ray diffraction. Energy-dispersive analysis analyses show that the elemental percentage of Mg and Ni is 67.8 % and 32.2 %, respectively. Those which are suitable for use as hydrogen storage alloy in nickel metal hydride batteries.

Keywords: Mechanical Alloying, Mg$_2$Ni alloy, X-ray diffraction.
PACS: 81.20.Ev

INTRODUCTION

The activity of Mg$_2$Ni with respect to hydrogen depends on the kind of preparation method, sintering and melting processes [1, 2]. Arc melting method leads to poor kinetics and are very complex [3-5]. In the ingot process, the control of the chemical composition of this alloy is very difficult because magnesium evaporates easily due to its high vapor pressure. Thus, a repeated remelting process with an additional supply of magnesium is needed to prepare Mg$_2$Ni alloy with a specific chemical composition. The alloy obtained by this method must be activated by many hydriding-dehydriding cycles before it can be put for practical use [6]. Lei *et al.* [7] reported that the Mg-Ni alloy can be prepared by mechanical alloying and can absorb-desorb hydrogen even at room temperature. This presents a new possibility for Mg$_2$Ni alloy application in Ni-MH batteries and reduces the production cost of nickel metal hydride (Ni-MH) batteries substantially [8].

In this work, the characteristics of Mg$_2$Ni hydrogen storage alloys prepared by the ball milling method are presented. The alloy is characterized for their structural, morphology, elemental composition and size. X-ray diffraction (XRD), scanning electron microscopy (SEM), energy-dispersive analysis of X-rays (EDX) and particle size distribution analysis were carried out to characterize the Mg$_2$Ni hydrogen storage alloy.

EXPERIMENTAL

Mg$_2$Ni alloy powders were synthesized by mechanical alloying. Pure Mg powder (R & M Chemical, ~ 100 μm, 99.9% purity) and Ni powder (Aldrich, ~100 μm, 99.9% purity) were mixed to give the desired composition. These starting materials were put into a stainless steel bowl and sealed under argon atmosphere with stainless steel balls. Mechanical alloying was carried out with a ball to powder ratio of 15:1, using a PASCAL 9VS horizontal ball mill at a milling speed of 200 rpm for 20 days.

RESULTS AND DISCUSSION

The XRD patterns of pure magnesium and nickel powders together with the mixtures of Mg and Ni ball-milled for 5, 7 and 20 days are shown in Fig. 1. The 2θ angles of Mg and Ni agree well with that contained in ASTM (4-0770) and ASTM (4-0850), respectively. The XRD pattern of Mg$_2$Ni sample obtained after ball milled for 20 days compared quite well with values from ASTM (1-1268) and the literature [9,10]. For comparison, only peaks of sample ball milled after 20 days match well with data from the literature. The peaks at 2θ = 32.5 and 34.5° are due to Mg.

It can be observed from the Fig. 1 that almost all peaks for the Mg$_2$Ni that are listed in the literature can be found in the XRD pattern of the present work although the peaks may differ in intensity. However, the most intense peak at 2θ = 45.2° agrees with the literature. The peaks at 2θ = 23.8, 32.2, 39.9 and 41.0°

CP989, Neutron and X-ray Scattering in Materials Science and Biology, International Conference on Neutron and X-ray Scattering 2007, edited by A. Ikram, A. Purwanto, Sutiarso, A. Zulfia, S. Hendrana, and Z. Nurachman

reported in the literature are not present in the XRD pattern of the present work.

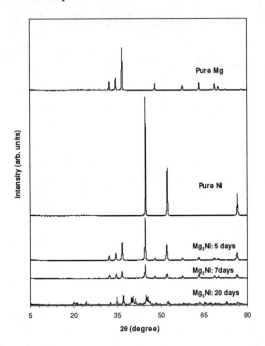

FIGURE 1. The XRD patterns of the powder: pure Mg, pure Ni and Mg_2Ni ball milled for various days. Peaks for the sample ball milled for 20 days have shifted to higher 2θ degree. Additional peaks can be observed around $2\theta = 20, 24$ and $45°$.

SEM micrographs were used to study the surface morphology of mechanically alloyed Mg_2Ni sample after ball milled for 5, 7 and 20 days. The images were scanned at a magnification of 100x and 700x.

The SEM micrographs after 5 days of milling are shown in Fig. 2 (a and b). The micrographs after 7 days milling are shown in Fig. 2 (c and d). The micrographs after milling for 20 days are shown in Fig. 2 (e and f). It can be observed that prolonged milling has produced very fine particles. Those which are suitable for use as hydrogen storage alloy. At this stage Mg_2Ni alloy phases was formed as proven by XRD.

After 20 days of milling, the sizes of the particles were smaller. Agglomerates formed could be as large as several hundreds of micrometers [11]. According to Orimo *et al.,* [12] the inter-grain region of Mg_2Ni increased after prolong milling time. This allows a larger amount of hydrogen to dissolve in the alloy, which can lead to improved kinetics of hydrogen diffusion when the alloy absorbs and desorbs hydrogen [13, 14].

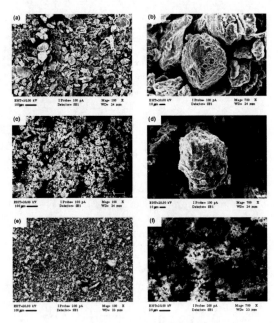

FIGURE 2. SEM micrographs of Mg_2Ni milled for 5 days (a and b), 7 days (c and d) and 20 days (e and f) at different magnifications.

EDX spectrometry was used to determine the major and minor elements present in Mg_2Ni alloy powder. Fig. 3 shows the EDX spectrum of Mg_2Ni, after milling for 20 days. EDX analyses show that the elemental percentage of Mg and Ni is 67.8 % and 32.2 %, respectively. The starting compositions were 66.8 % Mg and 33.2 % Ni. This result helps to confirm the formation of Mg_2Ni after 20 days ball milling.

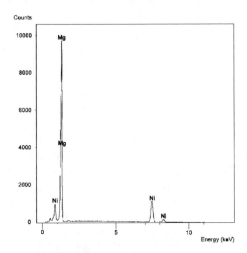

FIGURE 3. EDX spectrum for Mg_2Ni.

It is important to remember that the powder particles are usually agglomerated and therefore care has to be exercised in determining the accurate particle size. A powder particle may consist of several individual particles. Furthermore, an individual powder particle may contain a number of crystallites defined as coherently diffracting domains. Microscopic examination normally gives the particle size (or even grain size if sufficient resolution is available), whereas diffraction techniques (i.e. X-ray) give the crystallite size. The particle size distribution was carried out in order to know the effect of milling on Mg_2Ni powder.

Table 1 shows the average diameter and Fig 4 illustrated the diameter size upon milling. From the analysis of particle distribution, it can be found that milling the mixed Mg and Ni powder lead to gradual reduction in the lamellar spacing from about 100 μm for the mixed powder to about 36 μm after milling for 20 days. According to Chen et al. [14] and Abdellaoui et al. [15], there are many honeycomb cracks on the surface of mechanically alloyed powders. These crack resulted from the long time collision between the balls and the particles. This can explain why the mechanically alloyed powder has a larger specific surface area and the occurrence of defects on the surface as well as in the volume of the material.

TABLE 1. Average diameter for the samples and standard deviations (S.D.)

Sample	Mean ± S.D. (μm)
Magnesium pure	102.90 ± 47.7
Nickel pure	92.06 ± 31.8
5 days ball mill Mg_2Ni	68.53 ± 48.7
7 days ball mill Mg_2Ni	67.85 ± 52.7
20 days ball mill Mg_2Ni	35.77 ± 25.5

CONCLUSIONS

It can be seen from XRD, SEM and particle size distribution results, the forming of Mg_2Ni took a long time i.e. 20 days by using PASCAL 9VS horizontal ball mill (200 rpm). The Mg_2Ni phase with large surface area and 36 μm particle size was formed after milling for 20 days. It is hoped that mechanically alloyed Mg_2Ni can be a suitable active material for metal hydride electrodes in fabrication of Ni-MH batteries.

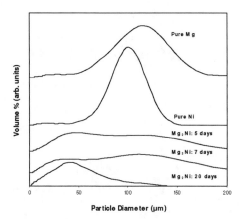

FIGURE 4. Particle distribution analysis of the powder: pure Mg, pure Ni and Mg_2Ni ball milled for various days.

REFERENCES

1. Y. Zhang, H. Yang, H. Yuan, E. Yang, Z. Zhou and D. Song *J. Alloys Compd.* 269, 278 (1998).
2. X. Jianshe, L. Guoxun, H. Yaoqin, D. Jun, W. Chaoqun and H. Guangyong *J. Alloys Compd.* 307, 240 (2000).
3. L. Zaluski, A. Zaluska and J.O. Strom-Olsen *J. Alloys Compd.* 217, 245 (1995).
4. G. Liang, S. Boily, J. Huot, A.V. Neste and R. Schulz, *J. Alloys Compd.* 268, 302 (1998).
5. G. Liang, S. Boily, J. Huot, A. Van Neste and R. Schulz, *J. Alloys Compd.* 267, 302 (1998).
6. L. Li, T. Akiyama, T. Kabutomori, K. Terao and J. Yagi, *J. Alloys Compd.* 281, 175 (1998).
7. Y.Q. Lei, Y.M. Wu, Q.M. Yang, J. Wu and Q.D. Wang, *Z. Phys. Chem.* 183, 379 (1994).
8. C.-H. Han, Y.-S. Hong, C.M. Park and K. Kim, *J. Power Sources* 92, 95 (2001).
9. H. Inoue, T. Ueda, S. Nohara, N. Fujita and I. Chiaki, *Electrochim. Acta* 43, 2215 (1998).
10. H. Inoue, S. Hazui, S. Nohara and C. Iwakura, *Electrochim. Acta* 43, 2221 (1998).
11. D. Cracco and A. Percheron-Guegan, *J. Alloys Compd.* 268, 248 (1998).
12. S. Orimo, H. Fujii and K. Ikeda, *Acta Mater.* 45, 331 (1997).
13. C. Iwakura, S. Nohara, S.G. Zhang and H. Inoue, *J. Alloys Compd.* 285, 246 (1999).
14. J. Chen, D.H. Bradhurst, S.X. Dou and H.K. Liu, *Electrochim. Acta* 44, 353 (1998).
15. M. Abdellaoui, D. Cracco and A. Percheron-Guegan, *J. Alloys Compd.* 268, 233 (1998).

Zircon Supported Copper Catalysts for the Steam Reforming of Methanol

M. Widiastri, Fendy, and I N. Marsih

*Inorganic and Physical Chemistry Group, Faculty of Mathematics and Natural Sciences,
Institut Teknologi Bandung, Jl. Ganesha 10 Bandung 40132, Indonesia*

Abstract. Steam reforming of methanol (SRM) is known as one of the most favorable catalytic processes for producing hydrogen. Current research on zirconia, ZrO_2 supported copper catalyst revealed that CuO/ZrO_2 as an active catalyst for the SRM. Zircon, $ZrSiO_4$ is available from the by-product of tin mining. In the work presented here, the catalytic properties of $CuO/ZrSiO_4$ with various copper oxide compositions ranging from 2.70% (catalyst I), 4.12% (catalyst II), and 7.12 %-mass (catalyst III), synthesized by an incipient wetness impregnation technique, were investigated to methanol conversion, selectivity towards CO formation, and effect of ZnO addition ($7.83\%CuO/8.01\%ZnO/ZrSiO_4$ = catalyst V). The catalytic activity was obtained using a fixed bed reactor and the zircon supported catalyst activity was compared to those of $CuO/ZnO/Al_2O_3$ catalyst (catalyst IV) and commercial Kujang LTSC catalyst. An X-ray powder diffraction (XRD) analysis was done to identify the abundant phases of the catalysts. The catalysts topography and particle diameter were measured with scanning electron microscopy (SEM) and composition of the catalysts was measured by SEM-EDX, scanning electron microscope-energy dispersive using X-ray analysis. The results of this research provide information on the possibility of using zircon ($ZrSiO_4$) as solid support for SRM catalysts.

Keywords: Steam reforming of methanol; incipient wetness impregnation technique; $CuO/ZrSiO_4$; CO formation
PACS: 61.05.cp

INTRODUCTION

Fuel cell is one of the alternative energy because it is more efficient than internal combustion engines in terms of converting fuel to power on the vehicle and low-emission. The primary fuel for fuel cells is hydrogen (H_2). Hydrogen can be produced from electrolysis of water, metal-hydride, gasoline, biomass, methanol, etc.

Methanol is the most favorable source of hydrogen onboard of a vehicle because as a liquid methanol is easy to transport. Furthermore, methanol is readily available and can be produced from biomass [1]. High H/C ratio in methanol makes great quantities hydrogen formed from methanol [2,3]. Moreover, methanol can be converted to hydrogen at relatively low temperature (200-400°C) [4].

There are three ways to produce hydrogen from methanol. They are decompositions of methanol (DM), partial oxidation of methanol (POM), and steam reforming of methanol (SRM). SRM is more favorable process because SRM forms the highest concentration of hydrogen and the lowest concentration of carbon monoxide (CO) than the other two processes [4].

The SRM is described by the following equation:

$$CH_3OH + H_2O \rightleftharpoons 3H_2 + CO_2 \qquad (1)$$

$CuO/ZnO/Al_2O_3$ is common catalyst for the SRM reaction. Unfortunately, $CuO/ZnO/Al_2O_3$ catalysts have short duration of life and poor thermal stability, yield high concentration of CO, and deactivated during the reaction [3,5].

As the solution, zirconia (ZrO_2) is chosen as support for Cu-based catalysts because it has long term and thermal stability and the catalysts, CuO/ZrO_2, yield the lowest concentration of CO than other catalysts for SRM [3,4]. But, ZrO_2 is not appropriate for industry application because of its high price.

In the work presented here, zircon ($ZrSiO_4$ or $ZrO_2.SiO_2$) from the by-product of tin mining were used as support for Cu-based catalysts. As a by product of tin mining, zircon is easy to get and cheap.

The catalytic properties of $CuO/ZrSiO_4$ with various copper compositions ranging from 2.70 to 7.12 %-mass, synthesized by an incipient wetness impregnation technique, were investigated with respect to activity, and CO formation. The catalytic activity of the catalysts were obtained using a fixed bed reactor and compared to commercial catalysts of $CuO/ZnO/Al_2O_3$.

CP989, Neutron and X-ray Scattering in Materials Science and Biology, International Conference on Neutron and X-ray Scattering 2007, edited by A. Ikram, A. Purwanto, Sutiarso, A. Zulfia, S. Hendrana, and Z. Nurachman

Powder X-ray diffraction (XRD) analysis was carried out to identify the abundant phases of the catalysts. In addition, the catalysts topography and particle diameter were measured with scanning electron microscopy (SEM) and the chemical composition of the catalysts was measured with SEM-EDX, energy dispersive using X-ray analysis.

The results of this research provide information about the prospect for using zircon as solid support for SRM catalysts.

EXPERIMENT AND METHOD

Catalyst preparation

The precursors of $CuO/ZrSiO_4$ catalysts were prepared by incipient wetness impregnation technique. Before impregnation, $ZrSiO_4$, provided by Department of Metallurgy Engineering ITB, Indonesia was washed with HNO_3 1 M by stirring it for 5 h and then centrifuged with water for several time. Then, $ZrSiO_4$ was impregnated with $Cu(NO_3)_2$ in the desired proportion, i.e. 2.70%$CuO/ZrSiO_4$ is referred to as catalyst I, 4.12%$CuO/ZrSiO_4$ is referred to as catalyst II, 7.12%$CuO/ZrSiO_4$ is referred to as catalyst III, and 7.83%CuO/8.01%$ZnO/ZrSiO_4$ is referred to as catalyst V. When the $ZrSiO_4$ was started to get wetted, the mixture were dried before further impregnation. This method was repeated until all of the $Cu(NO_3)_2$ were used up. The impregnated solid was dried at 100°C, and then calcined at 500°C for 12 h to form $CuO/ZrSiO_4$.

$CuO/ZnO/Al_2O_3$ catalyst was prepared by co-precipitation method (catalyst IV). Cu(II), Zn(II) and Al(III)-nitrate salts were precipitated by the addition of sodium carbonate solution. The mole ratio of Cu:Zn:Al was set to be 1:2:3. During the addition of carbonate solution, temperature of the solution was kept at 60-70°C. Precipitate formed was washed with water and then dried at 110°C for 16 h and finally calcined at 470°C for 12 h.

Characterizations and catalyst activity measurements

The XRD analysis was carried out on a PW1710 BASED Philips Analytical X-Ray B.V. diffractometer using K radiation at 40 kV and 30 mA. XRD analysis to identify the abundant phases of the catalysts was carried out at the Metallurgy Department at Institut Teknologi Bandung and at the Lembaga Ilmu Pengetahuan Indonesia (LIPI) Bandung. All of the diffractograms of the catalysts have the step size of 0.020 and time per step of 0.400s. SEM analysis to measure catalysts topography and particle diameter was carried out at the LIPI Bandung on a JSM-6360 instrument with magnitude 750x, 2000x, and 2500x with voltage of 20V. A SEM-EDX analysis to measure composition of the catalysts was carried out at Geology Museum, Bandung, Indonesia.

Catalytic activity measurement was carried out in a stainless steel tube reactor with internal diameter of 8 mm. The SRM were performed using 1 g of catalyst at temperature range of 150 – 300°C. Prior the SRM, the catalyst was reduced at 150°C for 2 h and at 300°C for 4 h by H_2/N_2 stream (40/30 mL/minute). After the reduction, the temperature was set to the reaction temperature and the H_2/N_2 stream was replaced by mixture of methanol/water/N_2. Gas mixture was sampled using a syringe from the reactor outlet for GC analysis after the reaction taken place for 2 h.

Activity of the catalysts synthesized in this research were compared to that of commercial catalyst, which is Kujang Low Temperature Shift Conversion (LTSC) catalyst type C18-7/Cu/Zn provided by SUD-CHEMIE.

RESULTS AND DISCUSSIONS

Catalyst characterizations

Several sharp peaks at the diffractogram show that $CuO/ZrSiO_4$ catalysts are crystalline (Figure 1). Base on the PCPDFWin database, $ZrSiO_4$ peaks at the diffractograms appear at 2θ of 20.080°, 27.045°, 33.875°, 35.670°, 38.585°, 40.765°, 44.530°, 47.645°, 52.210°, 53.475°, 55.660°, 62.920°, 67.805°, and 80.845°. The CuO peaks appear at 2θ of 35.580°, 38.565°, 48.775°, 53.595°, 67.820°, and 80.775° [7], except for diffractogram of catalyst I (Figure 1a) which peak for CuO does not appear. This case could be happened because of the concentration of $Cu(NO_3)_2$ is too low, or because of imperfect dispersion of $Cu(NO_3)_2$, thus the X-ray did not hit the CuO site.

Although almost all of the CuO peaks are too close with the $ZrSiO_4$ peaks, it can be concluded that those catalysts are $CuO/ZrSiO_4$. However, diffractogram of catalyst IV shows the amorphous phase (Figure 2). Base on the PCPDFWin database, CuO peaks at the diffractogram appear at 2θ of 32.476°, 38.686°, 48.659°, 56.642°, and 68.009°. A ZnO peak appears at 2θ of 36.207° while Al_2O_3 peak appears at 2θ of 37.780° [7].

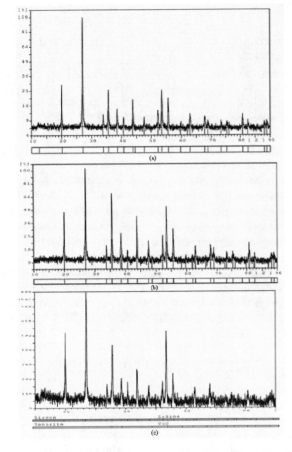

FIGURE 1. XRD Diffractograms of catalyst I (a); catalyst II (b); and catalyst III (c)

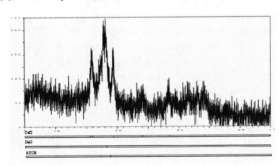

FIGURE 2. XRD Diffractogram of catalyst IV

The SEM images (Figure 3) show that particles diameter are between 0.5 – 10 μm for catalyst I, 10 – 40 μm for catalyst II, 0.5 – 8 μm for catalyst III, and 2-10 μm for catalyst IV. Those photographs have magnification of 2500 for catalyst I and III, 2000 for catalyst IV, and 750 for catalyst II.

The SEM-EDX result is shown on Table . Catalyst I contains 6.94% carbon (C), which could inhibit the catalytic activity [8]. As expected from the amount of $Cu(NO_3)_2$ added in the catalyst preparation, catalyst I

has the lowest CuO content, in the other hand, catalyst III has the highest (without catalyst V result).

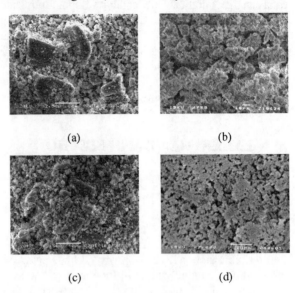

FIGURE 3. SEM Photograph of catalyst I (a); catalyst II (b); catalyst III (c); catalyst IV (d)

Table 1. SEM-EDX results of the CuO/ZrSiO₄ catalysts

Compound	%-mass on the CuO/ZrSiO₄ catalysts		
	I	**II**	**III**
C	6.94	-	-
Al₂O₃	1.11	-	1.66
SiO₂	25.06	25.33	21.89
TiO₂	4.44	4.45	4.33
CuO	5.30	15.41	26.07
ZrO₂	57.15	54.82	46.06

Catalytic activities of the catalysts

Figure 4 shows the effect of temperature on catalytic performance in the SRM over various catalysts and the methanol conversion is shown in Table 2.

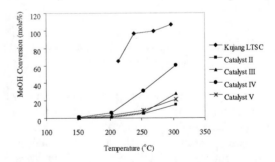

FIGURE 4. Effect of temperature on catalytic performance in the SRM over various catalysts. Reaction conditions: catalyst weight = 1g; CH_3OH: H_2O = 1:1.2; liquid flow rate of feed = 0,033 mL/min; flow rate of Ar = 15 mL/min

TABLE 2. Methanol conversion and selectivity towards CO formation (SCO) of the various catalysts at various reaction temperatures

Catalyst	T (°C)	MeOH Conversion (mole%)	S_{CO} (%)
Kujang LTSC	213	65.40	
	238	96.50	
	269	99.60	
	297	106.81	0.02
4.12%CuO/ZrSiO₄	151	0.03	
	202	0.60	
	252	4.72	
	303	15.46	
7.12%CuO/ZrSiO₄	151	0.03	
	202	1.73	
	252	6.12	
	303	27.76	0.12
CuO/ZnO/Al₂O₃	151	1.09	
	202	5.71	
	252	30.90	
	303	60.14	
7.83%CuO/8.01%ZnO/ZrSiO₄	151	0.28	
	202	3.02	
	252	8.73	
	303	21.08	

Figure 4 and Table 2 show that all of the catalysts activities are increased along with the temperature and Cu content. ZnO is a catalyst promoter which can increase surface area of Cu, thus it can increase catalyst activity. The addition of ZnO aimed to increase the catalyst activity, but the effect is not significant. In general, all of the CuO/ZrSiO₄ catalysts performed low activities. CO was appearing at 297°C on the Kujang LTSC catalyst and at 303°C on the catalyst III. These results agree with Breen et al who found that CO was formed at high methanol conversion condition or at long time contact [9]. At catalyst III, CO appears although the methanol conversion is still very low. The formation of CO is likely happened because of long time contact.

Reducing condition at 150°C for 2 h and at 300°C for 4 h by H_2/N_2 stream (40/30 mL/minute) was chosen because CuO was reduced at 150-300°C [6]. Reduction condition cannot be too hard because can damage the catalyst, thus the H_2/N_2 stream was chosen at 40/30 mL/minute. The reaction is:

$$CuO + H_2 \rightleftharpoons Cu + H_2O \qquad (2)$$

Because this reaction is reversible, after catalysts are reduced at 150°C for 2 h to prepare the next reduction, the temperature was quickly increased to 300°C to prevent reverse reaction.

CONCLUSIONS

Diffractograms of X-ray powder diffraction (XRD) indicated that all of the catalysts are crystalline except catalyst IV. Base on SEM-EDX result, CuO/ZrSiO₄ has the lowest CuO content, in the other hand, CuO/ZrSiO₄ III has the highest. All of the catalysts activities are increased along with temperature and Cu content. Using ZirSiO₄ as support, the addition of ZnO does not significantly increase the catalyst activity. In general, all of the CuO/ZrSiO₄ catalysts performed low activities. CO appears at high methanol conversion condition or long time contact.

ACKNOWLEDGEMENTS

The authors would like to thank Drs. I.G.B.N Makertihartha, Subagjo and M. Laniwati at the Chemical Engineering Department ITB for the discussion and support of this work, Chemical Engineering Department ITB for the permission of working at their laboratory, Dr. S. Soepriyanto at the Metallurgy Department ITB for the zircon, discussion, and support, Chemistry Department ITB for the grant and support, DIKTI for the grant.

REFERENCES

1. S. Velu, K. Suzuki. *Topics in Catalysis* **22**, 235 (2003).
2. D. Firmansyah, *Sintesis dan Uji Aktivitas Katalis Cu/ZnO/Al₂O₃ pada Reaksi Reformasi Kukus Metanol,* Master Thesis, Prog. Studi Kimia, Fakultas Matematika dan Ilmu Pengetahuan Alam, Institut Teknologi Bandung (2006).
3. A. Szizybalski, *Zirconium dioxide supported Copper Catalysts for the Methanol Steam Reforming*, PhD Dissertation, Mathematik und Naturwissenschaften der Technischen Universität Berlin (2005).
4. H. Purnama, *Catalytic Study of Copper based Catalysts for Steam Reforming of Methanol*, PhD Dissertation, Mathematik und Naturwissenschaften der Technischen Universität Berlin (2003).
5. Okada *et al. Method for Preparing Catalyst for Steam Reforming*, United States Patents 6, 844, 292; 1-8, 2005.
6. J. Rodriguez, J. Kim, J. Hanson, M. Pe´rez, and A. Frenkel, *Catal. Lett.* **85,** 247 (2003).
7. S. Asbrink, A. Waskowska, *J. Phys. Condens. Matter.* **3**, 8173 (1991).
8. E. Jatnika, S. Sianturi. *Sintesa dan Uji Aktivitas Katalis Cu/ZnO/Al₂O₃ untuk Reformasi Kukus Metanol Penyedia Hidrogen Sebagai Bahan Bakar Sel Tunam.* Skripsi. Facultas Teknologi Industri, Institut Teknologi Bandung, 2007.
9. J. P. Breen, Julian R.H. Ross, *Catal. Today* **51**, 521 (1999).

POLYMERS AND COLLOIDS

Preparation of Self-assembly Mesoporous TiO₂ Using Block Copolymer Pluronic PE6200 Template

W. Septina, B. Yuliarto, and Nugraha

Material Processing Laboratory, Engineering Physics Department, Institut Teknologi Bandung,
Jl. Ganesha No.10 Bandung 40132, Indonesia

Abstract. In this research, nanocrystal mesoporous TiO_2 powders were synthesized by sol-gel method, with $TiCl_4$ as a precursor in methanol solution. Block copolymer Pluronic PE 6200 was used as pores template. It was found that from the XRD measurements, both at 400°C and 450°C calcination temperatures, resulted in nanocrystal TiO_2 with anatase phase. Based on N_2 adsorption characterization (BET method), TiO_2 samples have surface area 108 m^2/g and 88 m^2/g for 400°C and 450°C calcination temperatures respectively. From Small-angle Neutron Scattering (SANS) patterns, it is investigated that TiO_2 samples have mesoporous structure where the pore order degree depend on the calcination temperature.

Keywords: Titanium dioxide, mesoporous material, sol-gel method, self-assembly, block copolymer, SANS.
PACS: 81.05.Rm

INTRODUCTION

Titanium dioxide (TiO_2) has attracted considerable interest in this recent years. In energy and environment field, research has been performed intensively on TiO_2 characteristics especially for its applications as a photocatalyst[1], sensor[2], and main component of dye-sensitized solar cell[3]. For these applications, high surface area material is preferred thus will increase contact area. Indeed nanostructure of TiO_2 become important characteristic as well as nanosized of pore. Class of nanomaterial was emerge from the requirements which called mesoporous material (2 – 50 nm of pore diameter)[4].

Mesoporous TiO_2 have been developed with various methods, most importantly using organic template to build porous structure. Synthesis of mesoporous TiO_2 first reported by Ying et al. at 1995 using ionic alkyl phosphat surfactan as a template with sol-gel method[5], however the resulted material was not pure TiO_2 because phosphat molecules was strongly bonded on TiO_2 structure. One of the alternative solution is using non-ionic surfactant, particularly block copolymer to replace ionic surfactant. Block copolymer have advantages on its easily to be removed from inorganic frameworks by calcination or solvent extraction since the block copolymer only involving H-bonding type rather than the electrostatic interactions found using ionic

surfactant[6]. And the properties of block copolymer can be continuously tuned by adjusting solvent composition, molecular weight or polymer architectures of block copolymer[7].

Since succesful synthesis of mesoporous material using block copolymer as a template, publication related to mesoporous TiO_2 continuously increasing, but mostly used titanium alkoxide as Ti source at aqueous solution[8]. Large amount of water inside solution caused hydrolysis and condensation process as well as mesostructure formation difficult to control because alkoxide highly reactive with water. And also most of reported mesoporous have amorphous wall which is unpreferred for most of its applications.

In this research, we prepared mesoporous TiO_2 at using triblock copolymer Pluronic PE 6200 as a template and $TiCl_4$ as a precursor at alcoholic solution. The resulted mesoporous TiO_2 have nanocrystalline wall which is necessary to ensure good electron diffusion[9].

EXPERIMENTAL METHOD

Chemicals

Triblock copolymer PEO-PPO-PEO Pluronic PE 6200 (PEO_8-PPO_{30}-PEO_8, M_{av} = 2450 g/mol) was received from BASF. Titanium tetrachloride ($TiCl_4$) as

CP989, *Neutron and X-ray Scattering in Materials Science and Biology, International Conference on Neutron and X-ray Scattering 2007*, edited by A. Ikram, A. Purwanto, Sutiarso, A. Zulfia, S. Hendrana, and Z. Nurachman

a precursor was purchased from Merck. Methanol was used as a solvent. All chemicals were used without further purification.

Synthesis

An initial solution was made by adding 1 gram Pluronic PE 6200 to 10 gram methanol and stirred for 30 minutes to initiate micelle structure. Then 1.9 gram $TiCl_4$ was slowly added to the solution under stirring for 30 minutes so then the molar ratio of $TiCl_4$:methanol:Pluronic PE6200 was 1:21.7:0.0408. $TiCl_4$ reaction with alcoholic solution is exothermic and produces large amont of HCl gas and a collection of chloro-alkoxide $(TiCl_2(OCH_3)_2)$[10][11]. The solutions were then aged for 4 days at 40-45°C in a furnace until dry-gel was formed. TiO_2 dry-gels were then calcined at 400°-450°C for 4 hours with heating rate 5-6°C/minute to remove the block copolymer and promote crystallization.

Characterization

X-Ray Diffractometer (XRD) patterns were obtained using Philips Analytical X-Ray with Cu radiation (λ= 0,154056 nm) and crystallite sizes were calculated from Scherrer equation. Nitrogen adsorption measurements were carried out on a NOVA 1000 High Speed Gas Sorption Analyzer to determine Brunau-Emmett-Teller (BET) surface area. Before measurements, samples were degassed at 200°C for 2 hour in vacuum condition to eliminate water and oil. Scanning Electron Microscope (SEM) images were obtained from JEOL operated at 20 kV. Thermogravimetric-Differential Thermal Analysis (TG-DTA) was performed with temperature range 50°-600°C with heating rate 10°C/minute. Small-angle Neutron Scattering (SANS) measurements were carried out at Neutron Scattering Laboratory, BATAN Indonesia using SMARTer. The neutron wavelength of SANS was λ = 0,566 nm. Sample to detector distances were varied at 2, 8, and, 18 meter to obtained Q range about 0.02 – 1.5 nm^{-1}.

RESULTS AND DISCUSSION

Figure 1 shows TG-DTA result for TiO_2 dry-gel after 4 days aging. From TGA curve, which show sample weight loss as the function of temperature, the sample weight was decrease about 75% when heated until 460°C. TGA curve also shows three weight-loss regions. First region at temperature below 200°C resulted from –OH groups evaporation. Second region between 200°- 325°C was atrributed to organic

decomposition. Third region from 325°C to 460°C due to residual block copolymer oxidation and decomposition of chlor bonded to Ti-OH. TGA curve also show that block copolymer completely removed at calcination temperature above 450°C.

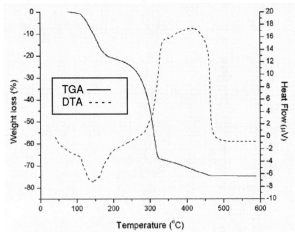

FIGURE 1. TG-DTA results for TiO2 dry-gel.

DTA curve clearly shown two peaks, endothermic peak at region 100°C - 200°C was attributed to solvent evaporation, and broad exothermic peak at 325°C - 460°C was atrributed to block copolymer and chlor decomposition, and phase transformation from amorphous to anatase TiO_2 which started at 350°C[12]. As shown at full TG-DTA curve, data from TGA curve corellated directly with DTA curve since weight-losses always followed by endothermic or exothermic processes.

TABLE 1. Properties of Mesoporous TiO_2

Sample	Calcination Temperature	Crystallite size[a]	Surface Area (m^2/g)
Ti-a	400°C	11 nm	108
Ti-b	450°C	14 nm	88

[a]Calculated from Scherrer equation at miller indices (101)

Wide-angle XRD patterns shown at Fig. 2 indicate both samples were well-crystallized TiO_2 with pure anatase phase according to JCPDS no. 21-1272. As shown at Table 1, higher calcination temperature resulted higher crystallite size. More intense and well-resolved peak were observed at higher calcination temperature indicate increasing the crystalline degree. This result correlated with the fact that full crystallization of TiO_2 occur at 550°C[13]. In accordance with TG-DTA result, TiO_2 sample calcined below 450°C still contain some residual carbon which observed from black colour at powder sample which may contribute to amorphous peak nature at XRD patterns.

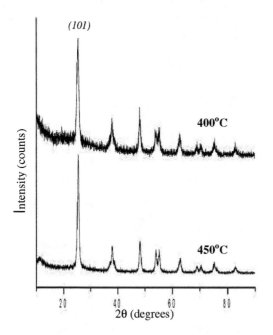

FIGURE 2. Wide-angle XRD Patterns.

Pore structure of TiO$_2$ samples was analyzed by Small-angle Neutron Scattering (SANS). SANS patterns (Fig. 3) show that correlation peak formed at both samples indicate mesoporous structures were preserved. Peak formed at low-angle can be attributed to distance between pore center[14]. Average pore distance ($d = 2\pi/Q_{peak}$) were about 22 nm (Ti-a) and 24 nm (Ti-b) which is typical the pore to pore distance of mesoporous materials. Average pore distance was increase with increase in calcination temperature because of structure contraction which is common observed from heat treatment.

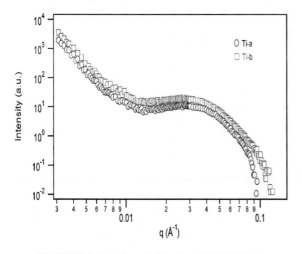

FIGURE 3. Small-angle Neutron (SANS) Patterns

SANS patterns also shown that peak intensity of lower calcination temperature sample (Ti-a) was slightly larger. This is indicating that sample have higher pore ordering than sample with higher calcination temperature (Ti-b).

FIGURE 4. SEM Images of Mesoporous TiO$_2$
(a) 400 °C (b) 450 °C

Samples morphology were observed with SEM (Scanning Electron Microscopy), images shown at Fig. 4. The structure morphology contains large amount of close packed particles which is typical of porous material. Mesoporosity of sample was partially due to intraparticle porosity and interparticle porosity.

According to N$_2$ adsorption measurements (the measurement result is not shown in this paper), Ti-a sample has BET surface area about 108 m^2/g and Ti-b about 88 m^2/g (Table 1). Higher calcination temperature always resulted in lower surface area since crystallization of TiO$_2$ always followed by some collapse and destruction of mesotructure[15]. These surface area are larger than commercial Degussa P-25 which is mainly use as a nano TiO$_2$ reference, with BET surface area about 50 m^2/g[16].

CONCLUSIONS

Mesoporous TiO_2 with nanocrystalline anatase wall have been succesfully prepared using combination of sol-gel method and block copolymer as a template. $TiCl_4$ was used as a Ti precursor and Pluronic PE6200 acted well as mesoporous template although short-range pore ordering was observed. It was also observed that calcination temperature is important parameter for mesostructure formation as well as crystallization process.

ACKNOWLEDGMENTS

Authors would like to thanks BASF Indonesia for providing block copolymer PE6200 and Dr. Edy Giri R. Putra from BATAN-Indonesia for his assistance in SANS characterization.

REFERENCES

1. K. Hashimoto, H. Irie, A. Fujishima, *Jpn J. Appl. Phys.* **44,** 8269-8285 (2005)
2. B. Yuliarto, I. Honma, Y. Katsumura, H. Zhou, *Sensors and Actuators B: Chemical* **114,** 109-111 (2006)
3. M. Gratzel, *J. Photochemistry and Photobiology C: Photochemi. Rev.* **4,** 145-153 (2003)
4. C.T. Kresge, M.E. Leonowicz, W.J. Roth, J.C. Vartuli, J.S. Beck, *Nature* **359,** 710-712 (1992)
5. D.M. Antonelli and J.Y. Ying, *Angew. Chem., Intl. Ed. Engl.* **34,** 2014-2017 (1995)
6. G.J. de A.A. Soler-Illia, E. L. Crepaldi , D. Grosso , C. Sanchez, *Curr. Opin. Colloid Interface Sci.* **8,** 109-126 (2003)
7. F.S. Bates, G.H. Fredrickson, *Physics Today* **52,** 32-38. (1999)
8. P. Yang, D. Zhao, D.I. Margolese, B.F. Chmelka, G.D. Stucky, *Chem. Mater* **11,** 2813-2826 (1999)
9. M. C. Fuertes and G. J. D. A. A. Soler-Illia, *Chem. Mater.* **18,** 2109-2117 (2006)
10. G. J. D. A. A. Soler-Illia, C. Sanchez, B. Lebeau, and J. Patarin, *Chem. Rev.* **102,** 4093-4138 (2002)
11. H. Luo, C. Wang, and Y. Yan, *Chem. Mater.* **15,** (3841-3846 (2003)
12. E.L. Crepaldi, G.J.D.A.A. Soler-Illia, D. Grosso, F. Cagnol, F. Ribot, and C. Sanchez, *J. Am. Chem. Soc.* **25,** 9770-9786 (2003)
13. E. L. Beltran, P. Prene, C. Boscher, P. Belleville, P. Buvat, S. Lambert, F. Guillet, C. Boissie`re, D. Grosso, and C. Sanchez, *Chem. Mater* **18,** 6152-6156 (2006)
14. M. S. Wong, E. S. Jeng, and J. Y. Ying, *Nano Letters* **1,** 637-642 (2001)
15. H. Yun, K. Miyazawa, I. Honma, H. Zhou, M. Kuwabara, *Mater. Sci. Eng.* **23,** 487-494 (2003)
16. L. Saadoun, J.A. Ayllo´n, J. Jime´nez-Becerril, J. Peral, X. Dome`nech, R. Rodrı´guez-Clemente, *Appl. Catal. B* **21,** 269-277 (1999)

SANS Study of Micellar Structures on Oil Solubilization

J. V. Joshi[1], V. K. Aswal[2] and P. S. Goyal[1]

[1]*UGC-DAE CSR, Mumbai Centre, Bhabha Atomic Research Centre, Mumbai 400 085, India*
[2]*Solid State Physics Division, Bhabha Atomic Research Centre, Mumbai 400 085, India*

Abstract. Micellar structures of cationic alkyltrimethylammonium bromide (C_nTAB) surfactants for different chain lengths (n=12, 14 and 16) in aqueous solution on the addition of two oils benzene and hexane have been studied by small-angle neutron scattering (SANS). Measurements have been performed for fixed 0.1 M concentration of surfactant and oil. It is found that the size and the aggregation number of the micelles increase on oil solubilization with increase in oil concentration. The effect is more pronounced for larger size of the micelles having longer chain length and also for the oil benzene than hexane. The axial ratio of the micelles shows a large increase and the effective charge on the micelles decreases on addition of oil. There is only a small increase in semiminor axis on addition of oil, which suggests that distribution of oil is uniform inside the micellar core.

Keywords: Small-angle neutron scattering, micelles, oil solubilization.
PACS: 61.12.Ex, 61.25.Hq, 82.70.Dd

INTRODUCTION

Surfactant molecules consist of a polar head group and a long hydrophobic tail connected to the head group. These molecules in aqueous solution above the critical micellar concentration (CMC) are known to self-aggregate to form micelles [1-3]. It has long been recognized that micelles have the ability to incorporate hydrophobic compounds such as oil and thus to enhance greatly the water solubility of these materials. This property of solubilization of a hydrophobic component is useful in many industrial applications which range from enhanced oil recovery, and cleanup of ground water contaminated with hydrophobic pollutants, to stabilizing cosmetics and foodstuffs. Life science applications range from drug delivery to providing understanding of basic biological processes.

The propensity of the oil solubilization by a micelle depends on its shape and size. It is expected that micelles with larger size can solubilize greater amount of oil. The use of non-spherical micelles, such as ellipsoidal or cylindrical, is of great interest for this purpose as they can enhance the oil solubilization by many folds. The formation of non-spherical micelles depends on the chemical structure of the surfactant as well as on the solution conditions such as temperature, concentration, ionic strength, etc [4, 5]. The packing parameter (p) is a useful quantity to determine the micellar structure in surfactant solutions [6]. This packing parameter is defined in terms of three quantities: (i) volume *v*, (ii) effective head group area *A* and (iii) effective chain length *l* of the surfactant molecules. It is given as $p=v/Al$. It can be shown that surfactant molecule with $p < 0.33$ form spherical micelles whereas ellipsoidal or cylindrical micelles are formed for $0.33 < p < 0.5$. For higher values of packing parameters, $p > 0.5$, the surfactant molecules aggregate to form disk-like micelles. While the solubility of oil in different shapes and sizes of micelles has been studied in considerable detail [7, 8], there are no such studies on the structural changes of the micelles during the oil solubilization.

Herein, we report studies on the structural changes in micelles of different sizes on oil solubilization as characterized by small-angle neutron scattering. The size of the micelle is controlled by varying the chain length of the surfactant. Cationic alkyltrimethyl-ammonium bromide surfactants [$C_nH_{2n+1}N(CH_3)_3Br$] with n = 12, 14 and 16 have been used in the present study. The aggregation number and the axial ratio of the micelles of these surfactants increase significantly with increase in the value of n [9-11]. The solubilization of two different oils, benzene and hexane, has been examined for the above micelles. SANS is one of the best suited and well established techniques for such studies [12].

CP989, Neutron and X-ray Scattering in Materials Science and Biology, International Conference on Neutron and X-ray Scattering 2007, edited by A. Ikram, A. Purwanto, Sutiarso, A. Zulfia, S. Hendrana, and Z. Nurachman

EXPERIMENTAL

All the chemicals used, surfactants [dodecyltrimethylammonium bromide (C_{12}TAB), tetradecytrimethylammonium bromide (C_{14}TAB), cetyltrimethylammonium bromide (C_{16}TAB)] and oils (hexane, benzene), were obtained from Aldrich. The samples were prepared by dissolving known amounts of surfactants and oils in aqueous solution. D_2O (99.9 % D atom) was used as solvent instead of H_2O, which provides better scattering contrast in neutron scattering experiments. Small-angle neutron scattering experiments were performed using a SANS diffractometer at BARC, Mumbai [13]. The incident neutron beam wavelength (λ) was 5.2 Å and the data were recorded in the wave vector transfer Q range of 0.018 to 0.20 Å^{-1}. The concentration of surfactants and oils were fixed at 0.1 M. During the experiments, samples were held in quartz sample holders having sample thickness 5 mm. The sample temperature was maintained at 30ºC for all the measurements. The data were corrected using standard procedures for background, empty cell contribution and sample transmission.

SANS ANALYSIS

In small-angle neutron scattering experiment, one measures differential scattering cross-section per unit volume ($d\Sigma/d\Omega$) as a function of Q, and for a micellar solution it can be expressed as [14]

$$\frac{d\Sigma}{d\Omega}(Q) = nP(Q)S(Q) + B \qquad (1)$$

where n is the number density of the particles. $P(Q)$ is the intra-particle structure factor and depends on the shape and size of the particles. $S(Q)$ is the inter-particle structure factor and is decided by the interaction between the particles. B is a constant that represents the incoherent scattering background, which occurs mainly due to hydrogen in the sample.

Data were analyzed using the method similar to that as discussed in an earlier paper [15]. Micelles are treated as prolate ellipsoidal. In general, micellar solutions of ionic surfactants show a correlation peak in the SANS distribution. The peak arises because of the corresponding peak in the interparticle structure factor $S(Q)$ and indicates the presence of electrostatic interactions between the micelles. $S(Q)$ specifies the correlation between the centers of different micelles and it is the Fourier transform of the radial distribution function $g(r)$ for the mass centers of the micelles. Unlike the calculation of $F(Q)$, it is quite complicated to calculate $S(Q)$ for any other shape than

spherical. This is because $S(Q)$ depends on the shape as well as relative orientation of the particles with respect to the beam direction and the relative orientation of neighboring particles. To simplify this, prolate ellipsoidal micelles are assumed to be equivalent spherical. We have calculated $S(Q)$ as derived by Hayter and Penfold from the Ornstein-Zernike equation and using the mean spherical approximation [16]. The micelle is assumed to be a rigid equivalent sphere interacting through a screened Coulomb potential.

The dimensions, aggregation number and the fractional charge on the micelles have been determined from the analysis of the SANS data. The semimajor axis (a), semiminor axises ($b = c$), and fractional charge (α) are the parameters in analysis. The aggregation number is calculated by the relation $N = 4\pi ab^2/3v$, where v is the volume of the surfactant chain. Throughout the data analysis corrections were made for instrumental smearing [13]. The parameters in the analysis were optimized by means of a nonlinear least-square fitting program.

RESULTS AND DISCUSSION

Strong correlation peaks of the structure factor $S(Q)$ are observed in the SANS distributions of pure 100 mM micellar solutions of C_{12}TAB, C_{14}TAB and C_{16}TAB at 30ºC (Fig. 1). These peaks are indication of interacting charged micelles in the solution [15]. The approximate position of the correlation peak occurs at $Q_{max} = 2\pi/d$, where Q_{max} is the value of Q at the peak position and d is the average distance between the micelles. The value of the differential scattering cross-section ($d\Sigma/d\Omega$) increases and the peak position shifts to lower Q values as the chain length of the surfactant is increased from C_{12}TAB to C_{16}TAB. The shifting of peak position towards low Q values suggests an increase in the value of d as a result of increase in size of the micelle. The higher intensity in case of C_{16}TAB surfactant is understood in terms of that this surfactant forms larger micelles as compared to C_{14}TAB and C_{12}TAB. The structural parameters of the micelles obtained after detailed analysis using eq. (1) are given in Table 1. It can be observed that the aggregation number (N) increases from 56 to160 and the fractional charge on the micelle (α) decreases from 0.26 to 0.14 when the hydrophobic chain length is increased from C_{12}TAB to C_{16}TAB. Both the semimajor and semiminor axes of the micelles increase with increase in chain length of the surfactant. The value of semiminor axis is determined by the chain length of the surfactant. On the other hand, the semimajor axis depends on the fractional charge on the micelle, which increases with decrease in the value of the fractional charge.

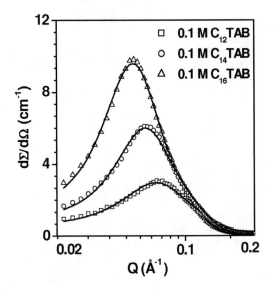

FIGURE 1. The SANS distributions of 0.1 M C_nTAB for different chain length of surfactants n = 12, 14 and 16.

TABLE 1. Micellar parameters of 0.1M C_nTAB surfactant solutions for n = 12, 14 and 16 for (a) pure, (b) with 0.1 M benzene and (c) with 0.1 M hexane.

n	Aggregation number N	Semimajor Axis a (Å)	Semi-minor axis $b = c$ (Å)	Fractional charge α
(a) Pure				
12	56	26.8	15.1	0.26
14	86	33.5	17.6	0.21
16	160	50.5	20.2	0.14
(b) with Benzene				
12	89	42.05	17.43	0.20
14	142	56.02	19.87	0.14
16	475	142.32	23.7	0.08
(c) with Hexane				
12	82	43.90	17.28	0.21
14	152	61.76	20.59	0.15
16	212	74.52	22.99	0.12

Fig. 2 shows the SANS distributions for 0.1M micellar solutions of C_{12}TAB, C_{14}TAB and C_{12}TAB with and without 0.1M benzene. Similar to the effect of increasing chain length of the surfactants (Fig.1), the effect of addition of benzene shifts the correlation peak to lower Q values, with an increase in the overall scattering for each surfactant. The effect of addition of benzene is observed to be more pronounced for the higher chain length surfactants. The analysis of data (Table 1) shows that the aggregation number

significantly increases with addition of benzene. There is an increase in value of the semimajor axis as well as in the value of the semiminor axis. However, the increase in value of the semiminor axis is much smaller than in the semimajor axis. The only marginal change in semi minor axis on addition of hydrocarbon oil suggests that oil can not be solubilized as a droplet surrounded by the surfactant molecules. The possibility is that oil is uniformly solubilized in the micellar core and a large increase in the size of the micelle or semimajor axis is expected because of increase in the packing parameter of the surfactant. The solubilization of oil increases the effective volume of the surfactant molecule (surfactant/oil complex), hence the increase in the packing parameter. The aggregation number of the micelles increases significantly because of the increase in the packing parameter. It is observed that the fractional charge for all the micelles, irrespective of the chain length, decreases on benzene solubilization. This can be understood in terms of increase in counterion condensation as the micellar size increases [15,17].

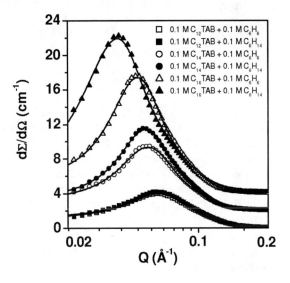

FIGURE 2. The SANS distributions of 0.1M C_nTAB for n = 12, 14, 16 with 0.1M benzene. Distributions for C_{14}TAB and C_{16}TAB are shifted by 2 and 4 units, respectively on the Y axis.

SANS data for comparing the effect of addition of hexane to that of benzene is shown in Fig. 3. The data are significantly different with the two oils. The correlation peak occurs at higher Q values and the scattering cross-section decreases with the hexane as compared to that for the addition of benzene. The differences in the effect for hexane and benzene increase with the chain length of the surfactant. These observations suggest the formation of smaller micelles for hexane amongst the two oils. The micellar parameters on addition of hexane in surfactant solutions of varying chain length are given in Table 1. It is clear from the Table 1 that smaller micelles are formed with hexane than benzene.

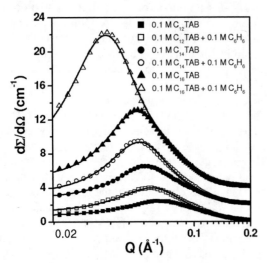

FIGURE 3. The comparison of SANS distributions of 0.1M CnTAB for n = 12, 14, 16 on addition of 0.1M benzene and 0.1M hexane. Distributions for C14TAB and C16TAB are shifted by 2 and 4 units, respectively, on the Y axis

CONCLUSIONS

SANS has been used to study the effect of oil solubilization for benzene and hexane on the sizes of micelles of C_nTAB (n = 12, 14 and 16) surfactants in aqueous solution. The size and aggregation number of the micelles increase on oil solubilization and this effect is more pronounced for the larger size of the micelles having longer chain length. The effective charge on the micelles decreases with increase in size of the micelles on addition of oil. The distribution of oil is found to be uniform inside the micellar core. In all these studies benzene has been found to be more

effective than hexane to induce changes in structure of micelles on oil solubilization.

REFERENCES

1. C. Tanford, *The Hydrophobic Effect: Formation of Micelles and Biological Membranes*, New York, Wiley, 1980.
2. V. Degiorgio and M. Corti (Eds.), *Physics of Amphiphiles: Micelles, Vesicles and Microemulsion*, Amsterdam, North Holland, 1985.
3. Y. Chevalier and T. Zemb, *Rep. Prog. Phys.* **53**, 279-371 (1990).
4. V.K. Aswal, P.S. Goyal and P. Thiyagarajan, *J. Phys. Chem.* **B102**, 2469-2473 (1998).
5. J.V. Joshi, V.K Aswal and P.S. Goyal, *J. Phys.: Condensed Matter* **19**, 196219-19627 (2007)
6. J.N. Israelachvili, *Intermolecular and Surface Forces*, New York, Academic Press, 1992.
7. S.D. Christian, J.F. Scamehorn (Eds.), *Solubilization in Surfactant Aggregates*, New York, Marcel Dekker, 1995.
8. R. Nagarajan, *Curr. Opin. Colloid Interface Sci.* **2**, 282-293 (1997).
9. S.S. Berr, *J. Phys.Chem.* **91**, 4760-4765 (1987).
10. V.K. Aswal and P.S. Goyal, *Chem. Phys. Lett.* **357**, 491-497 (2002).
11. J.V. Joshi, V.K. Aswal and P.S. Goyal, *Phys.* **B391**, 65-71 (2007).
12. V. K. Aswal and P. S. Goyal, *PRAMANA: J. Phys.* **63**, 65-72 (2004).
13. V.K. Aswal and P.S. Goyal, *Curr. Sci.* **79**, 947-953 (2000).
14. S.H. Chen and T.L. Lin, in *Methods of Experimental Physics*; D. L. Price and K. Skold, Eds.; New York, Academic Press, **23B**, 489-543, 1987.
15. V.K. Aswal and P.S. Goyal, *Phys. Rev.* **E 61**, 2947-2953 (2000).
16. J.B. Hayter and J. Penfold, *Mol. Phys.* **42**,109-118 (1981).
17. V.K. Aswal and P.S. Goyal, *Chem. Phys. Lett.* **368**, 59-65 (2003).

Polyblends of Poly(vinyl alcohol) and Poly(ε-caprolactone) and Their Properties

I M. Arcana and L. Alio

Inorganic and Physical Chemistry Research Groups, Faculty of Mathematic and Natural Sciences, Institut Teknologi Bandung, Indonesia

Abstract. The increasing volume of plastic has caused the serious problem in environment. One way to solve this problem is preparation of new plastic materials which can be decomposed by microorganisms in environment. These plastics may be prepared from non-biodegradable material by modification of theirs physical and chemical properties, preparation of theirs copolymers and polyblends. The main problem in preparation of polyblends is compatibility between polymers mixtures. In this work has focused on preparation of polyblends between poly(vinyl alcohol) (PVA) and poly(ε-caprolactone) (PCL) in various compositions by casting of polymers solution. Characterizations of polyblends were carried out by analysis of functional groups (FTIR), thermal property (DSC and TGA/DTA), mechanical properties (Tensile tester), and crystallinity (XRD). The results of polyblends showed that the compatible and homogeneous polyblends were obtained in solvent composition (dimethyl sulfoxide / tetrahydrofurane) (DMSO/THF) of 3:1 and PCL ratio in polyblends less than 15 % (w/w). The absorption intensity of carbonyl and alkyl groups observed in 1725 cm^{-1} and 2940 cm^{-1} increased with increasing PCL composition in polyblends. The melting point (Tm) and fusion enthalpy (ΔHm) for PCL region in polyblends decreased with decreasing PCL composition, but melting point (Tm) and fusion of enthalpy (ΔHm) for PVA region increased. The total fusion enthalpy value obtained by observation was smaller than that of calculation value, indicating the presence of interaction between PCL and PVA to form a part of compatible polyblends with more amorphous structure. The mechanical properties of polyblends tended to decrease with increasing PCL ratio in polyblends. These results were supported by analysis of crystallinity with using X-ray diffraction.

Keywords: Poly(vinyl alcohol), poly(ε-caproplactone), polyblends, and compatibility.
PACS: 61.25.Hq

INTRODUCTION

Increasing volumes of synthetic polymers are manufactured for various applications, mostly for packaging materials. Disposal of the used polymers has been accumulated under the land surface. Unlike natural polymers, the most synthetic polymers cannot be decomposed by microorganisms. So the landfill approach becomes inefficient, and other ways of plastics waste management should be found. New techniques have been developed to obtain biodegradable polymers, environmentally friendly polymers. A modification of non-biodegradable synthetic polymers with a biodegradable polymer producing environmentally friendly polymers is an established technique to solve this problem. The biodegradable plastic may be prepared from non-biodegradable material by modification of theirs physical and chemical properties, preparation of theirs copolymers and polyblends. The main problem in

preparation of polyblends is compatibility between polymers mixtures in order to obtain homogenous and compatible polyblends with better properties.

Poly(vinyl alcohol) (PVA) is a thermoplastic polymers produced commercially as plastic materials and slowly decomposed by microorganisms in environment (1). Poly(vinyl alcohol) is a water soluble polymer and has a important role in various industry mainly as emulsifier, coloid stability, and also adhesive materials. This polymer has high melting point, thermal stability, and mechanical property. These propeties cause the conventional processing of this polymer waste to become difficult. Poly(vinyl alcohol) is easily prepared by safonication of poly(vinyl ester) and poly(vinyl eter), or by partial hydrolysis of poly(vinyl acetate) in metanol solvent (2,3).

Among the biodegradable synthetic polymers that have been developed so far are aliphatic polyesters such as poly(ε-caprolactone) (PCL) (4,5). PCL has low

CP989, Neutron and X-ray Scattering in Materials Science and Biology, International Conference on Neutron and X-ray Scattering 2007, edited by A. Ikram, A. Purwanto, Sutiarso, A. Zulfia, S. Hendrana, and Z. Nurachman
© 2008 American Institute of Physics 978-0-7354-0508-0/08/$23.00

melting point (60°C) and brittleness, so it would appear to limit its practical application (6). But PCL is very attractive due to a valuable set of properties such as a high permeability, the lack of toxicity, biodegradability, and a capacity to be blended with various commercial polymers over a wide composition range (7,8). PCL can be easily prepared by ring-opening polymerization of ε-caprolactone monomer in the presence of distannoxane derivatives catalyst (5,9,10).

In previous papers, we prepared polyblends between modified polypropylene and synthetic polyesters such as poly-(R,S)-β-hydroxybutyrate (PHB) and poly-ε-caprolactone (PCL), and studied relationship between structure and properties of polyblends as well as their biodegradation behavior (11,12). The main problem in preparation of these polyblends is compatibility between polymers mixtures with different properties. In this paper, we report structure and properties of polyblends between poly(vinyl alcohol) (PVA) and PCL prepared by casting of polymers solution in various compositions followed by compression molding into thin film specimens.

EXPERIMENTAL

ε-Caprolactone monomer (ε-CL) was obtained commercially from Aldrich Chemical. Co., dried over CaH$_2$ and distilled under reduced pressure. Poly(vinyl alcohol), dibutyltin oxide (C$_4$H$_9$)$_2$SnO and dibutyltin dichloride (C$_4$H$_9$)$_2$SnCl$_2$ as catalyst precursors were also obtained from Aldrich Chemical Co.

Catalyst preparation

A mixture of dibutyltin oxide (60.0 mmol), dibutyltin dichloride (20.0 mmol), and 95 % ethanol was refluxed. After 6 h, the resulting transparent solution was concentrated to give a white powder, which was pulverized and then exposed to ambient atmosphere overnight to convert the partially formed 1-ethoxy-3-chlorotetrabutyldistannoxane into the corresponding 1-hydroxy-3-chlorotetrabutyl-distannoxane. Recrystallization of the crude product in hexane at 0°C produced crystals within the melting point range of 110-120 °C.

Poly-ε-caprolactone preparation

Preparation of PCL was done by polymerization of ε-CL monomer in presence of distannoxane catalyst under vacuum at 100°C for 4 h (11). 1-hydroxy-3-chlorotetrabutyl-distannoxane (1.2 x 10^{-4} mol) in hexane was first added to the reactor and then hexane

was evaporated by heating in silicone oil bath at 80°C under vacuum for 6 h. ε-CL monomer (0.06 mol) was added to the reactor at room temperature. The mixture was degassed during three freeze-thaw cycles and finally sealed under vacuum. Polymerization reactions were carried out at 100 °C for 4 h. The resulting mixture was dissolved in chloroform (100 mL g^{-1} of initial monomer), and polymer solution was stirred for 12 h at room temperature and refluxed for half an hour. The clear solution was then concentrated (10 mL g^{-1} of initial monomer), followed by precipitation in diethyl ether to give a white solid of PCL (6.50 g) in 95.0 % yield.

Polyblend preparation

Polyblends were prepared by casting of polymer solution in various compositions in dimethyl sulfoxide (DMSO) / tetrahydrofurane (THF) composition of 3/1 (v/v). Both polymers (PCL and PVA) were dissolved in mixture of DMSO and THF at 55°C for 45 minutes, and then polymer solution was placed in petri glass and their solvent was evaporated under vacuum at 55°C for 10 hours to produce polymer film. To obtain homogenous polyblends, this film was then compression molded with a hot press at 120°C for 2 minutes.

Polymers characterization

FTIR (Shimadzu 8501) was used to characterize the chemical structure of PVA, PCL and its polyblends. X-ray diffraction (Diano type 2100E) was used to determine the degree of crystallinity of the polymers formed. Thermal gravimetric analysis (TGA, Perkin Elmer, Pyris) was used to determine their thermal behavior. All the scans were carried out from 30 °C to 500 °C at a heating rate of 5°C min^{-1}. The tensile properties of the specimens were determined according to ASTM standard using a tensile tester machine /autograph (Shimadzu AGS 500D) at room temperature with single tension and a rate of 5 mm min^{-1}.

RESULTS AND DISCUSSION

Preparation of polyblends between PVA and PCL was carried out by casting of polymer solution in mixture of dimethyl sulfoxide (DMSO) / tetrahydrofurane (THF) with composition of 3/1 (v/v). In this solvent composition was obtained homogenous mixture of both polymers in various compositions.

FTIR spectra for polyblends of PVA and PCL shows the presence of superposition and displacement of characteristic peaks of polyblend components

(Fig.1c). The peak intensity ratio between -OH group at wave number about 3300 cm^{-1} to –CH– group at 2940 cm^{-1} for polyblends decreased compared to that for PVA (Fig. 1a). In addition, the peak intensity of carbonyl (–CO–) group at 1725 cm^{-1} and –CH– group at 2940 cm^{-1} for polyblends increased with the increase PCL composition in polyblends These results indicate the presence of an interaction between PVA and PCL as a result of van der Waals force or hydrogen bonding between carbonyl and hydroxyl groups of the components in the polyblends.

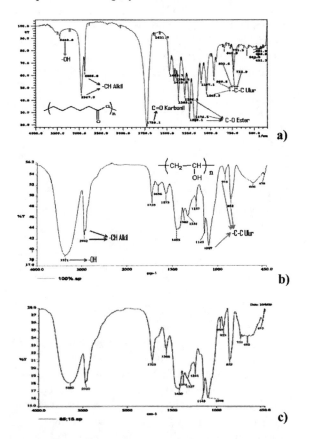

FIGURE 1. FTIR spectra. a). PCL, b). PVA, and c). Polyblends of PVA-PCL.

The thermal analysis with a differential scanning calorimeter (DSC) for polyblends of PVA-PCL in various proportions are described in Table 1. Thermograms of polyblends show two regions of fusion that represent the thermal properties of each component in the polyblends (Fig. 2). Furthermore, the fusion enthalpy of PVA component tends to decrease with increasing PCL content in the polyblends, whereas the fusion enthalpy of PCL component increases. These results are attributed the presence of conformity with the composition of the polyblends in each component. The separation of the melting transition regions in the polyblends of PVA and PCL

indicates that a part of the polyblends formed is non-compatible.

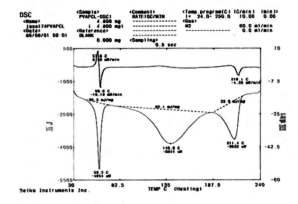

FIGURE 2. Thermogram of DSC for polyblend of PVA-PCL

In addition, we noted that the observed fusion enthalpy of PVA and PCL components for each proportion were lower than the calculated fusion enthalpy. These results might be caused by the presence of a part of PVA and PCL to form a homogenous or compatible polyblends. Based on the difference between calculated and observed fusion enthalpy, it might be predicted that the maximum compatible polyblends formed was obtained in PVA/PCL ratio of 90/10 (w/w) (Table 1). These results are supported by the analysis of crystallinity determined by X-ray diffraction which indicate that the crystallinity of polyblends decreases with addition of small PCL (10 % w/w) (Fig. 3).

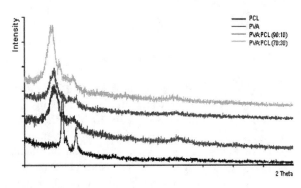

FIGURE 3. X-ray diffractogram of polyblends in various compositions

The decrease of crystallinity by addition of small PCL might be attributed the presence of interaction between PVA and PCL to form a homogenous or compatible polyblends. In previous study, the compatible polyblends of modified PP-PHB was found in a PP/PHB ratio of 90/10 (%w/w). The presence of some PHB in the amorphous parts of PP caused the structure of the PP component to become more regular and denser which was indicated by the maximum crystallinity and mechanical properties of the PP-PHB polyblends (11). Furthermore, the crystallinity of PVA-PCL polyblends tends to increase with the increase PCL content in the polyblends as a result of the high crystallinity and brittleness of PCL, and also the decrease of interaction or miscibility between PVA and PCL. Whereas the mechanical properties of polymers decrease with increasing PCL proportion in the polyblends (Table I). The decrease of mechanical properties might be caused by the high crystallinity and brittleness of PCL, and this property was approved by the difficulty of PCL to be formed polymer films.

The thermal degradation determined by analysis of TG/DTA thermogram was observed that pure PCL had a melting transition at about 59.2°C and it was decomposed thermally by one stage degradation started at 278°C and ended at 325°C (Fig. 4). Whereas in pure PVA was observed to have the melting transition at about 209.2°C and thermally degradation was occurred in three stages as characterized by three plateau regions.

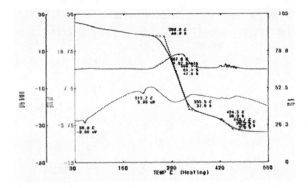

FIGURE 4. Thermogram of TGA for PVA-PCL polyblend

The first one at temperature lest than 240°C might be released a part of water and other impurity like solvent with a conversion of about 10 %. In the second between 240- 304°C was reached at conversion of about 60 % and pertains essentially to water formation and residual of polyene with low molecular weight. In the last stage at 407-435°C, the residual polymer containing a high concentration of conjugated C-C double bonds was cross-linked and converted to char. The mechanism of PVA pyrolysis can be seen in Fig. 5.

In thermogram of TG/DTA for PVA-PCL polyblend with PVA/PCL ratio of 80/20 (w/w) showed two melting transitions at about 56.8°C and 212.7°C attributed to melting transitions of pure PCL and PVA, respectively (Fig. 4). This result indicates that a part of PVA and PCL polyblend formed contains non-

compatible polyblend. In addition, thermal degradation of this polyblend was occurred in three stages as characterized by three plateau regions. The thermal behavior of this polyblend is similar to that of pure PVA, but the first stage occurred at lower temperature, lest than 230 °C might be released a part of water with a conversion of about 10 %. In the second stage at 230- 335°C might be attributed to the simultaneous decomposition of PVA and PCL, pertains essentially to water formation and residual of polyene with low molecular weight. In the last stage occurred at higher temperature (425-498°C), the high concentration of conjugated C-C double bonds of residual PVA was cross-linked and converted to char. Based on these thermal properties indicated that the increase PCL in polyblend could increase their thermal stability.

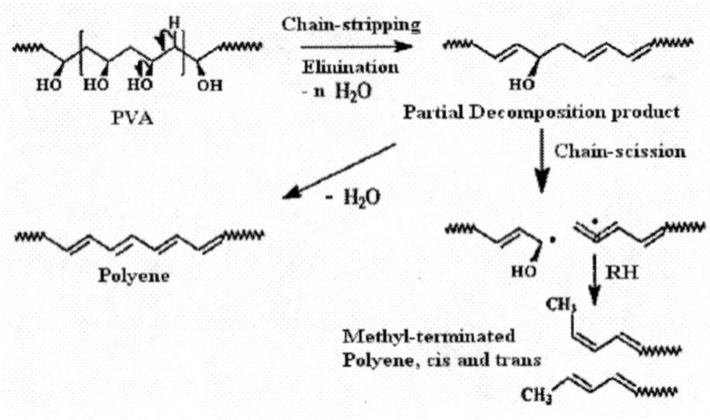

FIGURE 5. Pyrolysis mechanism of poli(vinil alkohol)

CONCLUSION

A part of the polyblends obtained between PVA and PCL were compatible polyblends, and the maximum compatible polyblends formed was obtained in PVA/PCL ratio of 90/10 (w/w). The melting point (Tm) and fusion enthalpy (ΔHm) for PCL region in polyblends decreased with decreasing PCL composition, but melting point (Tm) and fusion enthalpy (ΔHm) for PVA region increased. The total fusion enthalpy value obtained by observation was smaller than that of calculation value, indicating the presence of interaction between PCL and PVA to form compatible polyblends with more amorphous structure. The mechanical properties of polyblends decreased with increasing PCL proportion in the polyblends due to the high crystallinity and brittleness of PCL. These results were supported by analysis of crystallinity with using X-ray diffraction.

TABLE 1. Thermal properties of PVA-PCL polyblends in various compositions determined by differential scanning calorimetry (DSC)

Samples (PVA/PCL)	[a] Observed				[b] Calculated		[c] Compatible Component		[d] σ (MPa)
	PVA		PCL		PVA	PCL			
	Tm (°C)	ΔH (J/g)	Tm (°C)	ΔH (J/g)	ΔH (J/g)	ΔH (J/g)	ΔH (J/g)	(%)	
100/0	226.7	47.4	-	-	47.4	-	-	-	25.3
95/05	227.3	34.1	65.3	2.8	45.0	4.8	12.9	25.9	30.9
90/10	226.7	23.4	66.0	6.0	42.7	9.6	22.9	43.8	24.3
85/15	226.7	34.1	66.8	2.3	40.3	14.5	18.4	33.6	21.3
80/20	222.0	18.1	67.3	18.5	37.9	19.3	20.6	36.0	16.4
75/25	220.7	14.1	67.3	41.1	35.6	24.1	4.5	7.5	15.7
70/30	222.0	32.2	69.3	21.4	33.2	28.9	8.9	14.3	13.9
0/100	-	-	70.0	96.4	-	96.4	-	-	-

[a] Determined by observation of DSC thermogram, [b] Determined by calculation, [c] Determined by difference of fusion enthalpy between observed and calculated, [d] Determined by tensile tester

ACKNOWLEDGEMENTS

The authors gratefully acknowledge funding support from the Research Grant of Competition - Grant Program under the Directorate of Higher Education, Department of National Education, the Republic of Indonesia.

REFERENCES

1. A.B. Mathur, V. Kumar, A.K. Nagpal, and G.N. Mathur, *Ind J. Technol.* **19**, 89-91 (1981)
2. I. Palos, et. al., *J. Polym. Degrad. Stab.* **90**, 264-271 (2005)
3. Wonseoklyoo, et.al., *J. Polym. Degrad. Stab.* **83**, 117-125 (2000)
4. A. Lisuardi, A. Schoenber, M. Gada, R.A. Gross, and S.P. McCarthy, *Polym. Mater. Sci. Eng.* **67**, 298-199 (1992)
5. I.M. Arcana, O. Giani-Beaune, R. Schue, F. Schue, W. Amass, and A. Amass, *Polym. Int.* **51**, 859-866 (2002).
6. S. Akahori and Z. Osawa, *J. Polym. Degrad. Stab.* **45**, 261-265 (1994)
7. J.M. Vion, R. Jerome, Ph. Teyssie, M. Aubin, and R.E Prud'homme, *Macromolecules* **19**, 1828 (1986)
8. G.L. Brode, J.V. Koleske, *J. Macromol. Sci. Chem.* **A6**, 1109 (1972)
9. J. Otera, N. Dan-Oh, and H. Nosaki, *J. Organometal Chem.* **56**, 5307-5311 (1991).
10. R. Okawara and M. Wada, *J. Organometal Chem.* **1**, 81-88 (1963)
11. I.M. Arcana, A. Sulaeman, K.D. Pandiangan, A. Handoko, and M. Ledyastuti, *Polym. Int.* **55**, 435-440 (2006)
12. I.M. Arcana, B. Bundjali, I. Yudistira, B. Jariah, and L. Sukria, *J. Polym.* **39**, 12 (2007)

Physical Identification of Binary System of Gliclazide-Hydrophilic Polymers Using X-Ray Diffraction

H. Rachmawati, Yatinasari, Faizatun, S. A. Syarie

Research Group of Pharmaceutics, Institut Teknologi Bandung, Bandung 40132 Indonesia

Abstract. The formation of binary system in pharmaceutical solid state is aimed to improve the physicochemical characteristics of active compound, such as its solubility. To identify the physical change of the binary system including crystallinity or particle morphology, there are many methods can be applied. In present report, we study the physical interaction of the binary system of gliclazide and hydrophilic polymers. In this binary system, gliclazide was either dispersed or mixed with polyvinyl pirrolidone (PVP K30) or polyethylene glycol (PEG 6000). The dispersion system of gliclazide in the polymeric carriers was prepared by solvation-evaporation method, using dichloromethane/methylene chloride as an organic solvent. The physical characterization of both dispersed and mixed of gliclazide was studied using X-ray diffraction at interval 6-50°/2θ. As a comparison, the same procedure was performed for pure gliclazide. To confirm the diffractogram of this binary system, Fourier Transform Infrared (FT-IR) spectroscopy was carried out as well. Both diffarctogram and FT-IR spectra revealed that there was no new compound formed in the solid dispersion system of gliclazide:PEG 6000 and gliclazide:PVP K30. In contrast, the solubility as well as the dissolution rate of gliclazide in the presence of both hydrophilic polymers was increased as compared to pure gliclazide. We conclude therefore that solvatation followed by evaporation of gliclazide in the presence of either PEG 6000 or PVP K30 did not alter its crystalline characteristic. The improved of gliclazide solubility in the binary system might due to other mechanism such as increased in the wettability and the hydrophylicity effect of the polymers.

Keywords: Solid dispersion, gliclazide, binary system, X-ray diffractometry, FT-IR spectroscopy, solubility, dissolution
PACS: 61.05cp

INTRODUCTION

The term of solid dispersion has been utilized to describe a family of dosage forms whereby the drug is dispersed in a biologically inert matrix (hydrophilic polymers), usually with a view to enhancing oral bioavailability of the drug. The predicted mechanism of solid dispersion technique in increasing solubility of poorly soluble drugs are: formation of the amorphous state meaning that crystal lattice forces have already been overcome, formation of a solid solution or molecular dispersion, increase in wettability and generation of microenvironment conductive to dissolution. As a consequence of these combined effects, the solid dispersion/solution approach may result in enhanced bioavailability of water-insoluble drugs.[1-3]

Various hydrophilic carriers, such as polyethylene glycols (PEGs), polyvinylpyrrolidone (PVP), hydroxypropyl methylcellulose, gums, sugar, mannitol, and urea, have been investigated for improvement of dissolution characteristics and bioavailability of various water insoluble drugs.

Two common methods exist to prepare solid dispersion/solutions: the solvent method, and the hot melt method. With the solvent method, the drug and carrier are dissolved in a common organic solvent, followed by removal of the solvent by evaporation. The hot melt method, by contrast, consists of melting the carrier and drug resulting in formation of the solid dispersion upon cooling of the melt. In some cases, only the carrier is molten and the crystalline drug substance dissolves/disperses in the molten carrier.[4]

The aim of the present study was to improve the dissolution properties of gliclazide. The oral antidiabetic drug, gliclazide, was chosen as a model drug because of its low dose and poor solubility. One possible way to overcome this problem is to prepare solid dispersions of the drug with two different hydrophilic polymers: PVP K30 and PEG 6000.

To study the possible mechanism solubility, in particular amorphous formation as the most predicted mechanism involved in the improvement of gliclazide, X-ray diffraction method was performed. To confirm this predicted amorphous formation after solid dispersion procedure, other physical characterization

CP989, *Neutron and X-ray Scattering in Materials Science and Biology, International Conference on Neutron and X-ray Scattering 2007,* edited by A. Ikram, A. Purwanto, Sutiarso, A. Zulfia, S. Hendrana, and Z. Nurachman
© 2008 American Institute of Physics 978-0-7354-0508-0/08/$23.00

such as Fourier Transform Infrared (FT-IR) was also carried out.

EXPERIMENTAL METHOD

Preparation of solid dispersion

In our preliminary study, solid dispersion (SD) was prepared in various weight ratios of gliclazide:PEG 6000, ranging from ratio of 80:20, 60:40, 40:60, to 20:80. In present study, the solid dispersion was prepared in weight ratio of 20:80, that was the ratio resulting in maximum improved solubility of gliclazide (data not shown) and using PVP K30 at concentration of 83%. The solid dispersion was prepared by solvent evaporation method. Briefly, gliclazide and polymer were dissolved in proper solvent (dichloromethane for PEG and methylene chloride for PVP). The mixture was stirred and evaporated at room temperature until one-third of solvent was left. This mixture was then dried in oven at 40°C (PEG) or 70°C (PVP) until the weight was constant. The dried mass was pulverized and sieved defined mesh to obtain uniform size. The physical mixtures (PM) of gliclazide and PEG 6000 or PVP were prepared by mixing individual component that had previously been sieved. All the samples were stored in a desicator over silica gel till further use.

Powder X-ray Diffraction Analysis (XRD)

Powder X-ray diffraction patterns were recorded using a Powder X-ray diffractometer (Diano) under the following conditions: target/filter (monochro) Cu, voltage 40 kV, current 30 mA, receiving slit 0.3 inches. The data were collected in the continuous scan mode using a step scan of 0.02 deg/min. The scanned range was 5-60°.

Fourier Transform Infrared (FT-IR)

FT-IR spectroscopy was performed on fourier-transformed infrared spectrophotometer (8400S Shimadzu). The samples (gliclazide or solid dispersion or physical mixture) were previously ground and mixed thoroughly with potassium bromide, an infrared transparent matrix, at 1:5 (sample:KBr) ratio, respectively. The KBr discs were prepared by compressing the powders at a pressure of 20 psi for 10 min on KBr-press and the spectras were scanned over wave number range of 4500-500 cm^{-1}.

Dissolution studies

In vitro dissolution studies of gliclazide, physical mixture, and solid dispersion were carried out in two different mediums (phosphate buffer at pH 7.4 and HCl 0.1 N at pH 1.2), using USP paddle method (Hanson Research, SR6, Germany) by dispersed powder technique. Samples equivalent to 80 mg of gliclazide was added to 900 ml of dissolution medium at 37 ± 0.5°C and stirred at 100 rpm. An aliquot of 10 mL was withdrawn at different time intervals with a syringe filter (pore size of 0.45 μm). The withdrawn volume was replenished immediately with the same volume of the pre-warmed (37°C) dissolution medium in order to keep the total volume constant. The filtered samples were suitably diluted, if necessary, and assayed spectrophotometrically at 227 nm. Under these experimental conditions, nor PEG 6000 and PVP interfered with spectrophotometric assay.

RESULTS AND DISCUSSION

Powder X-ray Diffraction Analysis

The powder XRD patterns of various gliclazide, PEG 6000, PVP K30, and its binary systems were compared in figure 1. The diffraction pattern of the pure gliclazide showed its highly crystalline nature, as indicated by numerous distinctive peaks. The PEG 6000 alone exhibited two high intensity peaks at 19° and 23°, while PVP K25 showed no sharp peak indicating amorphous state was present. Both physical mixture and solid dispersion equivalent to the addition spectrum of polymer and the drug indicating no interaction occurred in the solid solution. This result is in accordance with FT-IR spectrograms (figure 2). The reduced intensity peaks were shown in both diffractograms of physical mixture and solid dispersion indicates lower concentration of the drug analyzed in those samples. This can be explained through determination of the crystallinity of gliclazide by comparing some representative peak heights in the XRD patterns of the binary system with those of reference.[1] The relationship used for the calculation of crystallinity was relative degree of crystallinity (RDC) = I_{sam}/I_{ref}, where I_{sam} is the peak height of the sample under investigation and I_{ref} is the peak height at the same angle for the reference with the highest intensity. Pure drug peak at 8° was used for calculating RDC of PM 20% and SD 20%. The RDC values of physical mixture and solid dispersion were 0.2 and 0.178, respectively. Suggesting, the gliclazide present in the solid dispersion would be mostly in the crystalline state and only with very few partially in amorphous molecules.

(1)

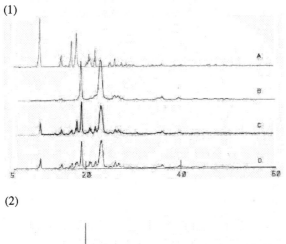

(2)

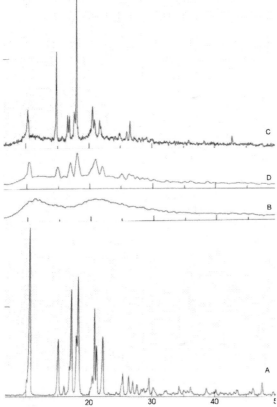

FIGURE 1. X-ray diffraction of gliclazide (A), PEG 6000 (B1), PVP K25 (B2), physical mixture (C), solid dispersion (D).

FT-IR spectroscopy

The interaction between the drug and the carrier often leads to identifiable changes in the FT-IR profiles of solid dispersion. FT-IR spectra for gliclazide, PEG 6000, physical mixture, and solid dispersion were depicted in figure 2.

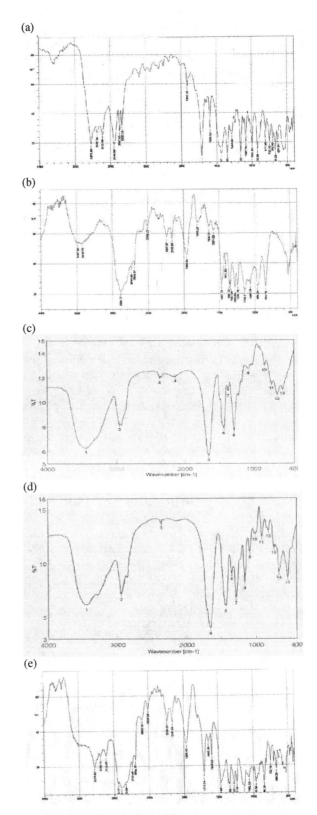

FIGURE 2. FT-IR spectra of gliclazide (A), PEG 6000 (B), PVP K30 (C), solid dispersion gliclazide:PVP K30 (D), solid dispersion gliclazide:PEG 6000 (E)

270

FT-IR spectrum of gliclazide exhibited characteristic signals at 3272.98, 3190.04, 3112.9 (N-H stretch), and a sharp peak at 1710.73 cm^{-1} for carbonyl group. Both spectra of solid dispersions using PVP and PEG were equivalent to the addition spectrum of PEG 6000 and gliclazide (spectra of physical mixtures were not shown). This result indicates absence of well-defined interaction between gliclazide and eithr PEG 6000 and PVP K30.

In vitro dissolution study

The dissolution profile of the gliclazide, the physical mixture, and the solid dispersion was shown in figure 3. The presence of PEG 6000 clearly improved the dissolution rate of gliclazide in both mediums. Although the dissolution profile of the physical mixture and the solid dispersion in phosphate buffer at pH 7.4 was not difference, we observed a better dissolution profile of the solid dispersion in HCl 0.1 N at pH 1.2 as compared to the corresponding physical mixture. This can be explained by the pH-solubility data of gliclazide at 37°C (table 1). Gliclazide, a weak acid with pKa of 5.8, showed better solubility in phosphate buffer (pH 7.4) than in HCl (pH 1.2). The presence of 80% PEG 6000 and 83% PVP K30 (data dissolution profile not shown) increased the solubility of gliclazide at both pHs, although the solubility at pH 7.4 remained superior. The pH-dependent solubility of gliclazide was also shown in dissolution profile of gliclazide, in which the solid dispersion resulted in increased dissolution rate at pH 1.2 as compared to the corresponding physical mixture, although the increased was not superior.

Possible mechanism of increased dissolution rates of solid dispersion has been proposed,[5-9] including reduction of crystallite size, a solubilization effect of the carrier, absence of aggregation of drug crystallities, improved wettability and dispersibility of a drug from the dispersion, dissolution of the drug in the hydrophilic carrier, conversion of drug to amorphous state, and finally, the combination of previously mentioned methods. Since DSC and XRD data demonstrated that there was no amorphous formation after the binary system preparation, the increased dissolution rate observed in present study can thus be contributed by solubilization effect of PEG 6000/PVP K30, improved wettability or combination of both. In addition to those factors, a slightly improved dissolution rate of the solid dispersion as compared to the physical mixture is suggested by improved dispersibility of the drug in the solid solution.

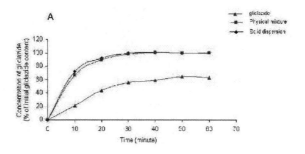

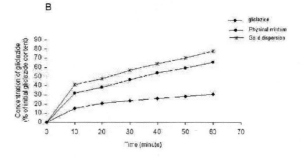

FIGURE 3. Dissolution profile of the preparations

X-ray diffraction analysis can be used to identify the interaction of two components after solid dispersion/solution procedure. Based on the X-ray diffractogram, it can be concluded that there is no amorphous formation after solid solution procedure of gliclazide in both polymers. Furthermore, no physical interaction between gliclazide and either PEG 6000 or PVP K30 was occurred after solid dispersion protocol. The increased solubility and dissolution rate of gliclazide in the presence of either PEG 6000 or PVP K25, therefore might be due to the solubilization effect of PEG and PVP, the improved wettability, and combination of both.

TABLE 1. pH-solubility data of gliclazide, physical mixture, and solid dispersion at 37°C for 30 min under stirring

Compound	Solubility (μg/mL)*	
	HCl 0.1N (pH 1.2)	Phosphate buffer (pH 7.4)
Gliclazide	67.25	229.54
Physical mixture	118.13	496.11
Solid dispersion	141.91	551.11

ACKNOWLEDGMENT

This research was partly funded by ITB (Institut Teknologi Bandung, Indonesia) research grant number 0004/K01.03.2/PL2.1.5/I/2006. We also thank Otto Pharmaceutical Industry for kindly providing gliclazide.

REFERENCES

1. V. S. G Kumar, D. Mishra, *Yakugaku Zasshi*, **126**, 657 (2006).
2. H. N. Joshi, *et al.*, *Int. J. Pharm.* **269**, 251 (2004)
3. C. M. Wassvik, A. G. Holmen, C. A. Bergstrom, I. Zamora, P. Artursson, *Eur. J. Pharm. Sci.* **29**, 294 (2006).
4. D. Q. M Craig, *Int. J. Pharm.* **231,** 131 (2002).
5. J. L. Ford, *Pharm Acta Helv.* **61**, 69 (1986)
6. A. A. Al-Angary, G. M. Al-Mahrouk, M. A. Al-Meshal, *Pharm. Ind.* **58**, 260 (1996)
7. E. Sjokvist Saers, D. Q. M. Craig, *Int. J. Pharm.* **83**, 211 (1992)
8. O. I. Corrigan, R. F. Timoney, *Pharm. Acta Helv.* **51**, 268 (1976)
9. A.T.M. Serajuddin, *J. Pharm. Sci.* **88**, 1859 (1999)

Effect of Varying Additives on Aqueous Solution of PEO-PPO-PEO Tri-block Copolymer

E. Giri Rachman Putra[1], A. Ikram[1], V. K. Aswal[2]

[1]Neutron Scattering Laboratory, BATAN, Kawasan Puspiptek Serpong, Tangerang 15314, Indonesia
[2]Solid State Physics Division, Bhabba Atomic Research Centre, Mumbai-400085, India

Abstract. The effect of addition of salt potassium chloride (KCl) and detergent sodium dodecyl sulfate (SDS) on the aqueous solution of PEO-PPO-PEO (polyethylene oxide-polypropylene oxide-polyethylene oxide) tri-block copolymer, i.e. $(EO)_{103}(PO)_{39}(EO)_{103}$ Pluronics F88 has been carried out. The concentration of F88 was fixed at 5 wt-% and the concentrations of KCl and SDS were varying in the range of 0.05 to 1.5 M and 0.001 to 0.125 M, respectively. Small-angle neutron scattering (SANS) experiments were carried out using 36m SANS spectrometer at Neutron Scattering Laboratory, Serpong (SMARTer), Indonesia. The wavelength of neutron beam was 3.90 Å and the experiments were performed at two different sample-to-detector distances of 2m and 8m. A ^{3}He of 128 x 128 two-dimensional position sensitive detector (2D-PSD) with a beam stopper of 60mm in diameter was used to detect the scattered beam in the range of $0.008 < Q (\text{Å}^{-1}) < 0.25$. All measurements were performed at room temperature. It is observed that increasing the concentration of the above additives induced the micellization of block copolymers to form a core-shell micelle structure.

Keywords: SANS, block copolymer, micellization, core-shell structure.
PACS: 25.40. Dn, 82.35. Jk, 82.70. Dd, 83.80. Qr, 83.85. Hf

INTRODUCTION

Block copolymers are macromolecules that compose of different blocks of different monomer types. The two or more different and incompatible parts provide unique solid state and solution properties that are useful for various applications. In solid state, microphase separation of microdomains rises to formation of different types.

Meanwhile, in aqueous solution which dissolves only one of the blocks, the molecules of polymer self-aggregate into specific structures. This self-aggregation or self-assembly exhibits a wide range of phase behavior, such as micelle formations of various sizes and shapes, i.e. spherical, cylindrical, lamellar, etc., complex structured, microemulsions, and liquid crystalline phases. The self-assembly of block copolymer in solution can be initiated either by changing the concentration at fixed temperature or by changing temperature at fixed concentration[1].

One of the most interesting block copolymers which are widely used for various industrial applications, mostly in nanotechnology, pharmaceutical, textile and detergent is PEO-PPO-PEO (polyethylene oxide-polypropylene oxide-polyethylene oxide) tri-block copolymers and produced commercially as Pluronics® [2]. Several studies in the series of PEO-PPO-PEO tri-block copolymers have been published for both static and dynamic properties using small-angle neutron scattering (SANS) technique involving temperature dependence experiments[3-5]. However, there are only few works in studying the presence of additives in initiating the self-assembly instead of changing the concentration and temperature for PEO-PPO-PEO tri-block copolymers[6-8].

It had been understood that the aggregation behavior of block copolymer in solution occurs with the differences of the chain block solubility in a specific solvent. In case of PEO-PPO-PEO tri-block copolymer in water, PEO is the water-soluble block meanwhile PPO is the hydrophobic or water-insoluble block. Thus, the dehydration in the PPO block will initiate to aggregate into micellar formation. However, this phenomenon can be driven thermodynamically or entropically[7]. Studies on the ionic surfactant such as SDS with PEO-PPO-PEO tri-block copolymer mixed system which is applied for many applications show that depending on their concentration and temperature, interaction between them lead to the formation of either mixed micelles or different kinds of mixed aggregates[8]. Copolymer with high PO/EO ratio shows a significantly higher stability of the micelles against addition of SDS than copolymers with low

CP989, *Neutron and X-ray Scattering in Materials Science and Biology, International Conference on Neutron and X-ray Scattering 2007,* edited by A. Ikram, A. Purwanto, Sutiarso, A. Zulfia, S. Hendrana, and Z. Nurachman

PO/EO ratio. Thus, in order to understand the effect of copolymer composition on the presence of additive such as SDS, thus series of PO/EO ratios were determined[8]. In this work we investigated the micellization of the lowest PO/EO ratio of tri-block copolymers by adding KCl and SDS with various concentrations to complete the whole figures of the effect, especially of SDS on the self-assembly behavior of the PEO-PPO-PEO tri-block copolymer.

EXPERIMENTAL METHOD

Tri-block copolymers of $(EO)_{103}$-$(PO)_{39}$-$(EO)_{103}$ F88 sample was provided by BASF. Potassium chloride KCl, Sodium Dodecyl Sulfate SDS, and D_2O (98% atom D) were obtained from Aldrich. The salt was dried at 100 °C in oven for overnight before used. A certain amount of F88 was dissolved in D_2O to obtain 5% (wt) solutions, and then KCl and SDS were added in various concentrations with concentration range of 0.05 - 1.5M and 0.002 - 0.125M, respectively.

SANS measurements were carried out using SANS spectrometer (SMARTer) at Neutron Scattering Laboratory (NSL) BATAN in Serpong, Indonesia. The detail of SANS spectrometer and its performance was described elsewhere[9,10]. The wavelength 1 of the neutron beam was 3.90 Å and the experiments were performed at two different sample to detector distances of 2m and 8m to cover a momentum transfer Q-range of 0.01 to 0.2 Å^{-1}. The scattered neutrons were detected using a two-dimensional position sensitive detector (2D-PSD) with 60 mm diameter of beam stop in the center.

The measurements were performed for both concentration and salt-type dependences using 5mm thickness of a quartz cell. Each sample has been exposed to the neutron beam for 1 hour. Meanwhile background, noise and detector efficiency measurements were carried out for 2, 16 and 12 hours, respectively. During the experiments, the temperature was maintained at room temperature.

The measurements on 5 wt-% F88 micellar solution as a function of temperature, from room temperature up to 100 °C were also carried out using 1 mm thickness of a quartz cell for 30 minute of exposure time at 2m and 8m sample to detector distances position in order to covering the Q-range experiment of 0.008 – 0.2 Å. These measurements were preformed as a comparative study on micellization of PEO-PPO-PEO tri-block copolymer which initiated by addition of additives.

RESULTS AND DISCUSSION

Figure 1 shows the SANS data from 5 wt% F88 solution as a function of temperature. At room temperature F88 in aqueous solution is a unimers which has Gaussian coil structure. The radius gyration of this structure was obtained by fitting the scattered data with data analysis program and the result is 22 Å. By increasing the temperature these unimers aggregate to form a micelle with core-shell structure which is PPO block as a core and PEO block as a corona or shell in aqueous solution. It is shown that at 35 °C 5 wt% F88 solution just started to form a micelle and once as the micellization has started, the number density of the micelles increases by further increasing the temperature.

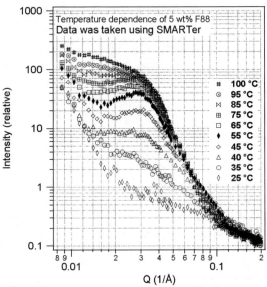

FIGURE 1. SANS data from 5 wt% tri-block copolymer as a function of temperature

Figure 2 shows the SANS data from 5 wt% F88 solution in presence of various KCl concentrations at room temperature. It is observed that the scattering intensity in the low Q value significantly increases with the addition of KCl after the concentration of 0.6 M. It is an indication of dissolved block copolymer as unimers transforms to micellar structures by increasing KCl concentration. It is also clear that there is a need of minimum amount of salt to induce the micellization. The number density of the micelles increases with the increase in the KCl concentration. Here we can also see that increasing the concentration of KCl in inducing micellization of F88 is similar to that of increasing temperature of the solution. The broad correlation peak in the SANS data in presence of KCl or increasing temperature is due to corresponding inter-particle structure factor.

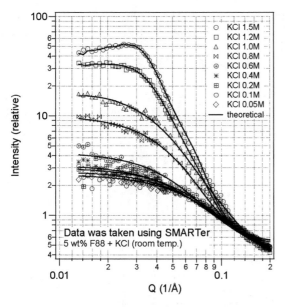

FIGURE 2. SANS data from 5 wt% block copolymer F88 in presence of various concentrations of KCl at room temperature

For structure analysis of the micelle, it is assumed that below 0.6M of KCl concentration, the Gaussian coil structure model was applied to fit the size of unimers. Meanwhile, from 0.6M and above, the spherical core-shell structure model was employed to fit the micelle sizes. These results are given in the table below for low and high KCl concentrations.

TABLE 1. Fitting results of 5 wt% tri-block copolymer F88 in presence of various concentrations of KCl.

KCl [M]	Radius of gyration (Å)	Core radius (Å)	Hard sphere radius (Å)	Aggregation Number
0.05	23.9	-	-	-
0.1	25.8	-	-	-
0.2	26.3	-	-	-
0.4	29.5	-	-	-
0.6	-	12.91	25.77	2.4
0.8	-	18.51	31.14	7.1
1.0	-	21.65	35.60	11.3
1.2	-	25.68	82.08	18.9
1.5	-	28.23	86.75	25.1

It is shown from Table 1 that at low KCl concentrations the unimers are slightly growing with the increase of KCl addition. Similar to that, from 0.6M of KCl concentration the core-shell micelles are also becoming bigger. Both models, Gaussian coil and spherical core-shell, have been applied and fitted very well to the experimental data as shown in Figure 2.

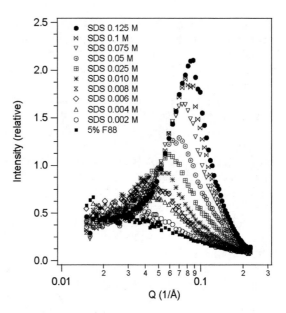

FIGURE 3. SANS data from 5 wt% block copolymer F88 in presence of varying concentration of SDS.

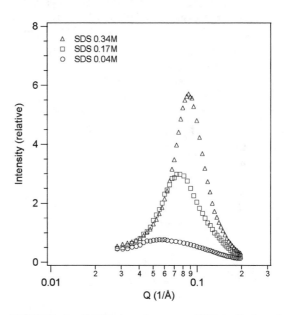

FIGURE 4. SANS data from pure SDS in D_2O at low concentration using SANS BATAN (SMARTer)

The scattering intensity increases significantly in the presence of additives as with the increase in temperature. At low temperature and without presence of any additive, the block copolymer is dissolved as unimers and thus the scattering from the solution is quite low. The scattering intensity increases since the unimers self-associate to form large micelles. The dehydration of water from PPO blocks with the increase in temperature results in transformation of unimers to the micelles.

On the other hand, Figure 3 shows that the micelle formation takes place even with the small addition of SDS detergent. It is observed that the increasing concentration of SDS in 5 wt% F88 results an increase in the intensity and a shift of the peak position to higher Q value. The micellization is also occurred by adding a minimum amount of SDS and the number density of these micelles increases with the SDS concentration.

The experiment data of 5wt% F88 with SDS does not fit to the model employing of spherical core-shell structure as it has been successful for the 5wt% F88 with KCl. Comparing to the SANS data from pure SDS in D2O at low concentration, it can be shown that the SANS profile of 5wt% F88 with SDS, Figure 3 is similar to pure SDS at low concentration, Figure 4. However, when we applied as same model as pure SDS in D_2O[11] for 5wt% F88 with SDS in structure analysis of the micelle, it also does not fit properly. It means that in the 5wt% F88 with SDS solutions form a mixed micelle where at room temperature the micellization occurs from tri-block copolymer as well as ionic surfactant molecule of SDS. The structure of the mixed micelles of PEO-PPO-PEO tri-block copolymer with SDS would be analysed further[12].

CONCLUSIONS

It is observed that increasing the concentration of the additives, KCl and SDS in inducing micellization of block copolymers is similar to that of increasing temperature of the solution. Presence of the KCl and SDS to the self-assembling is truly entropy driven.

ACKNOWLEDGMENT

The authors would like to tank Dr. P. S. Goyal from Bhabba Atomic Research Centre, India for beneficial discussions.

REFERENCES

1. P. Bahadur, *Current Science* **80**, 8, (2001)
2. S. Forster, et al. *J. Mater. Chem.* **13**, 2671 (2003)
3. K. Mortensen, *J. Phys. Condens. Matter* **8**, A103-A124 (1996)
4. K. Mortensen, *Colloid Surf.* **A 183**, 183-185, 277-299 (2001)
5. K. Mortensen, *Polym. Adv. Technol.* **12**, 2-22 (2001)
6. V. K. Aswal, P. S. Goyal., J. Kohlbrecher, P. Baduhur, *Chem. Phys. Lett.* **349**, 458 – 462 (2001)
7. V. K. Aswal, J. Kohlbrecher, E. Giri Rachman Putra, presented at the XIII International Conference on Small Angle Scattering (SAS2006), Kyoto July 9 – 13, 2006.
8. R. Ganguly, et al., *J. Phys. Chem.* **B 110**, 9843-9849 (2006)
9. E. Giri Rachman Putra, A. Ikram, E. Santoso, B. Bharoto, *J. Appl. Cryst.* **40**, s447-s452 (2007)
10. E. Giri Rachman Putra, B.Bharoto, E. Santoso, Y. A. Maulana, *Neutron News* **18.1**, 23 - 29 (2007)
11. E. Giri Rachman Putra, A. Ikram, *Indonesian J. Chem.* **6.2**, 117 – 120 (2006)
12. E. Giri Rachman Putra, paper in preparation.

Astrium GmbH
(formerly Dornier)
88039 Friedrichshafen
Germany

Hugo Betzold	Tel	+49-7545-8 25 96
	Fax	+49-7545-8 26 93
	@	Hugo.Betzold@astrium.eads.net
	www	http://www.astrium.eads.net

EQUIPMENT AND SOFTWARE FOR RESEARCH WITH NEUTRONS

Astrium GmbH, a company of the DaimlerChrysler Group is one of the leading high-tech companies in Germany.

In the field of advanced technologies developed Astrium GmbH neutron velocity selectors which reach the utmost limit of what is technically feasible today.

The unique feature of this SANS neutron velocity selector is the use of carbon fiber epoxy material for the very thin, helically twisted rotor lamellae which are coated with 10B or Gd203 as absorbing material.

This novel design allows rotor speeds of up to 28,300 rpm (= 430 m/s peripheral velocity).

Based on this technology Astrium is a partner for customer-specific:

- neutron velocity selectors (monochromators)
- higher-order neutron filters (harmonic filters)
- disk choppers (ball / active magnetic bearings)
- FERMI choppers (ball / active magnetic bearings)
- components for choppers (slats, multi-slit disks etc.)

as well as for software and peripheral instrumentation, e.g. as:

- control systems (PC based)
- monitoring systems for choppers and selectors
- interface to a host.

The successful use in many laboratories world-wide reflects the fact that Astrium's neutron research related instruments and components meet high experimental requirements and allows a broad versatility of applications.

References:
ANL, Argonne, IL; ANSTO, Lucas Heights, NSW; BNL, Upton, NY; CCLRC, Didcot; FZJ, Jülich; GKSS, Geesthacht; HMI, Berlin; IFE, Kjeller; ILL, Grenoble; IPC, Göttingen; ISSP, Tokyo; JAERI, Tokai; KAERI, Seoul; LANL, Los Alamos, NM; LLB, Saclay; MPI, Mainz, Stuttgart; NIST, Gaithersburg, MD; NRC, Chalk River, ON; PSI, Villigen; PSU, University Park, PA; PTB, Braunschweig; RISO, Roskilde; RMBI, Calgary, AB; TUM, Munich

General Description

GE Energy's Optimization and Control Business dedicates to help customers maximize the economic performance of their operating assets. GE Energy's portfolio of optimization and control solutions brings numerous complementary product lines and services together, addressing many of today's most pressing optimization and performance concerns:

- ◆ Asset lifecycle costs through proactive condition monitoring
- ◆ Operational flexibility, reliability, and efficiency through upgraded controls for machine, process, or plant
- ◆ Energy exploration through advanced drilling technologies
- ◆ Facility and homeland security through advanced radiation detection and cybersecurity technologies
- ◆ Reliability and maintenance practices through comprehensive consulting and outsourcing services
- ◆ Fuel costs, fuel efficiency, and dispatch decisions through intelligent software that maximizes revenue and minimizes costs

By bringing these solutions together, we've not only made it easier to combine our capabilities to meet your needs, we've made it easier for you to interact with us, putting a specialized team of optimization and control professionals at your disposal.

Products And Services

Bently Nevada™ Asset Condition Monitoring

Control Solutions

Oil & Gas Exploration & Production Technologies

Optimization and Control Services

Optimization and Diagnostic Software

Reliability Consulting and Implementation Services

Reuter Stokes Measurement Solutions

Author Index

K

Kalyanasundaram, P., 202
Kamiyama, T., 20
Kasiviswanathan, K. V., 202
Kasuda, K., 21
Kennedy, B. J., 60
Kennedy, S. J., 10
Khairurrijal, 117, 147
Knott, R. B., 40
Kohlbrecher, J., 53
Kurniawan, B., 151

L

Liu, F., 68
Lockman, Z., 155
Low, I. M., 168

M

Macdonald, J. E., 68
Mammou, L., 187
Manaf, A. , 77
Marsih, I. N., 248
Marsongkohadi, 4
Maryanto, S., 138
Mazumder, S., 29
Mohamad, A. A., 245
Mohamad, H., 161
Mohamed, A. A., 130, 194
Mokeddem, M. Y., 187
Muslih, M. R., 85, 92, 101

N

Nazri, I. M., 158
Nedjar, A., 187
Niimura, N., 47
Nishida, M., 21, 96, 101
Noor, A. F. M., 155
Nugraha, 255
Nurdin, W. B., 209
Nurhasanah, I., 147

O

O'Connor, B., 168
Onggo, D., 117
Othman, R., 158, 241
Özkendir, O. M., 233

P

Prajitno, D. H., 73, 77
Pratapa, S., 176
Prijamboedi, B., 164, 172
Priyanto, T. H., 85
Purwaningsih, S., 176
Purwanta, 92
Purwanto, A., 151, 214
Putra, E. G. R., 130, 176, 190, 273

R

Rachmawati, H., 268
Raghu, N., 202
Ramelan, A., 122
Ridwan, I., 180
Rosyidah, A., 117
Rusmiati, 172
Ryu, J. S., 198

S

Sairun, 92, 190
Saleh, J. M., 194
Santoso, E., 130, 190
Sanuddin, M., 221
Saragih, H., 237
Sari, N. R., 164
Sastry, P. U., 228
Schotte, K. D., 209
Septina, W., 255
Setiawan, A., 138
Setiawan, 85
Shin, J. W., 198
Shiro, A., 96
Soedarsono, J. W., 77
Sreekantan, S., 155
Subagio, A., 134, 224, 237
Sukirno, 134, 224, 237
Sumirat, I., 92

Suparno, N., 85
Supriyanto, E., 134, 224, 237
Sutanto, H., 134, 224, 237
Syarie, S. A., 268
Syarif, D. G., 122, 126, 143

T

Timmins, P., 35
Triwikantoro, 176
Tschierske, C., 68
Tun, Z., 28
Türker, A., 233
Tüzemen, 223

U

Ufuktepe, Y., 233
Umemoto, M., 241
Ungar, G., 68
Upadhyay, R. V., 228

V

Vavrin, R., 53
Vujičić-Žagar, A., 41

W

Wagh, A. G., 53, 107
Warikh, A. R. M., 89
West, A. R., 3
White, A. H., 59
Widiastri, M., 248
Wiendartun, 126

X

Xie, F., 68

Y

Yamin, B. M., 221
Yao, T., 138
Yatinasari, 268
Yuliarto, B., 255

Z

Zahirani, A. Z. A., 89
Zeng, X. B., 68
Zergoug, T., 187
Zuhailawati, H., 241